AF358475

Lithium-Ion Batteries and Li-Ion Capacitors: From Fundamentals to Practical Applications

Lithium-Ion Batteries and Li-Ion Capacitors: From Fundamentals to Practical Applications

Editor

Junsheng Zheng

Basel • Beijing • Wuhan • Barcelona • Belgrade • Novi Sad • Cluj • Manchester

Editor
Junsheng Zheng
Tongji University
Shanghai
China

Editorial Office
MDPI AG
Grosspeteranlage 5
4052 Basel, Switzerland

This is a reprint of articles from the Special Issue published online in the open access journal *Batteries* (ISSN 2313-0105) (available at: https://www.mdpi.com/journal/batteries/special_issues/5Z4W6K022F).

For citation purposes, cite each article independently as indicated on the article page online and as indicated below:

Lastname, A.A.; Lastname, B.B. Article Title. *Journal Name* **Year**, *Volume Number*, Page Range.

ISBN 978-3-7258-2419-9 (Hbk)
ISBN 978-3-7258-2420-5 (PDF)
doi.org/10.3390/books978-3-7258-2420-5

Preface

Lithium-ion batteries and lithium-ion capacitors, as representatives of energy and power devices, have been widely studied and rapidly developed in recent years. Lithium-ion battery is a kind of long-duration large-capacity energy storage device with a high conversion rate and a high energy density, and it has a wide range of applications and is in high demand. In fact, lithium-ion batteries are already widely used in consumer electronics, electric vehicles, and energy storage. Lithium-ion capacitors are a new type of capacitor combining the advantages of lithium-ion batteries and traditional supercapacitors, and they can effectively improve overall energy density on the basis of maintaining the advantages of supercapacitors. Lithium-ion capacitors are expected to solve the bottleneck problem of energy and power contradiction faced by power supply devices, and realize the balance and customizability of the energy, power and life of devices. Therefore, the best way to further improve the energy density of lithium-ion capacitors while maintaining high power density has become a focus of research on lithium-ion capacitors. Due to their high power densities, energy densities, and long cycle lives, the application of lithium-ion capacitors for automotive energy recovery, electrochemical energy storage and power assistance, fast charging, and multi-functional devices could be promising. However, there are still many problems that remain unsolved regarding the basic research and application of lithium-ion batteries and lithium-ion capacitors. This Reprint focuses on lithium-ion batteries and lithium-ion capacitors, including the rise of capacity, rate, and lifespan of electrode materials; the increase in the ion transmission and storage capacity of anodes and cathodes; and the improvement of the electrode/electrolyte interface and stability of the solid electrolyte interphase. On the other hand, improvements in the surface densities of electrodes and proportions of active substances have become key issues regarding research on lithium-ion batteries and lithium-ion capacitors. Furthermore, pack design, stacking technology, equalization technology, SOC estimation, operation, and monitoring are also crucial for the use of lithium-ion batteries and lithium-ion capacitors as power supply equipment.

I would like to thank all the authors who helped me to write this Reprint. It is your efforts and support that made this Reprint possible. I would also like to express my sincere thanks to my colleague at MEDPI for their guidance and help.

Junsheng Zheng
Editor

Article

Development of a Fusion Framework for Lithium-Ion Battery Capacity Estimation in Electric Vehicles

Bo Jiang [1], Xuezhe Wei [2] and Haifeng Dai [2,*]

[1] Postdoctoral Station of Mechanical Engineering, School of Automotive Studies, Tongji University, Shanghai 201804, China
[2] Clean Energy Automotive Engineering Center, School of Automotive Studies, Tongji University, Shanghai 201804, China
* Correspondence: tongjidai@tongji.edu.cn

Abstract: The performance of a battery system is critical to the development of electric vehicles (EVs). Battery capacity decays with the use of EVs and an advanced onboard battery management system is required to estimate battery capacity accurately. However, the acquired capacity suffers from poor accuracy caused by the inadequate utilization of battery information and the limitation of a single estimation method. This paper investigates an innovative fusion method based on the information fusion technique for battery capacity estimation, considering the actual working conditions of EVs. Firstly, a general framework for battery capacity estimation and fusion is proposed and two conventional capacity estimation methods running in different EV operating conditions are revisited. The error covariance of different estimations is deduced to evaluate the estimation uncertainties. Then, a fusion state–space function is constructed and realized through the Kalman filter to achieve the adaptive fusion of multi-dimensional capacity estimation. Several experiments simulating the actual battery operations in EVs are designed and performed to validate the proposed method. Experimental results show that the proposed method performs better than conventional methods, obtaining more accurate and stable capacity estimation under different aging statuses. Finally, a practical judgment criterion for the current deviation fault is proposed based on fusion capacity.

Keywords: lithium-ion battery; state estimation; battery capacity; adaptive fusion; estimation uncertainty

Citation: Jiang, B.; Wei, X.; Dai, H. Development of a Fusion Framework for Lithium-Ion Battery Capacity Estimation in Electric Vehicles. *Batteries* **2022**, *8*, 112. https://doi.org/10.3390/batteries8090112

Academic Editor: King Jet Tseng

Received: 22 July 2022
Accepted: 30 August 2022
Published: 5 September 2022

Publisher's Note: MDPI stays neutral with regard to jurisdictional claims in published maps and institutional affiliations.

1. Introduction

The rapid development of new energy vehicles benefits from the desire to reduce emissions and pollution [1]. Among different new energy vehicles, electric vehicles (EVs) are the most promising for mass commercialization, which gives credit to the excellent performance of lithium-ion batteries (LIBs) [2]. LIBs have the superiority of high power/energy density, light weight, and long life cycle [3]. However, their adaptability to the harsh working environment, poor thermal safety, and inevitable degradation need scientific improvements. The solution to the above challenges can be achieved through the battery design, for example, improvements in battery electrodes [4,5] and efficient battery management technologies for battery systems [6,7]. High-performance battery management technologies are an essential impetus for the development of EVs.

Among battery management technologies, LIB capacity is a vital parameter for characterizing battery performance and is deemed a direct indicator of battery health level [8]. Battery capacity affects the uninterrupted driving range of EVs and is a prerequisite for estimating some essential battery internal states. However, battery capacity is time varying and decays as the battery ages [9], which will challenge the capability of state estimation; hence, exact capacity information is essential for better management and utilization of LIBs. The direct measurement method is the most accurate to obtain battery capacity; however, this method is time consuming and usually limited by the operating conditions, which is unsuitable for real EV applications [10]. Therefore, accurate onboard estimation of battery capacity is a research hotspot in battery management technologies in EVs.

1.1. Review of Existing Battery Capacity Estimation Approaches

Many studies have investigated indirect estimation methods of battery capacity for EVs [11–13]. Generally, commonly used battery capacity estimation approaches categorize into three classes: the method based on the state of charge (SOC), the method based on incremental capacity analysis (ICA), and the data-driven method.

It can be found that SOC has a close relationship with battery capacity and the SOC-based methods utilize the change in SOC and charge accumulation over a while to estimate battery capacity [14]. Some researchers employed the measured or estimated open-circuit voltage to obtain SOC and achieve battery capacity [15,16]. Considering that battery capacity can be regarded as an unknown parameter, some joint estimation techniques for battery SOC and capacity are springing up. A single state vector containing battery SOC, capacity, and other parameters and corresponding state–space function was established to estimate battery capacity [17,18]. Using two or more estimators is more flexible and has been investigated by many researchers. Dual extended Kalman filter [19], dual sliding-mode observer [20], dual nonlinear predictive filter [21], and Kalman filter together with least squares [22] are common attempts to achieve the co-estimation. Battery SOC is fundamental in battery management technology, so the SOC-based capacity estimation method is accessible to implement. Some factors, including the SOC estimation error and the different time-varying characteristics between states, shall be considered when using this method [23].

The ICA method is another practical capacity estimation method, using the differential technology to transfer the battery charging capacity voltage (Q-V) curve into the incremental capacity (IC) curve. The features of the IC curve indicate the battery's internal electrode behavior; therefore, this method can analyze the battery aging mechanism [24]. A close relationship between IC curve features and battery health status can be established and further used for online capacity estimation. A necessary procedure in ICA technology is curve fitting/smoothing because the differential operation in ICA technology is sensitive to measurement noise. He et al. [25] compared six commonly used voltage curve fitting models in IC curve determination and the model in [26] was validated to be optimal for different types of batteries. After numerical differentiation, Gaussian filtering has an advantage over conventional moving average filtering in terms of IC curve smoothing [24,27], where low-frequency signals can be separated from high-frequency noise. An operating condition with a monotonic voltage change is necessary for ICA technology, limiting the application of this method.

The data-driven method has been widely employed in battery state estimation with the development of artificial intelligence technology, allowing knowledge related to battery aging to be learned from battery training datasets and further used for online capacity estimation [28,29]. Feature engineering is the first step in a data-driven estimation method. Compared with the discharging condition, the battery charging condition is more stable and regular and is usually used to extract battery-capacity-related features. Five charging-related features, including the initial and final charge voltage, the final charge current, the constant current charge capacity, and the constant voltage charge capacity, were extracted from charge curves [30]. After determining battery features, machine learning algorithms, including kernel techniques [31,32] and neural-network techniques [28,33], are employed to estimate capacity. The main concern during the online application of this method is the high computational effort.

Except for the above typical capacity estimation methods, fusion technologies have also been widely employed in battery state estimation. Zheng et al. [34] studied a novel capacity estimation method with the help of fusion estimation of battery charging curves and the Arrhenius aging model, which shows high estimation accuracy over the whole battery lifecycle. A multi-stage fusion method containing three battery models at different aging levels was proposed by Xiong et al. [35] and then the residual error of each model was employed to calculate the fusion weight. Moreover, Balasingam et al. [36] investigated a robust capacity estimation fusion method, which can optimally utilize the estimates from different approaches; however, only simulation data were used for validation in this work. Owing to the effective utilization of more information from different sources, the information fusion techniques can not only eliminate noise and outliers in the input information but also achieve a complete description of the observed object through information complementarity [37]. The multi-source information represents the data from different models [35], different approaches [34,36], and different operating scenes. Using the multi-source information, the fusion techniques promise to improve the estimation accuracy and enhance the estimation stability.

1.2. Existing Challenges and Original Contributions

The above overviews several typical capacity estimation methods and each approach shows better performance under specific conditions. However, the following issues remain to be addressed among current capacity estimation methods.

(1) **The inadequate utilization of battery information.** The attenuation of battery capacity affects the dynamic trajectories of other battery states, including SOC, terminal voltage, and open-circuit voltage. This inspires researchers to perform capacity estimation using the variation in the above states combined with the battery current. However, one primary shortage in the above approaches is that only a single battery information source (except battery current) is utilized in capacity estimation. For example, the SOC-based method only utilizes the battery SOC and current, while the ICA-based method only uses the battery voltage and current. The inadequate utilization of battery information could lead to difficulties in improving the estimation accuracy.

(2) **The limitation of EVs' operating conditions on estimation methods.** As introduced above, each method may only adapt to a specific operating condition. For the ICA-based and data-driven methods, the EV charging condition is stable and suitable for feature construction. In contrast, the EV discharging condition is random, not monotonous, and not convenient for the online application of these two methods. The adaptive filters offer advantages in SOC acquisition to the SOC-based method and battery model parameter determination is an important step. In this situation, dynamic discharging will be helpful because of adequate current excitation for parameter identification. Therefore, the battery operating conditions may limit the above capacity estimation methods.

(3) **The weak anti-interference ability to the system error.** Another deficiency in traditional capacity estimation is the weak anti-interference ability to system error because only one estimation approach is employed. Battery current is the critical information in capacity estimation and inaccurate current information affects the estimation accuracy. For example, a micro-internal short circuit (ISC) is a severe battery failure and may cause a thermal runaway of LIBs [38,39]. Supposing a battery has a micro-ISC fault (inaccurate current information), the SOC-based method may obtain a smaller capacity estimate, while the ICA-based method may get a larger capacity estimate. Hence, the weak anti-interference ability to system error may lead to capacity estimation deviations.

Considering the above problems, this study proposes an adaptive fusion method for battery capacity estimation under actual EV operation conditions, which takes advantage of information fusion technologies. Concretely, the SOC-based and ICA-based capacity estimation constitutes the fusion method, which can fully utilize battery information, including the battery current, SOC, and voltage. Moreover, the two base estimators operate in different working conditions, achieving the fusion framework during complete operating conditions. Here, the complete operating condition means the combination of discharging and charging conditions, which simulates the actual working conditions of EVs. The fusion of different capacity estimations enhances the anti-interference ability against system error because these two base estimators have different estimation characteristics under inaccurate battery current information. In our previous work, presented in VPPC 2020, Gijon, Spain [40], this framework was initially introduced and verified with a fresh battery cell. Significant extensions are provided in this study. Specifically, an improved estimation and fusion framework is investigated, in which the noise-compensating method is employed for capacity estimation during discharging. Moreover, the constructed fusion framework is validated by experimental data of aged battery cells and with inaccurate current information. The main work in this paper can be generalized as follows. Firstly, a general framework for battery capacity estimation and fusion is proposed, then the SOC-based and ICA-based capacity estimation approaches are revisited in different operating conditions and the corresponding error covariance of the two estimates is obtained separately. The acquisition of the fusion capacity is realized through the Kalman filter. Laboratory experiments simulating the EV operating conditions are designed and performed to validate the effectiveness of the proposed fusion method under different dynamic conditions and aging statuses. Furthermore, a practical judgment criterion for the current deviation fault is proposed based on fusion capacity.

1.3. Outline of the Paper

The remainder of this paper is organized as follows. Section 2 proposes the general estimation and fusion framework and derives the uncertainties of capacity estimation from two estimation approaches; further, the adaptive fusion of two estimations is constructed. The test bench and designed battery experiments are shown in Section 3. Section 4 shows the experimental results and discusses the accuracy and robustness of the fusion method. Finally, Section 5 concludes this paper.

2. Multi-Dimensional Capacity Estimation and Fusion

2.1. Development of the Adaptive Fusion Framework

Aiming to realize the adaptive fusion for capacity estimation, a general framework for battery capacity estimation and fusion is shown in Figure 1a. Three main procedures are included in the general framework: multi-dimensional capacity estimation, determination of estimation uncertainty, and fusion center.

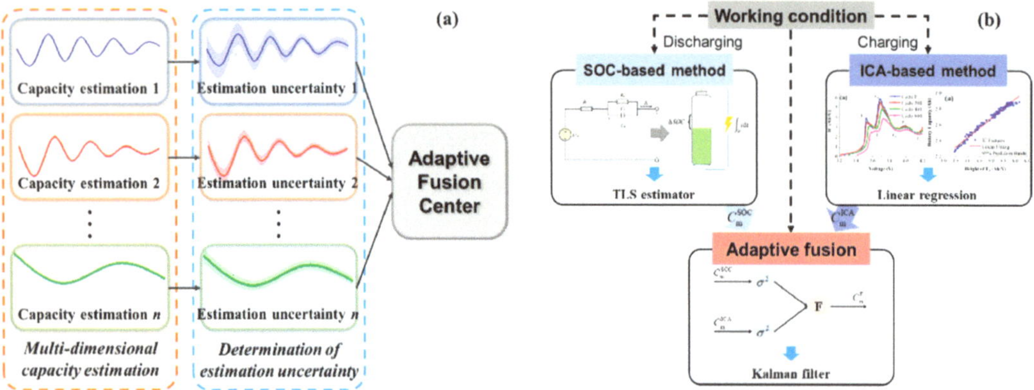

Figure 1. Battery capacity estimation and fusion. (**a**) A general framework of multi-dimensional capacity fusion. (**b**) The specific implementation in this study.

The multi-dimensional capacity estimation originates from the traditional methods listed in the literature review. With the increment of onboard chips' computing ability and improvement in estimation algorithms, multiple estimation methods can run simultaneously and export estimated capacity from different dimensions. Multi-dimensional capacity estimation brings opportunities for improving estimation accuracy because of the increase in useful information. The second step is the determination of estimation uncertainty.

Except for the estimated values, the estimated uncertainties are also essential knowledge in capacity fusion. The estimated uncertainties reflect the estimation error characteristics for estimated values and contribute to the adaptive filter-based fusion. The final is the adaptive fusion center, which receives information from all the capacity estimators and some adaptive fusion algorithms can be utilized for capacity fusion in the fusion center. The framework shown in Figure 1a is a centralized fusion system and the weighted average filter, least squares, and Kalman filter are common technologies in information fusion.

During the fusion process, a state–space function is established to describe the battery capacity characteristics. Battery capacity is a slowly varying state and can be further assumed to be disturbed by Gaussian white noise. The following equation is satisfied:

$$x_{l+1} = \phi x_l + w_l \tag{1}$$

where x_l represents the lth fusion state, ϕ is the system matrix, and w_l is the system noise during the fusion process and is considered a Gaussian white noise with covariance Q_l.

Two structures are usually employed in the fusion center: parallel structure and sequential structure. The parallel structure handles the independent estimated values from the transmitting sub-system simultaneously, while the sequential structure successively deals with the corresponding estimated values. The fusion measurement function can be expressed as:

$$z_l = Hx_l + v_l \tag{2}$$

where z_l estimated capacities and z_l^i comes from the ith sub-system, H is the output matrix, and v_l is corresponding measurement noise with covariance R_l^i.

It has been approved that parallel structure-based and sequential structure-based fusion have the same estimation accuracy [41]. Based on the constructed state–space function, some adaptive filters can obtain the optimal fusion state. Finally, the capacity fusion based on multi-dimensional estimation will be achieved.

According to the above general battery fusion methodology, together with the vision of utilizing more battery information effectively and taking advantage of battery operating conditions, a specific implementation is shown in Figure 1b. Inspired by Ref. [36], the proposed method will utilize more battery information effectively by combining the characteristics of battery operations. Generally, the implementation of capacity fusion is based on the sequential structure. It contains three main parts: (1) the SOC-based estimation during the dynamic discharging, (2) the ICA-based estimation under the stable charging condition, and (3) the final adaptive fusion under complete operating conditions. The first two parts have been preliminarily studied in our previous works [42,43]. This study will further revisit these two methods and obtain the estimation uncertainties, which will be used for adaptive fusion.

2.2. Revisiting Capacity Estimation at Different Working Conditions

2.2.1. SOC-Based Estimation during the Discharging Condition

The definition of battery SOC can be rewritten as

$$\sum_{k=k_1}^{k=k_2} i_k \Delta t = C_m \times (SOC_{k_1} - SOC_{k_2}) \tag{3}$$

where k_1 and k_2 are different sampling points, Δt is the sampling time, i_k is the battery current, and C_m is the battery capacity. Defining SOC change as $X = SOC_{k_1} - SOC_{k_2}$ and charge accumulation as $Y = \sum_{k=k_1}^{k=k_2} i_k \Delta t$, Equation (3) can be transformed to $Y = C_m X$. If N observations are performed for X and Y, the following equation in vector form can be obtained:

$$Y = C_m X \tag{4}$$

where $X = [X_1, X_2, \ldots, X_N]^T$ and $Y = [Y_1, Y_2, \ldots, Y_N]^T$.

The traditional least-squares method can be used for the solution of Equation (4), if the system data are accurate enough. However, the system input X and output Y may be noisy, for example, disturbed by the measurement and estimation noise. The noisy data can be expressed as

$$\begin{cases} X = \tilde{X} + \varepsilon_X \\ Y = \tilde{Y} + \varepsilon_Y \end{cases} \tag{5}$$

where $\tilde{X}$ and $\tilde{Y}$ are real values and ε_X and ε_Y are supposed to be zero-mean Gaussian white noises, respectively. The former noise mainly comes from the SOC estimation uncertainty, while the current measurement noise and numerical computation error will lead to the noise on Y. Under these circumstances, the conventional least-squares method is biased and unsuitable for online capacity estimation.

The total least squares (TLS) is a useful technique for such a system with noisy data [43–45]. In our previous study [43], co-estimation of battery SOC and capacity is realized. The joint estimation consists of an adaptive extended Kalman filter (AEKF) and a TLS estimator and uses a Thevenin model to describe the battery dynamic characteristic. When the AEKF estimator outputs the battery SOC, the TLS algorithm can perform the capacity estimation according to Equation (4). A Rayleigh-Quotient based TLS is employed to realize the recursive running of capacity estimation. The minimum of the following Rayleigh-Quotient function is the solution of the TLS estimator.

$$J(\theta_n, \overline{R}_n) = \frac{\theta_n^T \overline{R}_n \theta_n}{\theta_n^T \theta_n} \tag{6}$$

where $\theta_n = [C_m \ -1]^T$, $\overline{R}_n$ is the autocorrelation matrix of the augmented data matrix $H = [X \ \ Y]^T$. A gradient-based method is employed for recursively updating the C_m [45], where $C_{m,n} = C_{m,n-1} + \alpha_n X_n$ and α_n is to minimize the gradient of Equation (6).

The above procedures realize the capacity estimation during EV discharging and will be considered one of the essential inputs for capacity fusion.

2.2.2. ICA-Based Estimation during the Charging Condition

Employing differential technology, the ICA method can distinguish the voltage characteristics influenced by battery degradation. The ICA technique is usually adopted to investigate the battery aging mechanism [46]. For online capacity estimation, the ICA equation can be rewritten in a discrete form as

$$IC = \frac{dQ}{dV} \approx \frac{\Delta Q}{\Delta V} = \frac{Q_2 - Q_1}{\Delta V} = \frac{i\Delta T}{V_2 - V_1} \tag{7}$$

where ΔV and ΔT are the voltage interval and time interval to calculate the IC value, separately. The problem is that the selection of ΔV and ΔT is not adaptive and the IC curve will be susceptible to measurement noise. In our previous work [42], a filter-based method was proposed to generate the IC curves and this method showed good effectiveness and less computational cost.

Figure 2 shows the collected IC curves at different cycles under different aging conditions. Therefore, "0.5C-0.5C aging condition" means the accelerated cyclic aging test uses a 0.5C current for charging, followed by a 0.5C current for discharging. The IC curves show similar shapes, even at different aging cycles with different aging conditions. During battery charging, six noticeable features (noted as $F_1 \sim F_6$, including peaks and valleys) on the IC curves can be found. It should be noted that this is only one typical method for extracting features from IC curves. Dubarry et al. [47,48] presented a comprehensive summary of features of interest with battery SOH for different kinds of batteries, including the area, position, and intensity of the IC curves. In this study, with the increase in aging cycles, the battery IC curves shift significantly downwards and slightly towards higher voltage, which means that the IC curve features have a close relationship with battery aging and can estimate battery capacity.

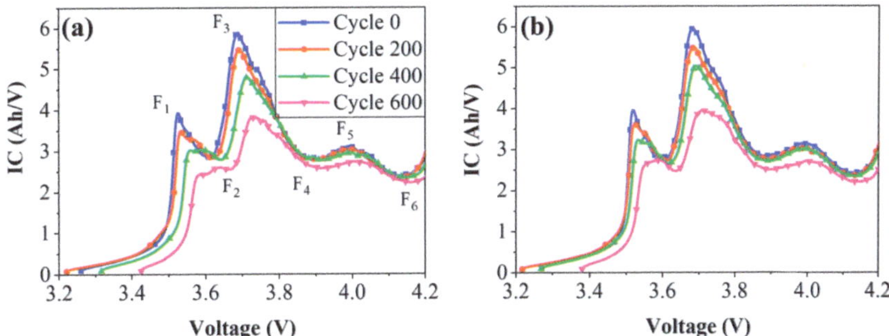

Figure 2. Battery IC curves at different aging cycles. (**a**) 0.5C-0.5C aging condition. (**b**) 0.5C-1C aging condition.

In this study, the height of these features is used for relationship construction because battery aging has less impact on the position of these features. Considering that when the battery is seriously aged (aging cycle $\geq$ 600), F_1 and F_2 blend together, which is difficult to identify and unsuitable for capacity, only F_3 to F_6 are taken into account in the following discussion.

It has been confirmed that the height of the IC features is linearly correlated to battery aging (Figure 3). The relationship between these features and battery capacity is constructed based on linear regression (LR), as shown in the following equation

$$C_m^{ICA} = \alpha_i \times H_{F_i} + \beta_i + \varepsilon_i \tag{8}$$

where H_{F_i} is the height of ith IC features; α_i and β_i are the regression coefficients; ε_i is the error term with zero-mean and variance σ_i^2.

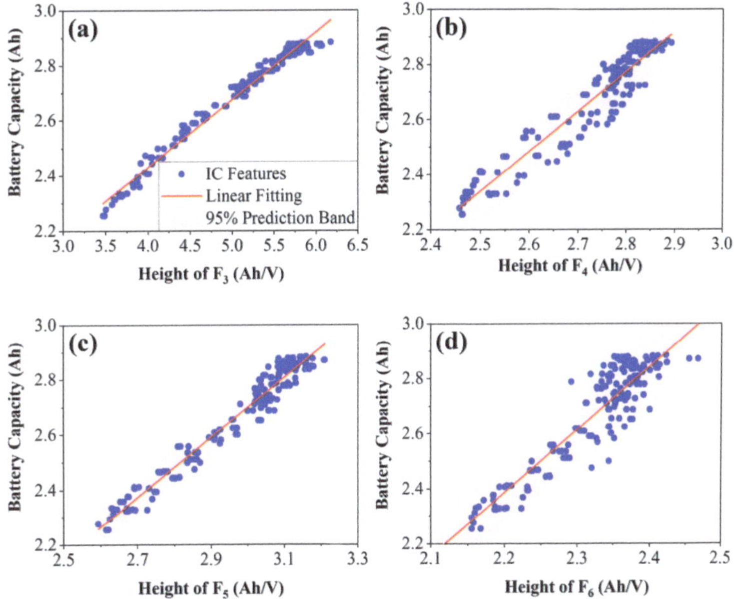

Figure 3. The linear relationship and prediction bands of the ICA method. (**a**–**d**) are the diagrams of F_3~F_6, separately.

2.3. Adaptive Capacity Fusion during Complete Operating Conditions

The above content discusses two capacity estimation methods, which mainly use different information and operate under different EV operating conditions. Inspired by the multi-source information fusion theory, this study investigates capacity fusion under complete operating conditions to acquire a more accurate and reliable capacity estimate.

Equations (1) and (2) show the state–space function of the sequential structure-based fusion. Either the estimated value from the SOC-based method or the ICA-based method can be regarded as an "observation" of the battery's actual capacity. Hence, two estimates (C_m^{SOC} and C_m^{ICA}) can be considered as the actual capacity disturbed by noise. As shown in Figure 1b, during the discharging process, $z_l = C_m^{SOC}$ and, during the charging process, $z_l = C_m^{ICA}$. Correspondingly, R_l is selected from the uncertainties according to the operating conditions.

For SOC-based estimation, C_m^{SOC} is realized through the TLS method. Crassidis et al. [49] derived that the error covariance matrix P of estimated parameters in the TLS problem is the inverse of the Fisher information matrix F ($P = F^{-1}$) and gave the approximate error covariance. Hence, in this study, the uncertainty of the SOC-based estimation is shown as follows

$$R^{SOC} = \left(\frac{1}{\theta_n^T \overline{R}_n \theta_n} \sum_{i=1}^{n} X_i X_i^T \right)^{-1} \tag{9}$$

where R^{SOC} is the error covariance of estimated C_m^{SOC}; other variables have the same meanings as in Equation (6).

For the ICA-based method, C_m^{ICA} is realized through the LR method. Equation (8) can be rewritten in a general form as $y = \alpha x + \beta + \varepsilon$, where x and y are the regression input and output. As shown in [50], the estimation of error variance σ^2 can be calculated as

$$\hat{\sigma}^2 = \frac{1}{N_{LR} - 2} (S_{yy} - \hat{\alpha} S_{xy}) \tag{10}$$

where N_{LR} represents the number of points used for regression, $\hat{\alpha}$ is the estimation of α; $S_{xy} = \sum (x_i - \overline{x})(y_i - \overline{y})$ and $S_{yy} = \sum (y_i - \overline{y})^2$, where S_{xy} is the sum of the deviation product of x and y and S_{yy} is the sum of the deviation square of y.

The estimation error covariance of the LR method can be expressed as

$$R^{ICA} = [1 + \frac{1}{N_{LR}} + \frac{(x' - \overline{x})^2}{S_{xx}}] \hat{\sigma}^2 \tag{11}$$

where $S_{xx} = \sum (x_i - \overline{x})^2$ is the sum of the deviation square of x. In the ICA-based method, $x = H_F$ and x' means the features obtained online.

Figure 3 shows the linear correlation between actual capacity and heights of IC features and the 95% prediction band of the LR method is also shown. Here, the prediction band is calculated based on Equation (11). It can be seen that the prediction band of F_4 and F_6 is broader than that of F_3 and F_5, which means that the error covariance of F_4 and F_6 will be larger than others.

The above obtains the estimation uncertainties of two methods and the complete state–space function is constructed. Due to the simplicity of the state–space function, the Kalman filter is utilized for state fusion, which has superior performance and is easy to implement. For detailed procedures of the Kalman filter, refer to [8,22].

3. Experimental Setups and Design

The battery test bench includes a battery test system to implement charging and discharging procedures, a thermal chamber to provide uniform environmental temperature and humidity, and a host computer for data logging. A kind of commercial 18650-type battery (the cathode is $LiNi_{0.8}Mn_{0.1}Co_{0.1}O_2$ and the anode is graphite) is used in this study. The nominal capacity is 2.9 Ah and the charge/discharge cut-off voltages are 4.2 V and

2.5 V, respectively. During the experiments, the battery temperature is maintained at 25 °C. The battery experiment is designed to simulate the actual operation of the onboard battery as much as possible and validate the fusion method, as shown in Figure 4.

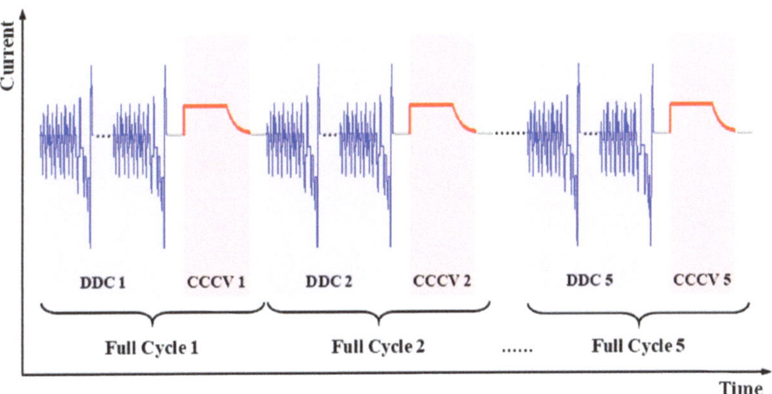

Figure 4. The current profiles of the designed experiment.

The battery will discharge to cut-off voltage with a dynamic discharging condition (DDC), followed by a constant current-constant voltage (CCCV) charging to 100% SOC, where the constant charging current is 1.375 A and the cut-off current for the CCCV protocol is about 60 mA. A DDC and a CCCV form a complete operating cycle and the whole experiment contains five complete cycles (notes as complete cycle 1–5). To verify the adaptability of the proposed method, three cells with different aging statuses are employed in this study, as shown in Table 1. During the experiments, Cell 1 and 2 will discharge with the new European driving cycle (NEDC) and Cell 3 will discharge with the Urban Dynamometer Driving Schedule (UDDS). The detailed current profiles of NEDC and UDDS can be found in Ref. [51]. Table 1 also shows the battery's actual capacity. The battery SOH is determined as the ratio of residual capacity to the nominal capacity [32,52].

Table 1. Cells used in the experiments.

Cell No.	Capacity	SOH	DDC
Cell 1	2.881 Ah	99.34%	NEDC
Cell 2	2.729 Ah	94.11%	NEDC
Cell 3	2.537 Ah	87.48%	UDDS

4. Results and Discussion

To evaluate the performance of the capacity fusion technique effectively, two general error criteria, including the maximum absolute percentage error (MaxAPE) and the root mean square error (RMSE), are adopted. The definition of these two criteria is shown as follows. The percentage error is used when calculating the RMSE.

$$\begin{cases} \text{MaxAPE} = \max\left(\left|\dfrac{\hat{y}_i - y_i}{y_i}\right|\right) \times 100\% \\ \text{RMSE} = \sqrt{\dfrac{1}{m}\sum_{i=1}^{m}\left(\dfrac{\hat{y}_i - y_i}{y_i}\right)^2} \times 100\% \end{cases} \tag{12}$$

where $\hat{y}$ and y are the estimated and actual battery capacity and m represents the number of capacity estimates used for assessment.

4.1. Effectiveness of the Proposed Capacity Fusion

The effectiveness of the capacity fusion technique will be first verified based on the experimental data for Cell 1. Figure 5 shows the estimation and fusion results.

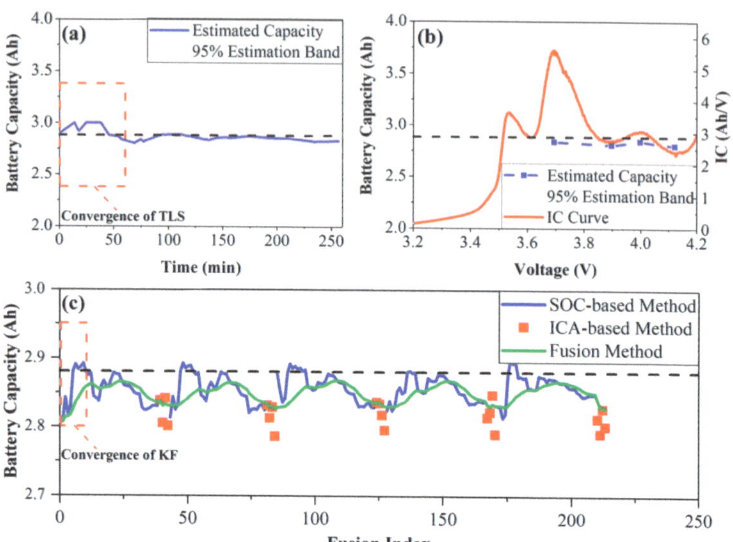

Figure 5. The capacity estimation and fusion results of Cell 1. (**a**) The estimated capacity and uncertainty from SOC-based method during DDC 1. (**b**) The estimated capacity and uncertainty from ICA-based method during CCCV 1. (**c**) The fusion results during 5 complete cycles.

When using the TLS estimator during DDC, the initial capacity is set to 2.9 Ah. Actually, this initial value has little influence on capacity estimation. Moreover, to prevent the influence of outliers on capacity estimation, the maximum value during recursion is set to 3.0 Ah. The estimation result during DDC 1 is shown in Figure 5a. The first several values have large deviations from the actual capacity (the dashed line represents the actual battery capacity). After convergence (about 2500 s for this condition), the estimated capacity becomes stable and close to the actual capacity. In addition, the 95% estimation band becomes narrower after convergence, which means the estimation uncertainties gradually decrease. Specifically, the convergence time of TLS is set to 1h and the estimated values before convergence will not be considered during fusion. After convergence, the MaxAPE and RMSE during DDC 1 are 2.558% and 1.195%, separately, indicating the effectiveness of the SOC-based method.

The estimation result of the ICA method during CCCV 1 is shown in Figure 5b. The capacity estimated through the last feature (F_6) has the biggest uncertainty. The MaxAPE and RMSE are 2.747% and 2.135%, which shows insufficient accuracy compared to the SOC-based method. Another shortcoming is that too few capacity values are updated based on IC curve features.

For the capacity fusion, the initial parameters of the Kalman filter during fusion are set to: $P_0 = 5 \times 10^{-3}$, $Q_l = 10^{-4}$, and R_l is calculated according to Equations (9) and (11). The capacity fusion result during the five complete cycles is shown in Figure 5c. It can be found that most of the estimated values of the SOC-based and ICA-based methods are smaller than the actual capacity. Through the Kalman filter, the trajectory of fusion capacity has fewer fluctuations. In most cases, the fusion value follows the estimates from the SOC-based method. However, when the uncertainty of the SOC-based method is considerable or the observations come from the ICA-based method, the trajectory of fusion capacity will change. Moreover, the ICA-based method has a larger estimation deviation and the fusion method can endure these errors. The first ten values during fusion are considered the Kalman filter's convergence process and will be ignored when assessing the performance. During five complete cycles, the MaxAPEs of SOC-based and ICA-based methods are 2.328% and 3.278%, while the MaxAPE of the fusion method is 1.862%, which is reduced by 0.466% and 1.416%, separately, compared with the above single method. It is noted that the ICA-based method has few estimates, which influences the performance of the fusion method.

4.2. Adaptability to Different Aging Statuses

To validate the adaptability of the fusion method, another two cells with larger capacity attenuation are employed to perform the capacity fusion, as shown in Figure 6. Figure 6a shows the experimental results of Cell 2. The fusion capacity is closest to the battery's actual capacity. At the beginning of each DDC, the TLS estimator exports some outliers, which results in a significant estimation error. On the contrary, the fusion method utilizes the information from the ICA-based method and the uncertainty from the SOC-based method and obtains a more accurate and stable estimation result. In addition, some inaccurate estimates from the ICA-based method were also corrected. The MaxAPE is reduced by 1.478% and 0.985%, separately, compared with the SOC-based and ICA-based methods, while the RMSE is reduced by 0.307% and 1.100%.

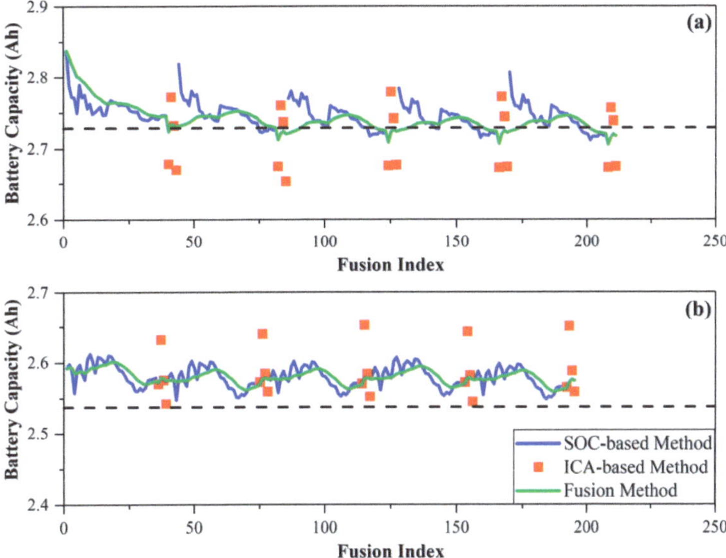

Figure 6. The estimation and fusion results of different cells. (**a**,**b**) are the results of Cell 2 and Cell 3.

The estimation results of Cell 3 are shown in Figure 6b. Unlike previous results for the other two cells, most of the estimated values for the SOC-based and ICA-based methods are larger than the actual capacity. Under such circumstances, the fusion method filters some outliers generated from the ICA-based method and obtains a smoother capacity estimation curve. During the five complete operating cycles, the MaxAPEs of the SOC-based, ICA-based, and fusion methods are 2.876%, 4.542%, and 2.547%. Compared with the single ICA-based method, the accuracy improvement in the fusion method is considerable.

Table 2 shows the estimation performance of different methods for cells with different aging statuses. For these three cells, the maximum MaxAPE and RMSE of the SOC-based method are 3.298% and 1.780% and those of the ICA-based method are 4.542% and 2.401%, while for the proposed fusion method, the maximum MaxAPE and RMSE are 2.547% and 1.703%, separately. The proposed fusion methods show the highest estimation accuracy from the capacity estimation results, indicating the adaptability to different aging statuses. It has been mentioned in Section 2.1 that the average filter is also a common technology in information fusion. Therefore, we compare the proposed fusion using the Kalman filter and traditional fusion using the moving average filter, as shown in Table 2. The window size for the moving average filter is 5. From Table 2, it can be found that moving average fusion also improves the estimation performance compared with SOC-based and ICA-based methods. Nevertheless, it only uses the estimation value while ignoring the estimation uncertainty. Comparatively, the proposed fusion method using the Kalman filter still has a minor estimation error, indicating the effectiveness of the proposed method.

Table 2. The performance of different methods for different cells.

Method	Criterion	Cell 1	Cell 2	Cell 3
SOC-based	MaxAPE (%)	2.328	3.298	2.876
method	RMSE (%)	1.061	0.882	1.780
ICA-based method	MaxAPE (%)	3.278	2.805	4.542
	RMSE (%)	2.300	1.675	2.401
Fusion method:	MaxAPE (%)	1.862	1.820	2.547
Kalman filter	RMSE (%)	1.144	0.575	1.703
Fusion method:	MaxAPE (%)	2.239	1.823	2.698
Moving average	RMSE (%)	1.141	0.751	1.788

4.3. Application with Inaccurate Battery Current Information

As discussed above, inaccurate current information will result in errors in capacity estimation and different estimation methods may have different error scenarios. For a battery with an early ISC fault, the measured current is larger than the battery's effective current during charging. In contrast, during discharging, the measured current is less than the battery's effective current. In this situation, for the SOC-based method, although the SOC estimator may obtain relatively accurate SOC based on the feedback of measurement voltage, the calculated charge accumulation is less than the actual value, which will further result in a smaller capacity estimate. For the ICA-based method during charging, the calculated IC value according to the measured current will be larger than its actual value; thus, the capacity estimation based on IC features will be larger.

This section simulates two fault scenarios; one is with a 10 mA current deviation (noted as Fault 1) and the other is with a 20 mA current deviation (noted as Fault 2). The experimental data for Cell 1 are used for fault simulation. Figure 7 shows the capacity estimation and fusion results.

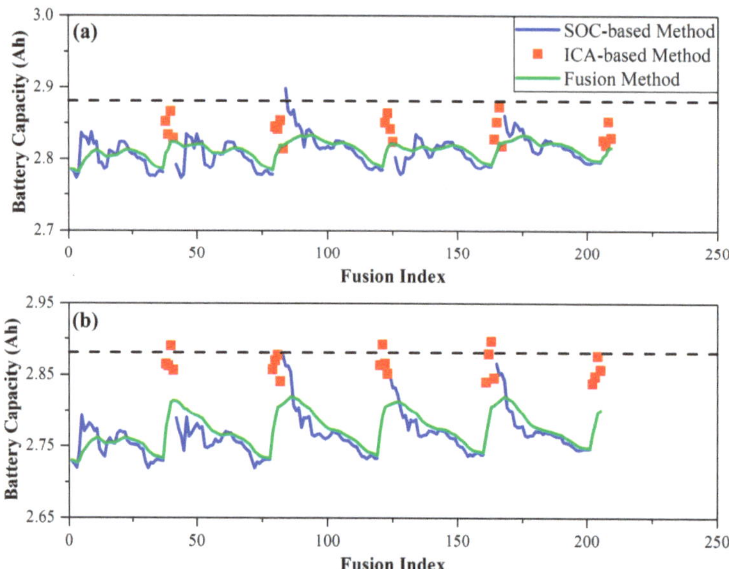

Figure 7. The estimation and fusion results of different fault scenarios. (**a**,**b**) are the results of Fault 1 and Fault 2.

With the influence of the current deviation, the SOC-based method's estimates have a downward trend, while the ICA-based method's estimates have an upward trend, compared with the results shown in Figure 5c. It should be noted that if the current deviation becomes large, the estimates from the ICA-based method will continue to move upward and the estimation error will become larger.

Compared with the single method, the estimation accuracy of the fusion method is not much improved. The main reason is that the ICA method outputs too few capacity estimates, which cannot provide continuous correction to estimates coming from the SOC-based method. Compared with the SOC-based method, the MaxAPEs of the fusion method under Fault 1 and Fault 2 are reduced by 0.392% and 0.474%, respectively, while the RMSEs under two fault conditions are reduced by 0.181% and 0.356%.

It is noticed that the difference between the three methods' estimation averages under the above scenarios is different. The difference between the three methods' estimation averages is slight for the normal conditions and conditions with a small current deviation. In contrast, for conditions with a large current deviation, the estimation average of the SOC-based and fusion method is close, while there is a significant deviation from that of the ICA-based method. To quantitatively describe the deviation between capacity estimates caused by inaccurate current information, the following criterion is proposed:

$$D_C = \max\left(\left|\frac{\overline{C_m^{SOC}} - \overline{C_m^F}}{\overline{C_m^F}}\right|, \left|\frac{\overline{C_m^{ICA}} - \overline{C_m^F}}{\overline{C_m^F}}\right|\right) \times 100\% \tag{13}$$

where D_C represents the capacity estimation deviation; $\overline{C_m^F}$, $\overline{C_m^{SOC}}$ and $\overline{C_m^{ICA}}$ mean the average during capacity fusion, SOC-based estimation, and ICA-based estimation.

Figure 8 shows the D_C for different cells under different current deviations. Note that with an increase in current deviation, the capacity estimation deviations both show a rising trend. Moreover, the more severe the battery aging, the higher the rising rate. It can be found that a threshold can be set to diagnose the fault condition of current inaccuracy. Here, we set this threshold to $T_C = 2\%$. Although the normal conditions and the slight fault conditions (10 mA deviation) of Cell 1 and Cell 2 cannot be distinguished from each other, the D_C of Cell 3 at the slight fault condition is larger than the set threshold. In addition, all the cells' D_C values are large than the set threshold. It is noted that the difference between the estimation averages will become larger with an increase in current deviation. Hence, it can be confirmed that a fault of current deviation can be identified when $D_C > T_C$ happens during capacity estimation and fusion.

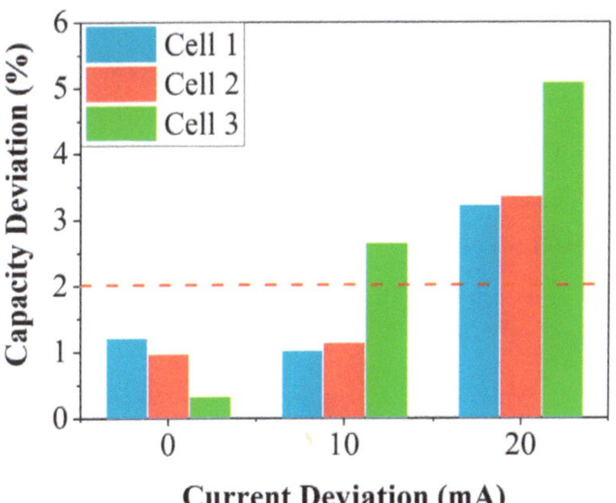

Figure 8. The capacity deviation for different cells.

5. Conclusion and Future Research Trends

Accurate and stable capacity acquisition has vital importance to the battery management technology in EVs. This study proposes an adaptive estimation–fusion method for battery capacity, utilizing more battery information and operating under complete battery operating conditions. The main conclusions of this study are drawn as follows:

(1) A general capacity fusion framework is proposed and then the SOC-based method and ICA-based method are revisited and employed during different operating conditions; further, the error covariance of different estimations is analyzed and derived.

(2) The adaptive battery fusion method is realized through the Kalman filter, which intelligently combines two estimates and takes advantage of estimation uncertainties.

(3) The fusion method outputs more accurate and stable capacity estimates. Owing to the utilization of more battery information, the maximum MaxAPE and RMSE are 2.547% and 1.703%, separately, under different aging statuses, which are smaller than traditional methods. Moreover, a judgment criterion based on capacity estimation and fusion for a sizeable current deviation fault is proposed.

Based on this research, there are several trends to be explored in future studies:

(1) More capacity estimates involved in fusion and more adaptive noise-matching methods in determining the process noise covariance will be helpful.

(2) Currently, the estimation during charging is the ICA-based method, which only updates four capacity estimates. A capacity estimation method with timely updating will continuously correct estimates from another method.

(3) The adaptability to different temperatures will be verified in further studies. Moreover, the estimation complexity caused by cell inconsistency shall be considered.

This study presents a detailed investigation of the application of information fusion technology in battery capacity estimation. For complex electrochemical systems, such as batteries, the information fusion technology will have more application scenarios, for example, battery multi-state estimation and remaining-life prediction. With the development of physical-based modeling and intelligent data-driven methods, as well as the availability of advanced-sensing techniques, information fusion will have diverse implementations and combinations to enhance the high-fidelity information acquisition from LIBs.

Author Contributions: Conceptualization, B.J. and H.D.; methodology, B.J.; software, B.J.; validation, B.J. and X.W.; formal analysis, B.J. and X.W.; investigation, B.J. and H.D.; data curation, B.J.; writing—original draft preparation, B.J. and X.W.; writing—review and editing, H.D.; visualization, B.J. and X.W.; supervision, H.D.; project administration, H.D.; funding acquisition, H.D. All authors have read and agreed to the published version of the manuscript.

Funding: This research was supported by the National Natural Science Foundation of China (NSFC, Grant No. U20A20310), Shanghai Sailing Program (Grant No. 22YF1450400), and China Postdoctoral Science Foundation (Grant No. 2022M712406).

Institutional Review Board Statement: Not applicable.

Informed Consent Statement: Not applicable.

Data Availability Statement: Not applicable.

Conflicts of Interest: The authors declare no conflict of interest.

References

1. Dai, H.; Jiang, B.; Hu, X.; Lin, X.; Wei, X.; Pecht, M. Advanced battery management strategies for a sustainable energy future: Multilayer design concepts and research trends. *Renew. Sustain. Energy Rev.* **2021**, *138*, 110480. [CrossRef]
2. Wang, X.; Wei, X.; Zhu, J.; Dai, H.; Zheng, Y.; Xu, X.; Chen, Q. A review of modeling, acquisition, and application of lithium-ion battery impedance for onboard battery management. *eTransportation* **2021**, *7*, 100093. [CrossRef]
3. Fasahat, M.; Manthouri, M. State of charge estimation of lithium-ion batteries using hybrid autoencoder and Long Short Term Memory neural networks. *J. Power Sources* **2020**, *469*, 228375. [CrossRef]
4. Moustafa, M.G.; Sanad, M.M.S. Green fabrication of $ZnAl_2O_4$-coated $LiFePO_4$ nanoparticles for enhanced electrochemical performance in Li-ion batteries. *J. Alloys Compd.* **2022**, *903*, 163910. [CrossRef]

5. Sanad, M.M.S.; Toghan, A. Unveiling the role of trivalent cation incorporation in Li-rich Mn-based layered cathode materials for low-cost lithium-ion batteries. *Appl. Phys. A* **2021**, *127*, 1–15. [CrossRef]
6. Hu, X.; Che, Y.; Lin, X.; Deng, Z. Health Prognosis for Electric Vehicle Battery Packs: A Data-Driven Approach. *IEEE/ASME Trans. Mechatron.* **2020**, *25*, 2622–2632. [CrossRef]
7. Wang, X.; Wei, X.; Dai, H. Estimation of state of health of lithium-ion batteries based on charge transfer resistance considering different temperature and state of charge. *J. Energy Storage* **2019**, *21*, 618–631. [CrossRef]
8. Li, X.; Wang, Z.; Zhang, L. Co-estimation of capacity and state-of-charge for lithium-ion batteries in electric vehicles. *Energy* **2019**, *174*, 33–44. [CrossRef]
9. Jiang, B.; Dai, H.; Wei, X.; Jiang, Z. Multi-kernel Relevance Vector Machine with Parameter Optimization for Cycling Aging Prediction of Lithium-ion Batteries. *IEEE J. Emerg. Sel. Top. Power Electron.* **2021**, 1–12. [CrossRef]
10. Goh, T.; Park, M.; Seo, M.; Kim, J.G.; Kim, S.W. Successive-approximation algorithm for estimating capacity of Li-ion batteries. *Energy* **2018**, *159*, 61–73. [CrossRef]
11. Farmann, A.; Waag, W.; Marongiu, A.; Sauer, D.U. Critical review of on-board capacity estimation techniques for lithium-ion batteries in electric and hybrid electric vehicles. *J. Power Sources* **2015**, *281*, 114–130. [CrossRef]
12. Berecibar, M.; Gandiaga, I.; Villarreal, I.; Omar, N.; Van Mierlo, J.; Van den Bossche, P. Critical review of state of health estimation methods of Li-ion batteries for real applications. *Renew. Sustain. Energy Rev.* **2016**, *56*, 572–587. [CrossRef]
13. Xiong, R.; Li, L.; Tian, J. Towards a smarter battery management system: A critical review on battery state of health monitoring methods. *J. Power Sources* **2018**, *405*, 18–29. [CrossRef]
14. Schwunk, S.; Armbruster, N.; Straub, S.; Kehl, J.; Vetter, M. Particle filter for state of charge and state of health estimation for lithium–iron phosphate batteries. *J. Power Sources* **2013**, *239*, 705–710. [CrossRef]
15. Einhorn, M.; Conte, F.V.; Kral, C.; Fleig, J. A method for online capacity estimation of lithium ion battery cells using the state of charge and the transferred charge. *IEEE Trans. Ind. Appl.* **2012**, *48*, 736–741. [CrossRef]
16. Zhou, Z.; Cui, Y.; Kong, X.; Li, J.; Zheng, Y. A fast capacity estimation method based on open circuit voltage estimation for LiNixCoyMn1-x-y battery assessing in electric vehicles. *J. Energy Storage* **2020**, *32*, 101830. [CrossRef]
17. Zou, Y.; Hu, X.; Ma, H.; Li, S.E. Combined state of charge and state of health estimation over lithium-ion battery cell cycle lifespan for electric vehicles. *J. Power Sources* **2015**, *273*, 793–803. [CrossRef]
18. Wei, Z.; Leng, F.; He, Z.; Zhang, W.; Li, K. Online state of charge and state of health estimation for a lithium-ion battery based on a data–model fusion method. *Energies* **2018**, *11*, 1810. [CrossRef]
19. Xiong, R.; Sun, F.; Chen, Z.; He, H. A data-driven multi-scale extended Kalman filtering based parameter and state estimation approach of lithium-ion polymer battery in electric vehicles. *Appl. Energy* **2014**, *113*, 463–476. [CrossRef]
20. Kim, I.L.S. A technique for estimating the state of health of lithium batteries through a dual-sliding-mode observer. *IEEE Trans. Power Electron.* **2010**, *25*, 1013–1022. [CrossRef]
21. Hua, Y.; Cordoba-Arenas, A.; Warner, N.; Rizzoni, G. A multi time-scale state-of-charge and state-of-health estimation framework using nonlinear predictive filter for lithium-ion battery pack with passive balance control. *J. Power Sources* **2015**, *280*, 293–312. [CrossRef]
22. Wei, Z.; Zhao, J.; Ji, D.; Tseng, K.J. A multi-timescale estimator for battery state of charge and capacity dual estimation based on an online identified model. *Appl. Energy* **2017**, *204*, 1264–1274. [CrossRef]
23. Jiang, B.; Dai, H.; Wei, X. A Cell-to-Pack State Estimation Extension Method Based on a Multilayer Difference Model for Series-Connected Battery Packs. *IEEE Trans. Transp. Electrif.* **2022**, *8*, 2037–2049. [CrossRef]
24. Li, Y.; Abdel-Monem, M.; Gopalakrishnan, R.; Berecibar, M.; Nanini-Maury, E.; Omar, N.; van den Bossche, P.; Van Mierlo, J. A quick on-line state of health estimation method for Li-ion battery with incremental capacity curves processed by Gaussian filter. *J. Power Sources* **2018**, *373*, 40–53. [CrossRef]
25. He, J.; Bian, X.; Liu, L.; Wei, Z.; Yan, F. Comparative study of curve determination methods for incremental capacity analysis and state of health estimation of lithium-ion battery. *J. Energy Storage* **2020**, *29*, 101400. [CrossRef]
26. Bian, X.; Liu, L.; Yan, J. A model for state-of-health estimation of lithium ion batteries based on charging profiles. *Energy* **2019**, *177*, 57–65. [CrossRef]
27. Zhang, S.; Zhai, B.; Guo, X.; Wang, K.; Peng, N.; Zhang, X. Synchronous estimation of state of health and remaining useful lifetime for lithium-ion battery using the incremental capacity and artificial neural networks. *J. Energy Storage* **2019**, *26*, 100951. [CrossRef]
28. Shen, S.; Sadoughi, M.; Chen, X.; Hong, M.; Hu, C. A deep learning method for online capacity estimation of lithium-ion batteries. *J. Energy Storage* **2019**, *25*, 100817. [CrossRef]
29. Deng, Z.; Hu, X.; Lin, X.; Xu, L.; Che, Y.; Hu, L. General Discharge Voltage Information Enabled Health Evaluation for Lithium-Ion Batteries. *IEEE/ASME Trans. Mechatron.* **2020**, *26*, 1295–1306. [CrossRef]
30. Hu, C.; Jain, G.; Zhang, P.; Schmidt, C.; Gomadam, P.; Gorka, T. Data-driven method based on particle swarm optimization and k-nearest neighbor regression for estimating capacity of lithium-ion battery. *Appl. Energy* **2014**, *129*, 49–55. [CrossRef]
31. Deng, Z.; Yang, L.; Cai, Y.; Deng, H.; Sun, L. Online available capacity prediction and state of charge estimation based on advanced data-driven algorithms for lithium iron phosphate battery. *Energy* **2016**, *112*, 469–480. [CrossRef]
32. Jiang, B.; Zhu, J.; Wang, X.; Wei, X.; Shang, W.; Dai, H. A comparative study of different features extracted from electrochemical impedance spectroscopy in state of health estimation for lithium-ion batteries. *Appl. Energy* **2022**, *322*, 119502. [CrossRef]

33. Tan, Y.; Zhao, G. Transfer learning with long short-term memory network for state-of-health prediction of lithium-ion batteries. *IEEE Trans. Ind. Electron.* **2020**, *67*, 8723–8731. [CrossRef]
34. Zheng, Y.; Qin, C.; Lai, X.; Han, X.; Xie, Y. A novel capacity estimation method for lithium-ion batteries using fusion estimation of charging curve sections and discrete Arrhenius aging model. *Appl. Energy* **2019**, *251*, 113327. [CrossRef]
35. Xiong, R.; Wang, J.; Shen, W.; Tian, J.; Mu, H. Co-Estimation of State of Charge and Capacity for Lithium-ion Batteries with Multi-Stage Model Fusion Method. *Engineering* **2021**, *7*, 1469–1482. [CrossRef]
36. Balasingam, B.; Avvari, G.V.; Pattipati, B.; Pattipati, K.R.; Bar-Shalom, Y. A robust approach to battery fuel gauging, part II: Real time capacity estimation. *J. Power Sources* **2014**, *269*, 949–961. [CrossRef]
37. Zhang, Y.; Jiang, C.; Yue, B.; Wan, J.; Guizani, M. Information fusion for edge intelligence: A survey. *Inf. Fusion* **2022**, *81*, 171–186. [CrossRef]
38. Qiao, D.; Wei, X.; Fan, W.; Jiang, B.; Lai, X.; Zheng, Y.; Tang, X.; Dai, H. Toward safe carbon–neutral transportation: Battery internal short circuit diagnosis based on cloud data for electric vehicles. *Appl. Energy* **2022**, *317*, 119168. [CrossRef]
39. Qiao, D.; Wang, X.; Lai, X.; Zheng, Y.; Wei, X.; Dai, H. Online quantitative diagnosis of internal short circuit for lithium-ion batteries using incremental capacity method. *Energy* **2021**, *243*, 123082. [CrossRef]
40. Jiang, B.; Dai, H.; Jiang, W.; Pei, F. A novel framework of multi-dimension capacity estimation and fusion for lithium-ion battery. In Proceedings of the 2020 IEEE Vehicle Power and Propulsion Conference (VPPC), Gijón, Spain, 26–29 October 2020.
41. Han, C.; Zhu, H.; Duan, Z. Estimation fusion. In *Multi-Source Information Fusion*, 2nd ed.; Tsinghua University Press: Beijing, China, 2010; pp. 250–319.
42. Jiang, B.; Dai, H.; Wei, X. Incremental capacity analysis based adaptive capacity estimation for lithium-ion battery considering charging condition. *Appl. Energy* **2020**, *269*, 115074. [CrossRef]
43. Jiang, B.; Dai, H.; Wei, X.; Xu, T. Joint estimation of lithium-ion battery state of charge and capacity within an adaptive variable multi-timescale framework considering current measurement offset. *Appl. Energy* **2019**, *253*, 13619. [CrossRef]
44. Wei, Z.; Zou, C.; Leng, F.; Soong, B.H.; Tseng, K.-J. Online model identification and state-of-charge estimate for lithium-ion battery with a recursive total least squares-based observer. *IEEE Trans. Ind. Electron.* **2018**, *65*, 1336–1346. [CrossRef]
45. Kim, T.; Wang, Y.; Sahinoglu, Z.; Wada, T.; Hara, S.; Qiao, W. A Rayleigh Quotient-based recursive total-least-squares online maximum capacity estimation for lithium-ion batteries. *IEEE Trans. Energy Convers.* **2015**, *30*, 842–851. [CrossRef]
46. Pastor-Fernández, C.; Uddin, K.; Chouchelamane, G.H.; Widanage, W.D.; Marco, J. A comparison between electrochemical impedance spectroscopy and incremental capacity-differential voltage as li-ion diagnostic techniques to identify and quantify the effects of degradation modes within battery management systems. *J. Power Sources* **2017**, *360*, 301–318. [CrossRef]
47. Dubarry, M.; Beck, D. Analysis of Synthetic Voltage vs. Capacity Datasets for Big Data Li-ion Diagnosis and Prognosis. *Energies* **2021**, *14*, 2371. [CrossRef]
48. Dubarry, M.; Berecibar, M.; Devie, A.; Anseán, D.; Omar, N.; Villarreal, I. State of health battery estimator enabling degradation diagnosis: Model and algorithm description. *J. Power Sources* **2017**, *360*, 59–69. [CrossRef]
49. Crassidis, J.L.; Cheng, Y. Error-covariance analysis of the total least-squares problem. *J. Guid. Control. Dyn.* **2014**, *37*, 1053–1063. [CrossRef]
50. Ramachandran, K.M.; Tsokos, C.P. Linear regression models. In *Mathematical Statistics with Applications in R*, 3rd ed.; Academic Press: Cambridge, MA, USA, 2021; pp. 301–341.
51. Fang, Q.; Wei, X.; Dai, H. A Remaining Discharge Energy Prediction Method for Lithium-Ion Battery Pack Considering SOC and Parameter Inconsistency. *Energies* **2019**, *12*, 987. [CrossRef]
52. You, H.; Zhu, J.; Wang, X.; Jiang, B.; Sun, H.; Liu, X.; Wei, X.; Han, G.; Ding, S.; Yu, H.; et al. Nonlinear health evaluation for lithium-ion battery within full-lifespan. *J. Energy Chem.* **2022**, *72*, 333–341. [CrossRef]

Review

A Review of the Application of Carbon Materials for Lithium Metal Batteries

Zeyu Wu, Kening Sun and Zhenhua Wang *

Beijing Key Laboratory of Chemical Power Source and Green Catalysis, School of Chemistry and Chemical Engineering, Beijing Institute of Technology, Beijing 100081, China
* Correspondence: wangzh@bit.edu.cn

Abstract: Lithium secondary batteries have been the most successful energy storage devices for nearly 30 years. Until now, graphite was the most mainstream anode material for lithium secondary batteries. However, the lithium storage mechanism of the graphite anode limits the further improvement of the specific capacity. The lithium metal anode, with the lowest electrochemical potential and extremely high specific capacity, is considered to be the optimal anode material for next-generation lithium batteries. However, the lifetime degradation and safety problems caused by dendrite growth have seriously hindered its commercialization. Carbon materials have good electrical conductivity and modifiability, and various carbon materials were designed and prepared for use in lithium metal batteries. Here, we will start by analyzing the problems and challenges faced by lithium metal. Then, the application progress and achievements of various carbon materials in lithium metal batteries are summarized. Finally, the research suggestions are given, and the application feasibility of carbon materials in metal lithium batteries is discussed.

Keywords: lithium metal batteries; carbon materials; composite anodes; current collectors

Citation: Wu, Z.; Sun, K.; Wang, Z. A Review of the Application of Carbon Materials for Lithium Metal Batteries. *Batteries* **2022**, *8*, 246. https://doi.org/10.3390/batteries8110246

Academic Editor: Junsheng Zheng

Received: 15 October 2022
Accepted: 15 November 2022
Published: 18 November 2022

Publisher's Note: MDPI stays neutral with regard to jurisdictional claims in published maps and institutional affiliations.

1. Introduction

The use of fossil fuels has made invaluable contributions to the development of human society. However, problems such as resource depletion and environmental pollution force human beings to develop new energy systems and adjust the energy structure [1]. New systems such as wind, hydro, solar, and fuel cells are expected to provide clean and sustainable energy for human society. As an effective way to store and transfer energy, electrochemical energy storage has gradually become an indispensable part of the transformation of energy structure [2,3]. Since the advent of lithium-ion batteries in the 19th century, they have become an irreplaceable energy storage device in various fields [4,5]. With the unremitting efforts of researchers, the safety performance, specific energy, and cycle life of lithium-ion batteries were greatly improved, but they are still far from meeting the development needs of electric vehicles and portable electronic devices [6,7]. The human pursuit of high specific energy, long life, and fast charging batteries has never stopped. In the current commercial lithium secondary batteries, high nickel cathode materials can provide a specific capacity of 220 mA h g^{-1} and an energy density of 800 W h kg^{-1} [8]. However, the intercalation cathode material is limited by the crystal volume and element quality. Its specific energy is difficult to further improve. In this case, research to improve the energy density of batteries has mainly focused on the anode side. Widely used graphite materials have low specific capacity and energy density (specific capacity 372 mA h g^{-1}, energy density 300–400 W h kg^{-1}). Lithium metal anodes with high specific capacity (3860 mA h g^{-1} or 2061 mA h cm^{-3}) are considered to be the best choice for next-generation lithium battery anodes [9–11]. In particular, Li-S and Li-O$_2$ batteries assembled with high specific energy cathodes such as S or O$_2$ can provide specific energies as high as 650 W h kg^{-1} and 950 W h kg^{-1}, respectively.

Lithium metal anodes were extensively studied long before graphite anodes, but were not effectively used until now. There are three main problems faced by Li metal anodes: (1) The depositing/stripping of Li metal during cycling is accompanied by a huge change in its volume, which leads to changes in the internal pressure and structure of the battery. (2) The side reactions that occur continuously with the battery cycle consume the effective components in the electrode and the electrolyte, resulting in a decrease in the coulombic efficiency of the battery or even failure. (3) Lithium metal tends to grow dendrites in liquid electrolytes. The formation of lithium dendrites has serious consequences: the larger specific surface area leads to more reactions with the electrolyte; the fracture of dendrites during cycling will cause part of the lithium metal to detach from the collector and become "dead lithium", resulting in a decrease in electrode capacity.; dendrites may also penetrate the separator and contact the positive electrode to short-circuit the battery and cause thermal runaway. The above problems have become the bottlenecks that limit their commercialization [12,13]. Paul Albertus et al. [14] believe that the application of a lithium battery needs to maintain an average Coulomb efficiency of 99.98% in 1000 cycles to reach the commercial level, and the current lithium metal anode is still far from this goal.

After more than 40 years of continuous research and deepening understanding of lithium metal electrodes, various strategies to control dendrite growth and improve the efficiency of lithium metal electrodes were proposed. These strategies have made various modifications and designs mainly from the perspectives of electrolytes, separators, SEIs, current collectors, composite electrodes, and solid electrolytes.

Electrolyte regulation For the electrolyte development of the lithium metal battery, the composition and structure of SEI are mainly regulated by changing the lithium salts and additives in the electrolyte [15]. In recent years, high-concentration and localized high-concentration electrolytes based on a solvated structure design have obtained lithium metal batteries with high Coulombic efficiency by promoting the decomposition of anions and reducing the reduction of free solvents [16–20]. However, expensive lithium salts and diluents hinder its further application. In addition to forming SEI, the electrolyte can also inhibit the growth of lithium dendrites by forming an electrostatic screen shielding layer on the surface of the lithium metal electrode. Cs^+, Rb^+, and other cations can suppress the formation of lithium dendrites by adsorbing on the surface of lithium metal to form an electrostatic shielding layer, thereby enabling the uniform deposition of lithium metal [21,22].

Separator modification: In the design of the lithium metal battery separator, the main purpose is to improve its Young's modulus to suppress the growth of lithium dendrites and to design the separator with uniform pores to obtain uniform lithium ion flux [23–26]. In addition, immobilizing inorganic materials or organic groups on the separator can obtain separators with specific functions, which can improve the safety and cycling stability of lithium metal batteries [27,28]. Utilizing the pore structure of the separator can also assist in adjusting the solvation structure of the electrolyte to obtain a functionalized separator with a wide electrochemical stability window and a stable SEI structure [29].

Artificial SEI The composition and structure of the SEI film on the surface of Li metal electrodes can be directly and effectively adjusted in situ by changing the composition of the electrolyte [30]. Artificial SEI layers can also be constructed on Li metal surfaces by ex situ methods [31,32]. Various organic polymers, inorganics, and organic–inorganic composite SEIs can provide high mechanical properties to suppress dendrite growth, high lithium ion conductivity to reduce polarization, and thus, improve the stability of lithium metal electrodes during battery cycling [33–38].

Current collector design Electrodeposition of lithium metal is closely related to its current density, and a large specific surface area can effectively reduce the local current density to inhibit dendrite growth [39]. Adding lithiophilic or nucleation-inducing substances to the current collector can also reduce the nucleation overpotential or uniform Li metal deposition. Metals and carbon materials such as Cu and Ni with excellent electronic conductivity are the two most common current collector materials [40–42]. In addition, carbon materials have the characteristics of light weight, high electronic conductivity, and easy modification and are one of the excellent current collectors for lithium metal anodes.

Solid-state electrolytes: Solid-state electrolytes mainly include organic polymer electrolytes, inorganic ceramic electrolytes, and organic–inorganic composite electrolytes [43–48]. Solid-state electrolytes are considered to be the technical direction with the most potential to enable lithium metal batteries to be applied. Solid-state electrolytes are believed to provide high modulus and migration numbers close to unity without causing lithium dendrite growth. However, in the research of organic and inorganic solid electrolytes, it is found that short circuit or thermal runaway caused by dendrite growth is still inevitable [49–51].

Various new materials and technologies were tried to solve the problems faced by lithium metal batteries, but the commercialization of lithium metal electrodes is still full of challenges. Compared to other materials, carbon materials have excellent electrical and thermal conductivity. Their structural plasticity and easy modification allow them to be designed in a variety of shapes and have specific functions. With the rise in materials science and nanotechnology, carbon materials of different dimensions (0–3D) and different scales (nano-millimeter) were developed and played a huge role in lithium metal batteries (Figure 1). In lithium metal batteries, carbon materials are mainly used as current collectors to disperse current and heat. In addition, carbon materials can also be used as additives or artificial SEI to participate in lithium metal electrodes to inhibit dendrite growth and improve battery life.

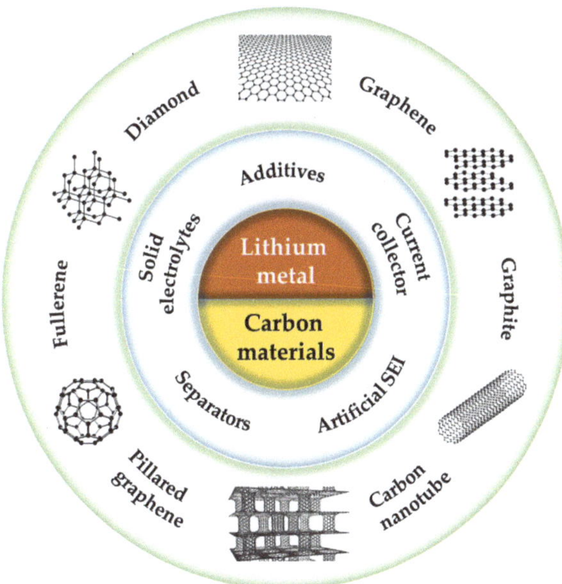

Figure 1. Various strategies applied by carbon materials in lithium metal batteries.

This review will start from the inherent scientific issues of metallic lithium anodes and introduce the problems and challenges of metallic lithium electrodes. Then, various applications of carbon and its derivatives/composites in lithium metal batteries are summarized. Finally, the research of carbon materials in lithium metal batteries is discussed and suggestions for its development direction are provided.

2. Issues and Challenges of Lithium Metal Anodes

2.1. Nucleation of Lithium Metal

Lithium metal anodes undergo reversible plating and stripping during battery cycling. In each charging process, lithium atoms go through a process from nucleation to growth. The nucleation state and growth pattern of Li metal have a huge effect on its deposition/dendritic morphology.

The nucleation process of Li metal is controlled by the properties of the deposition substrate, multiphysics, and SEI. Yan et al. [52] compared the nucleation process of lithium metal on different metal substrates and found that there is an obvious overpotential when lithium metal is deposited on the surface of a copper current collector, that is, the energy used to overcome the heterogeneous nucleation (Figure 2a). In contrast, the deposition of Li metal on the Au surface first forms a Li-Au alloy above 0 V, and then, there is almost no nucleation overpotential during the deposition process below 0 V (Figure 2b). They also verified the deposition process of lithium metal on more than ten current collectors and proved that the properties of the current collector have a significant impact on the lithium metal nucleation process (Figure 2c,d).

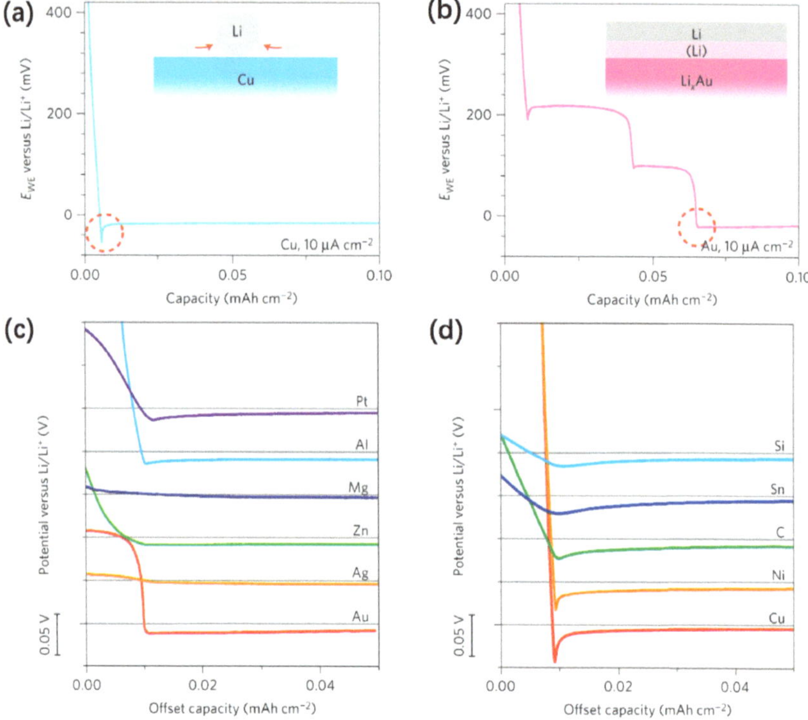

Figure 2. Voltage distributions of galvanostatic Li deposition on copper (**a**) and gold (**b**) substrates (EWE is the potential of the working electrode) (**c**) Voltage distributions of various materials with a certain solubility in Li during Li deposition, (**d**) Displacement voltage distributions for various materials with negligible solubility in Li [52].

In addition to the properties of the deposition substrate, the presence of various physical fields can have an effect on the nucleation of lithium metal. Han et al. [39] observed the nucleation state of Li metal on copper substrates at different temperatures and current densities by SEM (Figure 3a). At low temperature and high current density, the nucleation density of Li metal is higher and the particles are smaller. Further, Pei et al. [53] combined the analysis of the homogeneous nucleation equation and the SEM experimental results to conclude that the size of the lithium metal nuclei is inversely proportional to the overpotential, and the nucleus number density is proportional to the third power of the overpotential (Figure 3b–d).

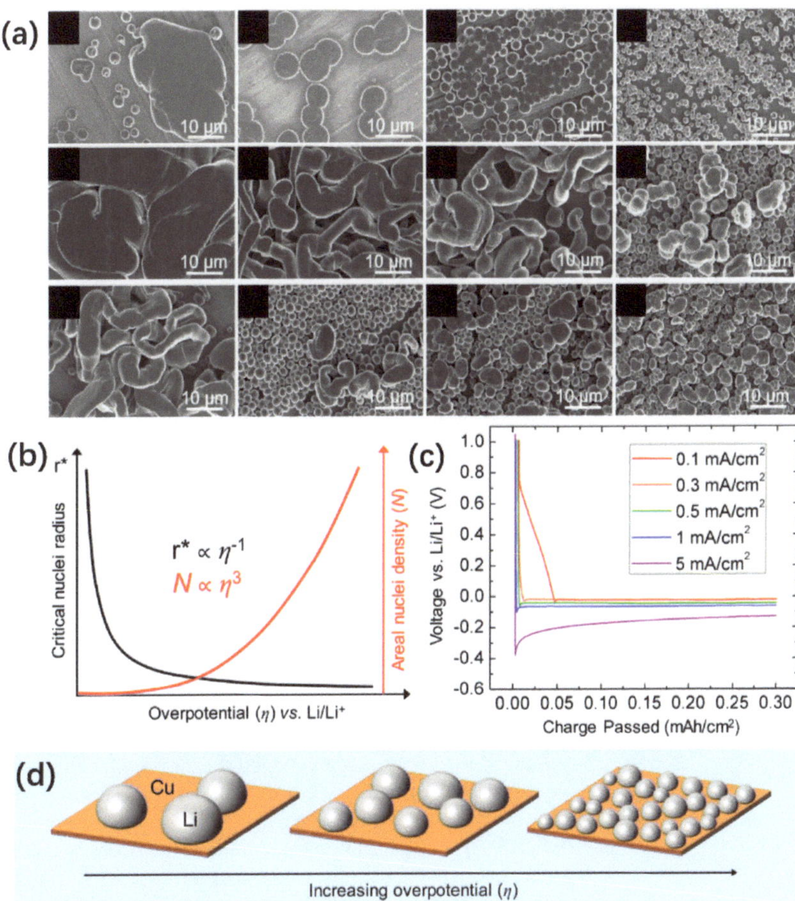

Figure 3. (**a**) Nucleation states of Li metal on copper substrates at different temperatures and current densities [39]. (**b**) Schematic diagram of the relationship between critical Li core radius and areal core density and Li deposition overpotential. (**c**) Experimental voltage distributions of Li deposition on Cu at different current densities. (**d**) Schematic illustration of the size and density of Li nuclei deposited on Cu at different overpotentials [53].

The use of lithium metal deposition materials with low nucleation overpotential and effective current density is an important way to regulate lithium metal nucleation. Carbon materials with high specific surface area and conductivity can effectively disperse the current density and, thus, make lithium metal deposition more uniform. Compared with metals such as Cu and Ni, carbon materials are lithiophilic. The lithiophilicity of carbon materials can be enhanced after doping or chemical treatment. Lithiophilic deposition substrates can induce more uniform nucleation and growth of lithium metals.

2.2. Growth of Dendrites and Formation of "Dead Lithium"

In terms of material properties, Li metal has a low surface energy and a high diffusion barrier (calculated about 0.14 eV [54]). During deposition, lithium metal tends to form a larger surface area to reach the lowest surface energy state. At the same time, lithium atoms have high diffusion energy on the bulk surface, which makes it difficult for atoms formed by lithium ions to obtain electrons to diffuse on the surface and form a uniform and flat structure. The natural properties of lithium metal determine that it tends to grow one-dimensionally, that is, to grow into dendrites [55,56]. The growth of lithium dendrites is mainly affected by factors such as electric field strength, lithium ion flux, current density, temperature, and pressure [56]. The widely accepted space charge model can well explain the growth of Li dendrites in the liquid electrolyte. Chazalviel et al. [57] believed that the space charge layer in dilute solution led to the formation of dendrites. Specifically, at the initial stage of deposition, lithium ions on the electrode surface gain electrons and become atoms. However, the lithium ions in the convection region cannot reach the electrode surface quickly due to the slow diffusion rate, and a concentration gradient of lithium ions is formed on the electrode surface at this time. The existence of the concentration gradient resulted in the formation of a space charge layer on the electrode surface. At this time, the uneven surface of the electrode leads to uneven distribution of electrons at the interface. Under the combined action of the electric field and ion field, lithium ions will preferentially deposit at the protrusions (lithium metal nucleation sites or dendrite tip), a phenomenon also known as the "tip effect". This growth mode is quantified by Sand and summarized as Sand's Time formula to predict the critical condition for dendrite growth [58]:

$$\tau_s = \pi D \left(\frac{C_0 e z_c}{2J}\right)^2 \left(\frac{\mu_a + \mu_c}{\mu_a}\right)^2 \tag{1}$$

$$D = \frac{\mu_a D_c + \mu_c D_a}{\mu_a + \mu_c} \tag{2}$$

where e is the electron charge, J is the effective electrode current density, z_c is the number of cationic charges, and C_0 is the initial concentration of Li salt. μ_c and μ_a are the anion and cation (Li+) mobilities, D is the bipolar diffusion coefficient, D_c and D_a are the cation and anion diffusion coefficients.

It can be seen that reducing the current density can increase the onset time of lithium dendrite growth so as to achieve the effect of inhibiting the growth of dendrites. Yoon et al. [59] analyzed the relationship between the current density and deposition capacity of lithium metal and the growth of lithium dendrites and showed that with the increase in current density, the critical discharge required to form dendrites first increased and then decreased. In addition to current density, compressive stress is also an important factor controlling dendrite growth. Monroe and Newman predicted that shear moduli over 10^9 Pa could inhibit dendrite growth [60]. Temperature can affect the growth process and morphology of dendrites through the diffusion and surface reaction of Li ions. The study by Hitoshi Ota et al. [61] found that dendrite growth is more serious at a lower temperature.

In different electrolyte systems, dendrites appear with different morphologies, indicating that the morphology of lithium dendrites is also controlled by the electrolyte and the SEI derived from the electrolyte. The presence of defects induces Li metal deposition on current collectors or SEIs at defects, dislocations, grain boundaries, and even contaminants. Under the combined action of many influencing factors, lithium dendrites can appear as needle-like, mossy-like or tree-like. The growth of dendrites mainly leads to three adverse consequences: penetrating the separator and contacting the positive electrode, resulting in a short circuit or even thermal runaway of the battery; the larger specific surface area aggravates the reaction between the electrode and the electrolyte, which consumes the electrolyte in the battery; lithium dendrites form "dead lithium" after fracture, resulting in the loss of lithium inventory.

"Dead lithium" formed after lithium dendrites break cannot continue to participate in the electrochemical process because they are separated from the current collector. The loss of active material will lead to the decrease in battery capacity, and the large amount of "dead lithium" wrapped by SEI on the electrode surface will also hinder the mass transfer process and cause the increase in polarization. The composition of SEI and "dead lithium" in pulverized lithium anodes can be quantitatively analyzed by titrimetric gas chromatography [62] and nuclear magnetic resonance spectroscopy [63]. This study found that the "dead lithium" that loses activity due to detachment from the current collector is an important factor leading to battery capacity decay.

In this case, the growth of lithium dendrites can be inhibited by reducing the effective current density. Carbon materials with high specific surface areas can disperse the current density, thereby reducing the local current density to obtain a dendrite-free lithium metal anode. At the same time, carbon materials with good mechanical strength as artificial SEI can also inhibit the growth of lithium dendrite.

2.3. Solid Electrolyte Interphase of Lithium Metal Anode

The solid electrolyte interphase (SEI) is a passivation interface layer formed by chemical and electrochemical reactions on the surface of the negative electrode. Similar to the solid-state electrolyte, the SEI acts to conduct lithium ions and block further reactions between the active material and the electrolyte. Lithium metal and the electrolyte will form an SEI film through a chemical reaction at the moment of contact. This is because the electrochemical window of general organic electrolytes is about 1–4.7 V (relative to Li metal potential), and the electrode potential of Li metal is lower than the reduction potential of electrolytes [64,65]. Goodenough et al. [2] explained this phenomenon using the molecular orbital theory. As shown, the thermodynamic stability of the electrolyte is determined by its lowest unoccupied orbital (LUMO) and highest occupied orbital (HOMO) (Figure 4). When the electrochemical potential of the electrode is higher than the LUMO or lower than the HOMO, the contact between the electrode and the electrolyte will no longer be a thermodynamically stable state. At this time, the electrode will react with the electrolyte to form SEI. The presence of SEI complies with the electrochemical potentials of the electrode and electrolyte so that no further reactions can take place.

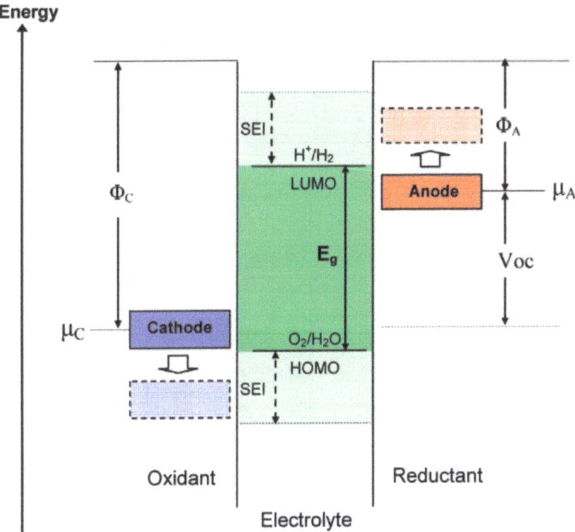

Figure 4. Schematic diagram of open circuit energy in liquid electrolytes. Φ_A and Φ_C are the anode and cathode working potentials, respectively. E$_g$ is the electrolyte thermodynamic stability window. μ_A > LUMO and/or μ_C < HOMO indicate that the SEI layer needs to be formed to reach a kinetically stable state [2].

The native SEI formed by chemical reaction is often not enough to maintain stability during the electrochemical process, so the lithium metal electrode will also reduce the electrolyte through an electrochemical reaction to form SEI. During the electrochemical process, not only the new electrolyte is reduced to increase and thicken the SEI but the structure and composition of the original SEI also change. In addition, continuous cracking and regeneration of SEI also occurs due to the volume change and non-uniformity of plating/stripping of Li metal. As shown in Figure 5, Cohen et al. [66] believed that during the deposition of lithium metal, the growth of lithium dendrites would also break the SEI, and the fresh lithium metal would continue to react with the electrolyte to form the SEI. When the lithium metal is peeled off, the brittle SEI film will rupture to expose the fresh lithium metal. At this time, the exposed lithium metal will continue to react with the electrolyte to form a new SEI.

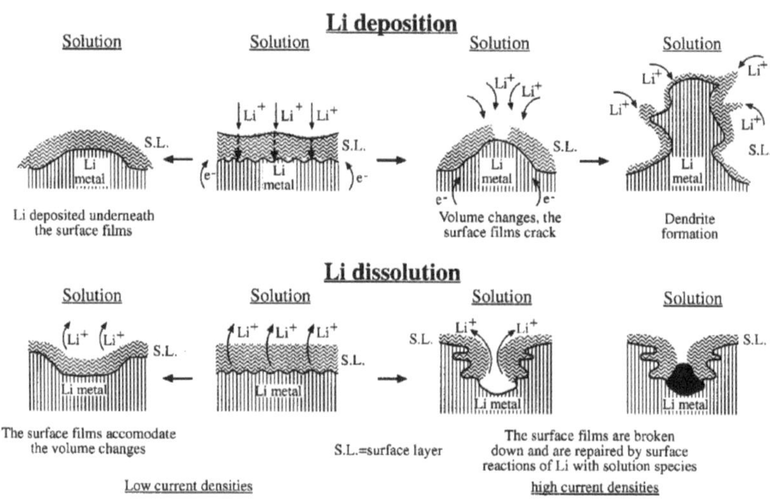

Figure 5. Description of the morphology and failure mechanism of Li electrodes during Li deposition and dissolution, with dendrite formation and heterogeneous Li dissolution accompanied by surface film destruction and repair [66].

The ideal SEI is believed to have high insulating properties, high lithium selectivity, and ionic conductivity; to be as thin as possible; with high strength and resistance to expansion and contraction stresses; to be insoluble in electrolytes; over a wide range of operating temperatures, and with stability under potential [67]. The active material and electrolyte components consumed to form the SEI lead to a decrease in battery capacity and even failure. The formation and evolution of SEI are affected by many factors, such as electrolyte composition, current collector material, temperature, electrolyte salt concentration, reduction current rate, side reactions, impurities, and uneven current distribution. The low thickness, complex structure, heterogeneous composition, and dynamic properties (spatial and temporal variations of morphology and composition) of SEI pose challenges to a comprehensive understanding of SEI [30].

With the development and innovation of characterization techniques, people's understanding of the structure and composition of SEI is constantly changing. Since SEI is an insoluble substance formed by the in situ reaction of lithium metal with the electrolyte, its chemical composition is dependent on the formulation of the electrolyte (salt anion, solvent, additives, concentration, and solvation structure). In addition, factors such as current density, cut-off voltage, capacity utilization, temperature, and pressure also have certain effects on the composition of SEI [30]. In the classic carbonate electrolyte with $LiPF_6$ as lithium salt, SEI is mainly composed of organic alkyl lithium carbonate ($ROCO_2Li$, ($ROCO_2Li)_2$), inorganic lithium salts (LiF, Li_2CO_3, Li_2O), and a small amount of fluorophosphate ($LiPO_xF_y$)) [68–70]. There are three main models for the structure of SEI: the bilayer model [71], the mosaic model [72], and the mosaic pudding model [30,73] (Figure 6). Aurbach's analysis based on Raman, FTIR and XPS found that the SEI is a mixture of various organic and inorganic substances, and that the inorganic-rich inner layer (in contact with Li) and the organic-rich outer layer (in contact with the electrolyte) constitute the two-layer structure of the SEI. The mosaic model believes that the SEI on the surface of lithium metal is composed of different lithium salts with mosaic-like morphology and stacking, and its inner layer is inorganic components, and the outer layer is mainly organic components. In recent years, based on the application of cryo-transmission electron microscopy, more researchers believe that the structure of SEI occurs when some single crystals of inorganic lithium salts are dispersed in an amorphous structure, and the first-principles density functional theory calculations and experiments have proven that amorphous regions and grain boundaries are the main routes for lithium ion transport [74].

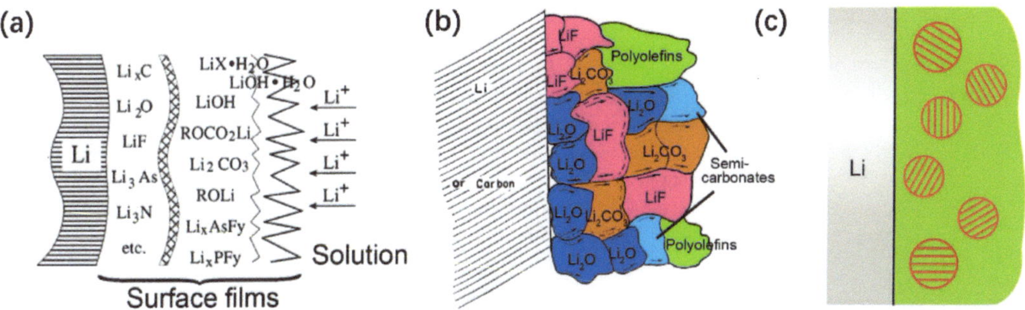

Figure 6. SEI model: (**a**) Multilayer model [71]; (**b**) Mosaic model [72]; (**c**) Raisin pudding model [30].

The regulation of the electrolyte directly affects the composition and structure of the anodes' SEI and is one of the simplest and most effective ways to improve the SEI. The design and use of an artificial SEI can compensate for the shortcomings of natural SEI and obtain a stable interface structure and dendrite-free lithium metal anode.

3. Recent Progress

3.1. Electrolyte Additives

The Fermi level of Li metal is lower than most common organic electrolytes, which leads to the inevitable reduction in Li salt and solvent molecules on the anode surface to form a solid electrolyte interphase (SEI). As an important component of the electrolyte, the use of additives can improve the various properties of anodes by forming SEI, changing the solvated structure, or changing the electric double layer structure [75–78].

Different from general electrolyte additives, carbon materials have electronic conductivity and can promote lithium metal nucleation when used as additives. Cheng et al. [79] used octadecylamine-treated nanodiamonds as additives in lithium metal batteries. Nanodiamonds with low diffusion barriers provide nucleation sites for Li metal, inducing the uniform deposition of Li metal. The nanodiamond-decorated electrolyte enables stable cycling of Li|Li symmetric cells at 2.0 mA cm^{-2} and 1.0 mA cm^{-2} for 150 h and 200 h, respectively. A Coulombic efficiency of 96% was obtained in Li|Cu cells. The addition of surfactants is bound to affect the battery. Hu et al. [80] added graphene quantum dots into the electrolyte to continuously control the growth morphology of lithium metal. Graphene quantum dots with a smaller size can be uniformly dispersed in the electrolyte without modification.

3.2. Separator Modification

In lithium batteries, the separator mainly functions to separate the electrodes and allow the electrolyte to pass through. It is a simple and valuable direction to improve lithium metal battery performance by modifying separators. Coating carbon materials on the separator surface is a simple and effective strategy. Carbon materials with higher mechanical strength has an inhibitory effect on dendrite growth. In addition, the porous carbon material with high specific surface area can control the lithium ions passing through the separator to be uniformly redistributed and, thus, deposit uniformly on the electrode surface. For example, Xu et al. [81] used carbon nanosheet coatings with cubic cavities to suppress dendrite growth (Figure 7a,b). Connected cubic carbon channels enable stable Li metal battery cycling by modulating Li deposition behavior. Li|Li symmetric cells cycled for over 2600 h at 6 mA cm^{-2} and 2 mA h cm^{-2}, while Li|Cu cells achieved an average Coulombic efficiency of 98.5% at 2 mA cm^{-2} and 2 mA h cm^{-2}. Li et al. [82] coated a thin layer of ultra-strong diamond-like carbon (DLC) on the negative side of the polypropylene (PP) separator (Figure 7c). The coating not only has a high modulus t to inhibit the growth of lithium dendrites but also undergoes in situ chemical lithiation with lithium metal in the battery, transforming into an excellent three-dimensional lithium ion conductor to redistribute lithium ion flux. The dual role of the DLC/PP separator enables the Li|Li symmetric cell to achieve stable cycling for over 4500 h at a current density of 3 mA cm^{-2}. Wang et al. [83] coated carbon fibers on the surface of separators for lithium metal batteries. The presence of carbon fibers improves the spatial electric field on the Li metal electrode surface and effectively suppresses the tip effect during dendrite growth. This study also provides new insights into the mechanism of action of carbon materials to modify the separator.

There are a large number of functional groups on the surface of carbon nanomaterials such as graphene oxide, and chemical reactions can be used to modify or modify these functional groups to obtain materials with specific functions. Li et al. [25] coated polyacrylamide-grafted graphene oxide nanosheets (GO-g-PAM) on one side of a commercial PP separator (Figure 7d). The robust GO backbone improves the mechanical strength, and the brush-like PAM chains on the graphene oxide surface contain a large number of polar groups such as C=O, N-H, etc., which provide functions for the efficient adhesion and uniform distribution of Li ions at the molecular level. Furthermore, the gaps between the stacked 2D molecular brushes provide a fast pathway for electrolyte diffusion. Liu et al. [84] coated the surface of the separator with functionalized nanocarbons modified with lithium p-benzenesulfonate groups and stabilized the deposition of lithium metal by inducing the opposite growth of lithium dendrites from the current collector and the separator (Figure 7e). In Li|LFP coin cells, this method can achieve long-term stable cycling (800 cycles with 80% initial capacity retention and 97% Coulombic efficiency).

3.3. Artificial SEI

The native SEI on the surface of Li metal electrodes is often difficult to adapt to the huge volume changes and electrochemical reactions of the electrodes during cycling. An artificial SEI used in situ and ex situ was designed to obtain a more stable interface structure. Carbon materials have good mechanical strength and good chemical/electrochemical stability. Carbon materials with different structures and their composites are designed as an artificial SEI to stabilize the electrode–electrolyte interface.

Cui et al. [37] used a monolayer of interconnected amorphous hollow carbon nanospheres as an artificial SEI layer to cover the Li metal surface (Figure 8a,b). The highly insulating top surface of the hollow carbon nanospheres promotes the deposition of metallic Li under the carbon nanospheres. The carbon layer as SEI can easily adapt to the volume change of Li metal during cycling. The Li|Cu half-cell assembled with ether electrolyte maintained a Coulombic efficiency of 99% for 150 cycles at a current density of 1 mA cm^{-2} and an areal capacity of 1 mA h cm^{-2}.

Graphene has excellent mechanical properties. The presence of defects and functional groups gives it excellent processability properties. Graphene and its derivatives or composites have received extensive attention as strategies for artificial SEI-stabilized lithium metal anodes [85–87]. Zhou et al. [88] covered the Li metal surface with several layers of parallel aligned graphene (Figure 8c). Flexible graphene films can adapt to the volume change of lithium metal during cycling. The Li|Li symmetric cell with this artificial SEI can operate for 1000 h at a current density of 5 mA cm^{-2} and a deposition capacity of 2.5 mA h cm^{-2}.

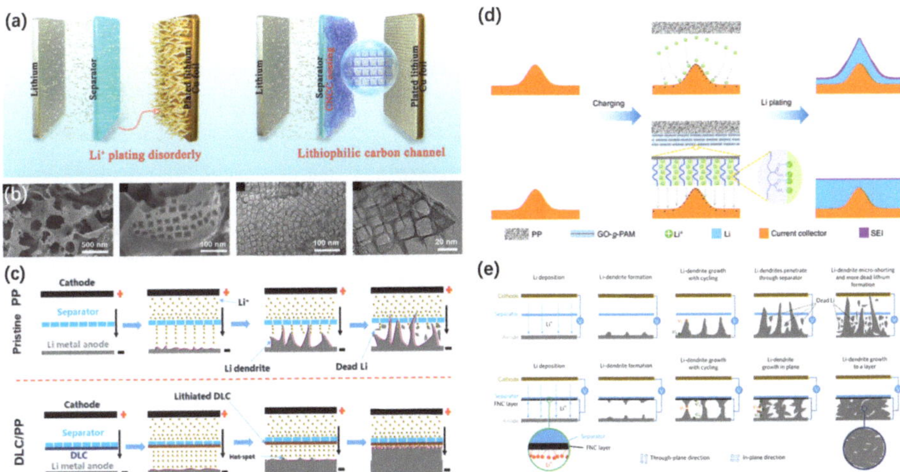

Figure 7. (**a**) Schematic illustration of Li-ion deposition in batteries with pristine separators or CNCC-coated separators. (**b**) SEM and TEM images of the CNCC coating [81]. (**c**) Schematic illustration of Li dendrite growth in pristine PP and LMB based on DLC/PP separator [82]. (**d**) Schematic illustration of Li ion deposition on electrodes with microscopic surface roughness [25]. (**e**) Growth of dendrites in blank cells and FNC cells [84].

Benefiting from the simple preparation process and good modification properties of graphene, a graphene artificial SEI combined with three-dimensional current collectors can provide higher Coulombic efficiency for lithium metal batteries. Xie et al. [89] grew graphene on the surface of nickel foam by chemical vapor deposition(CVD), and lithium metal was uniformly deposited between the nickel foam and graphene (Figure 8d). The graphene-based artificial SEI layer can inhibit the growth of dendrites and improve the cycling stability of the battery. On this basis, Wang et al. [35] composited graphene oxide and P(SF-DOL) to form an artificial SEI layer. The addition of polymers with Li-ion conductivity provides the artificial SEI with flexibility and Li-ion conductivity (Figure 8e). Combined with the three-dimensional copper foam current collector, the lithium metal battery protected by this artificial SEI maintains an average Coulombic efficiency of 99.1% over 300 cycles at a current density of 4.0 mA h cm^{-2} and a deposition capacity of 2.0 mA cm^{-2} (Figure 8f). In addition, graphene can also be composited with Prussian blue [90], LiF [91], etc. as artificial SEI layers to obtain dendrite-free Li metal batteries.

3.4. Current Collector Design

3.4.1. Lithium Metal Anode Using Carbon Material as Current Collector

As an important component of lithium batteries, current collectors not only play the role of transferring electrons between active materials and external circuits but also diffuse the heat generated inside the battery [92]. Meanwhile, the 3D current collector design can not only tolerate the huge volume change of Li metal during cycling but also achieve uniform Li deposition by reducing the current density. The properties of current collectors play an important role in the nucleation and deposition morphology of Li metal. Compared with metal materials, carbon materials have the advantages of low specific gravity and high abundance, as well as excellent electronic conductivity and lithiophilicity [93]. Thanks to their good plasticity and modifiability, various scales and various functionalized carbon materials were designed as current collectors for lithium metal batteries [94–98]. The large specific surface area can effectively reduce the local current density. At the same time, the lithiophilicity of carbon materials can be improved through surface modification to induce the uniform deposition of lithium metal. According to the morphological characteristics of the carbon material monomer, it can be divided into several categories from 0D to 3D.

Among them, 0D carbon materials mainly include carbon spheres, carbon nanoparticles, carbon quantum dots, etc.; 1D carbon materials mainly include carbon nanotubes, carbon nanowires, carbon fibers, etc.; 2D materials mainly include graphene, carbon nanosheets, etc.; 3D materials mainly include porous carbon, aerogel, and three-dimensional structures built from various carbon materials. Studies using carbon of different dimensions as current collectors are summarized and listed in Table 1.

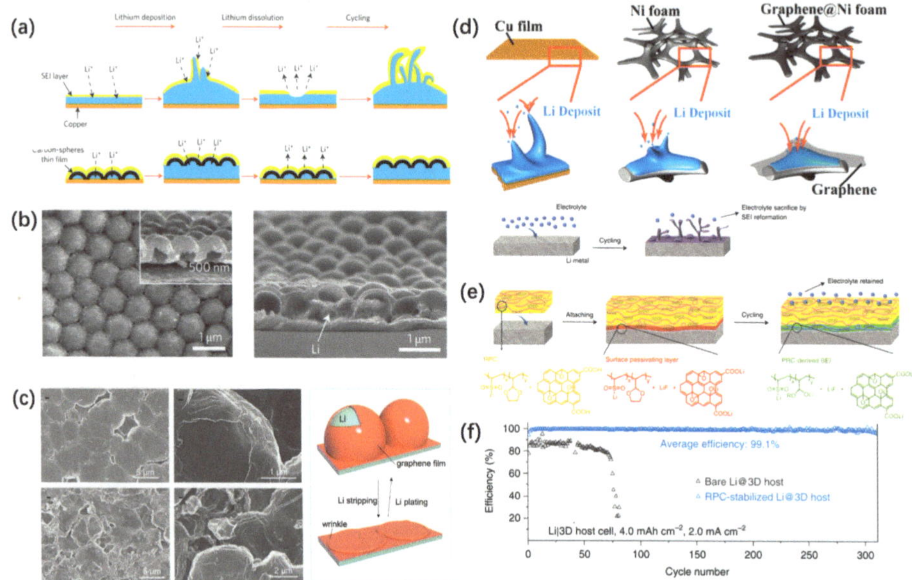

Figure 8. (**a**) Schematic illustration of the deposition of lithium metal in the presence of a stable SEI layer created by the hollow carbon nanosphere layer-modified copper substrate. (**b**) SEM image of hollow carbon nanospheres after initial SEI formation [37]. (**c**) Morphological changes and schematic diagrams of graphene films on Li metal anodes [88]. (**d**) Schematic illustration of Li deposition on bare copper foil, nickel foam, and 3D graphene@Ni foam. (**e**) Schematic illustration of molecular-level design of polymer-inorganic SEI using reactive polymer composites [89]. (**f**) Efficiency of Li | Cu 3D cells with RPC artificial SEI protection at a capacity of 4.0 mA h cm^{-2} [35].

Table 1. Performance of lithium metal batteries using carbon materials as current collectors.

	Current Collector	Half Cell Performance (Cycle Number/h, CE)	Operating Conditions (Current Density/mA·cm^{-2}, Areal Capacity/mA h·cm^{-2})	Reference
	Au@hollow carbon sphere	300, 98%	0.5, 1	[52]
	S-doped carbon nanospheres	220, −97.5 %	0.5, 1	[99]
0D	Nitrogen-doped hollow porous carbon spheres	270, 98.5%	1, 1	[100]
	hollow carbon spheres modified with evenly dispersed Ni$_2$P nanoparticles	400, 98.4%	1, 1	[101]
	graphitic carbon tubes	350, 99.3%	0.5, 1	[102]
	hollow carbon fiber	350, 99.5%	0.5, 2	[103]
1D	Lotus-root-like Ni–Co hollow prisms@carbon fibers	250, 98%	3, 1	[104]
	Li/carbon nanotube hybrid	150, 95%	1, 0.5	[105]
	hollow carbon fiber	350, 99.5	0.5, 2	[106]
	oxygen-rich carbon nanotube	200, 99%	2, 1	[107]

Table 1. *Cont.*

2D	layered reduced graphene oxide	-	-	[108]
	oxygen-codoped vertical carbon nanosheet arrays	325, 98%	0.5, 1	[109]
	S-doped graphene	180, 98.23%	1, 0.5	[110]
	3D fluorine-doped graphene	150, 98%	2, 1	[111]
3D	Nitrogen-doped amorphous Zn–carbon multichannel fibers decorated with carbon cages	800, 98%	1, 1	[112]
	Au nanoparticles@graphene hybrid aerogel	175, 91.2%	1, 1	[113]
	Carbon nanofiber-stabilized graphene aerogel film	100, 98.5%	3, 1	[114]

Although different morphologies of carbon materials can be used as current collectors for lithium metal electrodes, in order to obtain large specific surface area and porosity, most strategies are to design materials into 3D structures. The 3D carbon structure provides a larger specific surface area to reduce local current density, higher porosity, and mechanical strength to accommodate the volume change in Li metal during deposition and exfoliation. Infusion of molten lithium metal into 3D current collectors is the most common method, but this also requires the current collector itself to have a certain lithiophilicity [115]. Lin et al. [108] obtained graphene oxide films with good lithiophilicity through Li-assisted reduction vacuum filtration and then injected molten lithium into the uniform nano-gap of the graphene films (Figure 9a). The layered graphene can not only adapt to the huge volume change of Li metal but also stabilize the deposition and interface structure of Li metal. The mass fraction of graphene in the electrode is only 7%, which ensures the high specific capacity of the electrode.

In addition to the simple use of carbon materials to build 3D conductive frameworks, the modification of carbon materials and their surfaces can obtain current collector materials with special functions. Modification methods mainly include: doping, deposition, and chemical group modification.

Doping is a common means of modifying carbon materials. Elements such as N, O, and S can be doped into carbon materials to improve their lithiophilicity. Zhang et al. [116] designed a N, S co-doped ordered mesoporous carbon nanospheres as a deposition substrate for Li metal electrodes. The experimental and computational results show that the synergistic effect of N/S double doping enhances the surface electronegativity of the carbon spheres and lowers the nucleation energy barrier of Li-Au on the surface of the carbon spheres, enabling uniform nucleation in the initial stage, thereby inducing branch-free crystalline Li deposition. At the same time, N and S elements also help to form a more stable SEI layer, which prolongs the cycle life (400 h) of lithium metal symmetric batteries at high current density (20 mA h cm^{-2}).

In addition to doping, loading lithiophilic metal materials on the surface of carbon materials can induce nucleation and reduce overpotential [117,118]. Li et al. [119] and Tian et al. [120] coated the carbon cloth with Au and Ag layers, respectively, and then placed the metal-coated side away from the separator when assembling the battery. Lithium metal preferentially nucleates and grows at the metal coating during deposition. At the same time, the upper part of the porous skeleton of the carbon cloth also provides enough space to buffer the volume expansion of metallic lithium.

There are a large number of active functional groups on the surface of carbon materials such as graphene oxide and carbon nanotubes. Using these active sites to design chemical reactions can obtain carbon materials with specific functions. For example, Gao et al. [121] introduced benzenesulfonyl fluoride molecules on the surface of reduced graphene oxide aerogels (Figure 9b). During the metal deposition process, the labile molecules not only generate metal-coordinated benzenesulfonate anions to guide homogeneous metal deposition but also introduce lithium fluoride into the SEI to improve the SEI composition on the Li surface. High-efficiency lithium deposition with low nucleation overpotential is achieved at a current density of 6.0 mA cm^{-2}. Niu et al. [106] designed a lithium anode structure based on an amine-functionalized mesoporous carbon fiber framework (Figure 9c). The introduction of amine groups enhanced the wettability of carbon fibers to lithium metal, which enabled the smooth deposition of lithium metal on the surface of carbon fibers. The full cell assembled with this anode can maintain stable cycling for 200 cycles at a low N/P ratio (< 2).

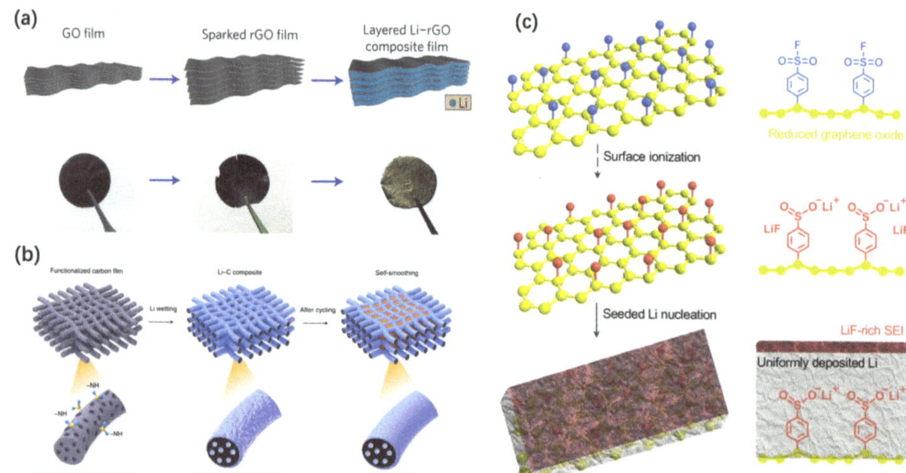

Figure 9. (**a**) Preparation of the layered Li-rGO composite film [108]. (**b**) Schematic illustration of the stabilized lithium deposition interface using the active organic molecule benzenesulfonyl fluoride (BSF) [121]. (**c**) Schematic illustration of the self-smoothing behavior of the lithium–carbon anode [106].

It is worth noting that either excess Li metal or excessively heavy current collectors will weaken or even offset the advantages of Li metal's high specific energy. Therefore, the design of thin and light and lithium-lean/lithium-free anodes has practical application value. The design of the 3D current collector is one of the most widely used and promising solutions for carbon materials in lithium metal batteries.

3.4.2. Graphite–Lithium Metal Composite Electrode

In recent years, a graphite-lithium metal composite electrode was proposed to simultaneously obtain the intercalation capacity of the graphite anode and the conversion capacity of the lithium metal by depositing a certain amount of lithium metal on the graphite electrode [122]. The use of 3D porous graphite hosts is expected to alleviate the volume expansion and dendrite growth problems of Li metal. Compared with the general lithium metal anodes using metal or carbon materials as current collectors, LiC_6 formed by graphite intercalation is considered to have good lithiophilicity [123]. Lithium metal can obtain lower nucleation overpotential on the surface of LiC_6, resulting in more uniform deposition. In addition, the current collector material occupies more mass and volume in the electrode, which weakens the advantage of the high specific volume of the metal lithium electrode. The use of graphitized carbon materials with lithium intercalation ability combined with lithium metal is expected to break the capacity limitation of graphite anodes and provide electrodes with higher effective capacity. It is worth noting that ordinary lithium metal batteries often use an excess of lithium metal as the negative electrode, thereby ignoring the volume/mass ratio of the negative electrode in the battery. The lithium-free design of the composite anode is expected to improve the specific capacity of the full cell. From a practical point of consideration, the graphite–lithium metal composite electrode uses a commercial graphite anode as the lithium deposition substrate without changing the existing production process. Graphite has the advantages of high abundance and low cost, and the use of graphite as the deposition substrate of lithium metal has high application feasibility. The design of the composite electrode is actually a compromise between the advantages and disadvantages of graphite and lithium metal. The specific capacity of graphite is improved while maintaining the stability and safety.

Although graphite-lithium metal composite anodes have many advantages compared with graphite anodes or lithium metal anodes, they also face many difficulties and challenges. Graphite was intensively researched and widely used as a mature lithium-ion battery anode. After Li metal is deposited on the graphite surface, the excess Li coating quickly fails in common carbonate-based electrolytes, resulting in a rapid decrease in battery capacity [124].

Graphitized carbon materials with various structures and functions have begun to be used as active substrates for lithium metal. These graphitic materials mainly function as 3D current collectors in electrodes [125]. A composite electrode with a higher capacity was obtained by depositing lithium metal into the voids of artificial graphite by Cui et al. (Figure 10a) [126]. Wan et al. [125] deposited Li metal on a 3D framework wrapped by graphitized carbon spheres, and the full cell assembled with LiFePO$_4$ achieved a lifespan of 1000 cycles using an anode with 5% Li pre-deposited by electrochemistry. Zuo et al. [127] reported that the graphitized carbon fiber electrode can be used as a multifunctional 3D current collector to enhance the lithium storage capacity. Intercalation and electrodeposition reactions can provide areal capacities up to 8 mA h cm^{-2} without significant dendrite formation.

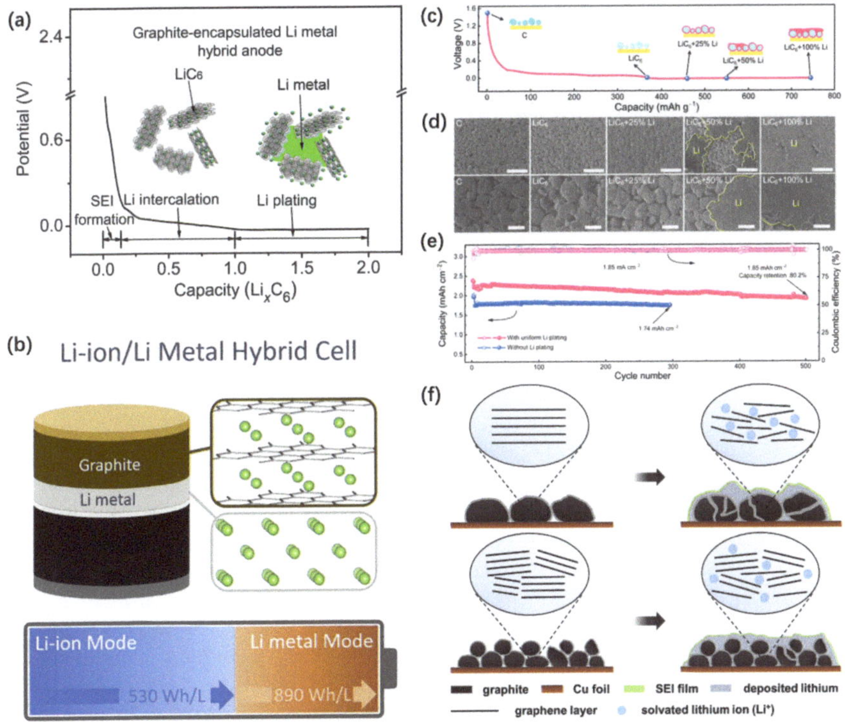

Figure 10. (**a**) Voltage variation of Li metal deposition in bulk artificial graphite [126]. (**b**) Schematic and specific energy of the hybrid Li-ion/Li-metal battery [122]. (**c**,**d**) Characterization of the morphological evolution of lithium intercalation and electroplated graphite electrodes. (**e**) Electrochemical performance of the lithium–graphite composite battery with localized highly concentrated electrolyte [128]. (**f**) Schematic illustration of the effects of NG and BMG structures on Li deposition and carbon structure evolution [129].

However, the research on graphite–lithium metal electrodes often ignores the capacity contribution of graphite itself, which also makes the volume-specific energy advantage of composite anodes not effectively utilized. In recent years, Dahn et al. [122] proposed the concept of lithium ion–lithium metal composite batteries. They believe that the use of graphite–lithium metal composite anode can increase the volume energy density of the anode from 530 W h L^{-1} to 890 W h L^{-1} (Figure 10b). However, the composite anode used by Dahn mainly deposits lithium on the surface of graphite to form a double-layer structure, and no further studies on the distribution of lithium were carried out. Zhang et al. [128] explored the boundary values for Li plating on graphite (Figure 10c–e). Combined with thermal monitoring, SEM, TOF-SIMS, and other characterizations, the properties of graphite–lithium metal electrodes with different lithium contents were tested. Their results show that the electrode surface has the most uniform lithium distribution when depositing lithium metal with a graphite capacity of 25%. Of course, the boundary value is affected by many conditions such as temperature, magnification, porosity, etc., and more work is needed to verify.

From a material point of view, reducing the particle size of graphite is considered to be more effective to obtain a more stable structure and a larger specific surface area (Figure 10f) [129]. The study of Chen et al. [124] showed that the capacity attenuation of graphite–lithium metal composite anodes mainly comes from the accumulation of dead lithium and the decrease in graphite capacity. The results of in situ X-ray microtomography analysis also confirmed this statement [130]: the main reason for the decrease in capacity after lithium deposition from graphite is that the graphite under the lithium metal layer is affected by mass transfer and cannot achieve the effective intercalation of lithium ions.

The focus of graphite–lithium metal composite anode research is on the construction of stable SEI and the maintenance of battery capacity. In order to construct a more robust SEI, Wu et al. [131] obtained a graphite–lithium metal composite anode with a longer cycle life by coating PVDF on the surface of the graphite electrode. By changing the carbon matrix or electrolyte, a uniform and stable in situ SEI can be effectively constructed. Wang et al. [132] fluorinated the edge of mesocarbon microspheres to obtain an LiF-rich stabilized SEI. Benefiting from the extensive research on lithium metal anodes in recent years, electrolyte systems suitable for lithium metal anodes were also used in graphite–lithium metal anodes. Lithium salts such as $LiBF_2(C_2O_4)$- $LiBF_4$ [122], LiFSI [124,133] were used in composite electrodes and obtained more stable SEI and higher Coulombic efficiency. Zhang et al. [128] used a localized highly concentrated electrolyte to promote more uniform Li deposition, and the full cell matching NCM532 achieved a capacity retention of 80.2% after 500 cycles.

3.5. Carbon Materials in Solid-State Batteries

In lithium metal solid-state batteries, especially inorganic ceramic solid-state batteries, the solid–solid contact between the electrolyte and the two electrodes is poor, and some electrolyte materials have poor compatibility and affinity with lithium metal. In order to obtain a stable structure, the use of carbon materials as interface layers or current collectors can improve the interface stability and affinity of lithium metal anodes with solid electrolytes. Feng et al. [134] obtained a pure air-stable surface on $Li_{6.75}La_3Zr_{1.75}Ta_{0.25}O_{12}$ (LLZTO) by thermal decomposition vapor deposition (TVD) (Figure 11a). Benefiting from the amorphous structure of low graphitized carbon (LGC), instantaneous lithiation is achieved, and the impedance of the Li/LLZTO interface is reduced to 9 Ω cm^{-2}. Chen et al. [135] carbonized a mixture of phenolic resin and polyvinyl butyral on the surface of LLZTO to obtain a porous hard carbon layer (Figure 11b). The multi-layered pore structure of the hard carbon layer provides capillary force and large specific surface area, which, coupled with the chemical reactivity of the carbon material with Li, facilitates the penetration of molten Li with the garnet electrolyte. The Li/LLZTO interface exhibits a low interfacial resistance of 4.7 Ω cm^{-2} and a higher critical current density at 40 °C. Lee et al. [136] mixed silver and carbon nanoparticles to make anodes, and during the deposition and exfoliation of Li metal, the silver and carbon nanoparticles moved away from the electrolyte and closer to the electrolyte, respectively (Figure 11c). The gradient electrode structure provides both nucleation sites and interfacial protection layers for Li metal deposition. The pouch cells assembled with silver pyroxene Li_6PS_5Cl exhibited high energy density (900 W h l^{-1}) and superior cycle life (1000 cycles, Coulombic efficiency 99.8%).

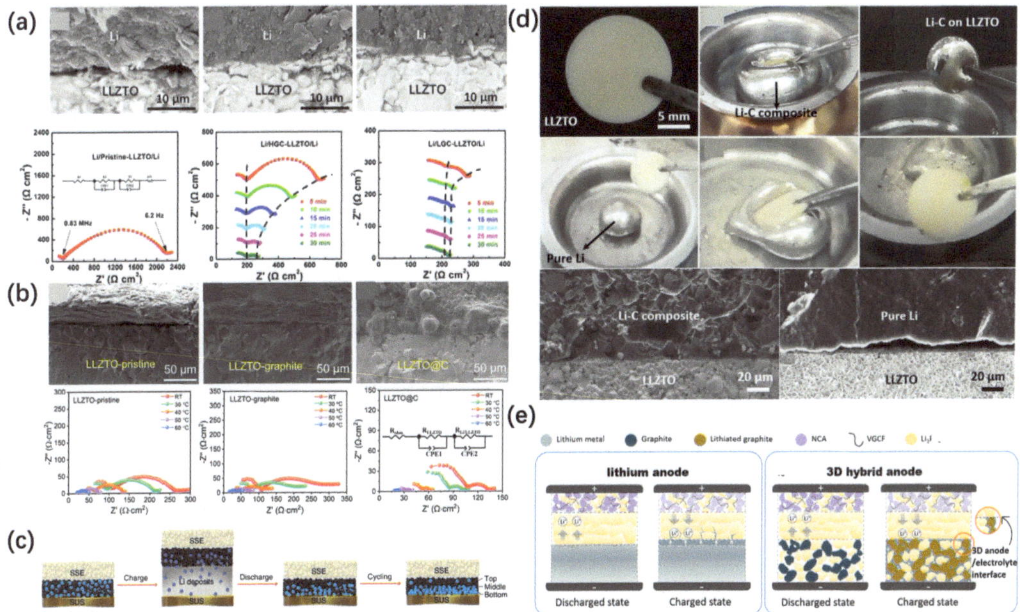

Figure 11. (**a**) Amorphous low graphitized carbon stabilized Li/LLZTO interface [134]. (**b**) Li/LLZTO interface stabilized by the porous hard carbon layer [135]. (**c**) Schematic illustration of the role of the Ag-C nanocomposite layer on the current collector to stabilize the Li coating during charging and discharging [136]. (**d**) Comparison of the interfacial behavior of LLZTO with pure Li and Li-C [137]. (**e**) Schematic illustration of the deposition process on Li metal and 3D composite anode in an all-solid-state Li metal battery [138].

In addition, graphite–lithium metal composite electrodes can also be designed using the lithiophilic properties of graphite in solid-state batteries. Duan et al. [137] cast a mixed slurry of lithium metal and graphite into a pole piece and applied it in an LLZO electrolyte battery (Figure 11d). The graphite–lithium metal composite electrode can effectively improve the affinity with the solid electrolyte and reduce the interfacial impedance.

Integrating graphite directly into solid-state electrolytes can utilize the interstitial spaces between graphite and ceramic particles to store lithium metal. Furthermore, the lithium-free negative electrode design can also obtain high specific energy batteries. Ping Liu et al. [138] mixed graphite into the sulfide solid electrolyte, and the resulting composite anode could effectively alleviate the infiltration of lithium metal in the lattice gap and prevent short circuits (Figure 11e). The critical current density of the electrode increases and the interface resistance decreases.

The main problem in organic polymer electrolytes is their low electrical conductivity. Adding fillers can effectively reduce the crystallinity of the electrolyte and improve the conductivity. Materials such as graphene [139,140] and carbon quantum dots [141] as fillers added to polymer electrolytes can simultaneously improve the mechanical properties and electrical conductivity of the electrolytes.

4. Summary and Outlook

Compared with other materials, carbon materials are unique in the field of energy storage due to their low cost, controllable microstructure, tunable electrical conductivity, and modifiable surface structure. Diverse carbon materials have played a huge role in lithium metal batteries. Lithium metal batteries using carbon materials as current collectors can effectively reduce the current density and disperse heat. For the modified carbon material, it will also have the effect of regulating the nucleation and growth of lithium metal. In particular, graphitized carbon materials can be used as a deposition substrate to effectively improve the coulombic efficiency of the anode. The use of carbon materials as additives or artificial SEI in lithium metal batteries can achieve the role of stabilizing the interface layer. In solid-state batteries, carbon materials as interface layers can improve the wettability of lithium metal and electrolyte and increase the ultimate exchange current density. We summarize the application and research of carbon materials in lithium metal batteries in recent years. These works explore the possibilities of carbon materials from various angles. Combined with our reflections on current research, we make some empirical recommendations:

1. When introducing carbon materials into the design of lithium metal batteries, the negative effects of carbon materials, such as chemical/electrochemical stability, structural stability, etc., should be considered at the same time.
2. When designing carbon-based three-dimensional current collectors, the effects of porosity and specific surface area should be considered at the same time. The size of porosity directly affects the mass transfer process of lithium ions: too large porosity will weaken the advantages brought by the 3D structure, while too small porosity will affect the mass transfer process of lithium ions in it. A large specific surface area can achieve more uniform deposition by dispersing the local current density, but at the same time, it will also increase the SEI film area and reduce the first effect and Coulomb efficiency of the battery.
3. The lithium metal foil used in the laboratory test is generally thick. The excessive lithium metal and electrolyte greatly prolong the failure time of the battery. When conducting battery tests, the experimental conditions should be scientifically controlled in order to truly reflect the role of materials in the battery.
4. Pay attention to the overall specific capacity of the battery. Excess lithium metal will reduce the actual specific capacity of the battery. The use of carbon materials can improve the cycle stability and battery life of lithium metal batteries to a certain extent. However, the mass and volume of carbon materials themselves are often overlooked. Controlling the lithium–carbon ratio is particularly important to ensure the specific capacity of the battery.
5. The experiment is established on the basis of the full cell, and its feasibility is verified with a pouch cell or a cylindrical cell.

With the advancement of materials science and the development of nanotechnology, carbon materials are increasingly incorporated into various battery systems and are successfully applied. With the unremitting efforts of mankind, carbon materials will also provide a strong boost to the development of lithium metal batteries.

Author Contributions: Z.W (Zeyu Wu) drafted the majority of this review paper. K.S. and Z.W. (Zhenhua Wang) were involved in the editing and literature-screening efforts. All authors participated in writing the manuscript and discussing the contents to be included in this review. All authors have read and agreed to the published version of the manuscript.

Funding: This research received no external funding.

Data Availability Statement: The data presented in this study are available on request from the corresponding authors.

Conflicts of Interest: The authors declare no conflict of interest.

References

1. Gür, T.M. Review of electrical energy storage technologies, materials and systems: Challenges and prospects for large-scale grid storage. *Energy Environ. Sci.* **2018**, *11*, 2696–2767. [CrossRef]
2. Goodenough, J.B.; Kim, Y. Challenges for Rechargeable Li Batteries. *Chem. Mater.* **2010**, *22*, 587–603. [CrossRef]
3. Dunn, B.; Kamath, H.; Tarascon, J.M. Electrical Energy Storage for the Grid: A Battery of Choices. *Science* **2011**, *334*, 928–935. [CrossRef] [PubMed]
4. Winter, M.; Barnett, B.; Xu, K. Before Li Ion Batteries. *Chem. Rev.* **2018**, *118*, 11433–11456. [CrossRef] [PubMed]
5. Li, M.; Lu, J.; Chen, Z.; Amine, K. 30 Years of Lithium-Ion Batteries. *Adv. Mater.* **2018**, *30*, 1800561. [CrossRef]
6. Armand, M.; Tarascon, J.M. Building better batteries. *Nature* **2008**, *451*, 652–657. [CrossRef]
7. Tarascon, J.M.; Armand, M. Issues and challenges facing rechargeable lithium batteries. *Nature* **2001**, *414*, 359–367. [CrossRef]
8. Liu, W.; Oh, P.; Liu, X.; Lee, M.J.; Cho, W.; Chae, S.; Kim, Y.; Cho, J. Nickel-rich layered lithium transition-metal oxide for high-energy lithium-ion batteries. *Angew. Chem. Int. Ed.* **2015**, *54*, 4440–4457. [CrossRef]
9. Choi, J.W.; Aurbach, D. Promise and reality of post-lithium-ion batteries with high energy densities. *Nat. Rev. Mater.* **2016**, *1*, 16013. [CrossRef]
10. Zhang, W.-J. A review of the electrochemical performance of alloy anodes for lithium-ion batteries. *J. Power Sources* **2011**, *196*, 13–24. [CrossRef]
11. Lin, D.; Liu, Y.; Cui, Y. Reviving the lithium metal anode for high-energy batteries. *Nat. Nanotechnol.* **2017**, *12*, 194–206. [CrossRef] [PubMed]
12. Guo, Y.; Li, H.; Zhai, T. Reviving Lithium-Metal Anodes for Next-Generation High-Energy Batteries. *Adv. Mater.* **2017**, *29*, 1700007. [CrossRef] [PubMed]
13. Zu, C.-X.; Li, H. Thermodynamic analysis on energy densities of batteries. *Energy Environ. Sci.* **2011**, *4*, 2614–2624. [CrossRef]
14. Albertus, P.; Babinec, S.; Litzelman, S.; Newman, A. Status and challenges in enabling the lithium metal electrode for high-energy and low-cost rechargeable batteries. *Nat. Energy* **2018**, *3*, 16–21. [CrossRef]
15. Zhang, H.; Eshetu, G.G.; Judez, X.; Li, C.M.; Rodriguez-Martinez, L.M.; Armand, M. Electrolyte Additives for Lithium Metal Anodes and Rechargeable Lithium Metal Batteries: Progress and Perspectives. *Angew. Chem. Int. Ed.* **2018**, *57*, 15002–15027. [CrossRef]
16. Chen, S.R.; Zheng, J.M.; Mei, D.H.; Han, K.S.; Engelhard, M.H.; Zhao, W.G.; Xu, W.; Liu, J.; Zhang, J.G. High-Voltage Lithium-Metal Batteries Enabled by Localized High-Concentration Electrolytes. *Adv. Mater.* **2018**, *30*, 1706102. [CrossRef]
17. Ren, X.D.; Zou, L.F.; Cao, X.; Engelhard, M.H.; Liu, W.; Burton, S.D.; Lee, H.; Niu, C.J.; Matthews, B.E.; Zhu, Z.H.; et al. Enabling High-Voltage Lithium-Metal Batteries under Practical Conditions. *Joule* **2019**, *3*, 1662–1676. [CrossRef]
18. Qian, J.F.; Henderson, W.A.; Xu, W.; Bhattacharya, P.; Engelhard, M.; Borodin, O.; Zhang, J.G. High rate and stable cycling of lithium metal anode. *Nat. Commun.* **2015**, *6*, 6362. [CrossRef]
19. Fan, X.L.; Chen, L.; Borodin, O.; Ji, X.; Chen, J.; Hou, S.; Deng, T.; Zheng, J.; Yang, C.Y.; Liou, S.C.; et al. Non-flammable electrolyte enables Li-metal batteries with aggressive cathode chemistries. *Nat. Nanotechnol.* **2018**, *13*, 715–722. [CrossRef]
20. Suo, L.M.; Xue, W.J.; Gobet, M.; Greenbaum, S.G.; Wang, C.; Chen, Y.M.; Yang, W.L.; Li, Y.X.; Li, J. Fluorine-donating electrolytes enable highly reversible 5-V-class Li metal batteries. *Proc. Natl. Acad. Sci. USA* **2018**, *115*, 1156–1161. [CrossRef]
21. Ding, F.; Xu, W.; Graff, G.L.; Zhang, J.; Sushko, M.L.; Chen, X.L.; Shao, Y.Y.; Engelhard, M.H.; Nie, Z.M.; Xiao, J.; et al. Dendrite-Free Lithium Deposition via Self-Healing Electrostatic Shield Mechanism. *J. Am. Chem. Soc.* **2013**, *135*, 4450–4456. [CrossRef]
22. Zhang, S.; Cheng, B.; Fang, Y.; Dang, D.; Shen, X.; Li, Z.; Wu, M.; Hong, Y.; Liu, Q. Inhibition of lithium dendrites and dead lithium by an ionic liquid additive toward safe and stable lithium metal anodes. *Chin. Chem. Lett.* **2022**, *33*, 3951–3954. [CrossRef]
23. Zhang, W.D.; Tu, Z.Y.; Qian, J.W.; Choudhury, S.; Archer, L.A.; Lu, Y.Y. Design Principles of Functional Polymer Separators for High-Energy, Metal-Based Batteries. *Small* **2018**, *14*, 1703001. [CrossRef] [PubMed]
24. Hao, X.M.; Zhu, J.; Jiang, X.; Wu, H.T.; Qiao, J.S.; Sun, W.; Wang, Z.H.; Sun, K.N. Ultrastrong Polyoxyzole Nanofiber Membranes for Dendrite-Proof and Heat-Resistant Battery Separators. *Nano Lett.* **2016**, *16*, 2981–2987. [CrossRef] [PubMed]
25. Li, C.F.; Liu, S.H.; Shi, C.G.; Liang, G.H.; Lu, Z.T.; Fu, R.W.; Wu, D.C. Two-dimensional molecular brush-functionalized porous bilayer composite separators toward ultrastable high-current density lithium metal anodes. *Nat. Commun.* **2019**, *10*, 1363. [CrossRef] [PubMed]
26. Jin, R.; Fu, L.X.; Zhou, H.L.; Wang, Z.Y.; Qiu, Z.F.; Shi, L.Y.; Zhu, J.F.; Yuan, S. High Li+ Ionic Flux Separator Enhancing Cycling Stability of Lithium Metal Anode. *ACS Sustain. Chem. Eng.* **2018**, *6*, 2961–2968. [CrossRef]
27. Woo, S.G.; Hwang, E.K.; Kang, H.K.; Lee, H.; Lee, J.N.; Kim, H.S.; Jeong, G.; Yoo, D.J.; Lee, J.; Kim, S.; et al. High transference number enabled by sulfated zirconia superacid for lithium metal batteries with carbonate electrolytes. *Energy Environ. Sci.* **2021**, *14*, 1420–1428. [CrossRef]
28. Liu, Y.J.; Tao, X.Y.; Wang, Y.; Jiang, C.; Ma, C.; Sheng, O.W.; Lu, G.X.; Lou, X.W. Self-assembled monolayers direct a LiF-rich interphase toward long-life lithium metal batteries. *Science* **2022**, *375*, 739–745. [CrossRef] [PubMed]
29. Yang, Y.; Yao, S.Y.; Liang, Z.W.; Wen, Y.C.; Liu, Z.B.; Wu, Y.W.; Liu, J.; Zhu, M. A Self-Supporting Covalent Organic Framework Separator with Desolvation Effect for High Energy Density Lithium Metal Batteries. *ACS Energy Lett.* **2022**, *7*, 885–896. [CrossRef]
30. Wu, H.; Jia, H.; Wang, C.; Zhang, J.G.; Xu, W. Recent Progress in Understanding Solid Electrolyte Interphase on Lithium Metal Anodes. *Adv. Energy Mater.* **2020**, *11*, 2003092. [CrossRef]

31. Meyerson, M.L.; Papa, P.E.; Heller, A.; Mullins, C.B. Recent Developments in Dendrite-Free Lithium-Metal Deposition through Tailoring of Micro- and Nanoscale Artificial Coatings. *ACS Nano* **2021**, *15*, 29–46. [CrossRef] [PubMed]
32. Fedorov, R.G.; Maletti, S.; Heubner, C.; Michaelis, A.; Ein-Eli, Y. Molecular Engineering Approaches to Fabricate Artificial Solid-Electrolyte Interphases on Anodes for Li-Ion Batteries: A Critical Review. *Adv. Energy Mater.* **2021**, *11*, 2101173. [CrossRef]
33. Liu, K.; Pei, A.; Lee, H.R.; Kong, B.; Liu, N.; Lin, D.C.; Liu, Y.Y.; Liu, C.; Hsu, P.C.; Bao, Z.A.; et al. Lithium Metal Anodes with an Adaptive "Solid-Liquid" Interfacial Protective Layer. *J. Am. Chem. Soc.* **2017**, *139*, 4815–4820. [CrossRef] [PubMed]
34. Tu, Z.Y.; Choudhury, S.; Zachman, M.J.; Wei, S.Y.; Zhang, K.H.; Kourkoutis, L.F.; Archer, L.A. Designing Artificial Solid-Electrolyte Interphases for Single-Ion and High-Efficiency Transport in Batteries. *Joule* **2017**, *1*, 394–406. [CrossRef]
35. Gao, Y.; Yan, Z.F.; Gray, J.L.; He, X.; Wang, D.W.; Chen, T.H.; Huang, Q.Q.; Li, Y.G.C.; Wang, H.Y.; Kim, S.H.; et al. Polymer-inorganic solid-electrolyte interphase for stable lithium metal batteries under lean electrolyte conditions. *Nat. Mater.* **2019**, *18*, 384–389. [CrossRef]
36. Cha, E.; Patel, M.D.; Park, J.; Hwang, J.; Prasad, V.; Cho, K.; Choi, W. 2D MoS2 as an efficient protective layer for lithium metal anodes in high-performance Li-S batteries. *Nat. Nanotechnol.* **2018**, *13*, 337–344. [CrossRef]
37. Zheng, G.; Lee, S.W.; Liang, Z.; Lee, H.-W.; Yan, K.; Yao, H.; Wang, H.; Li, W.; Chu, S.; Cui, Y. Interconnected hollow carbon nanospheres for stable lithium metal anodes. *Nat. Nanotechnol.* **2014**, *9*, 618–623. [CrossRef]
38. Lin, Z.; Ma, Y.; Wang, W.; He, Y.; Wang, M.; Tang, J.; Fan, C.; Sun, K. A homogeneous and mechanically stable artificial diffusion layer using rigid-flexible hybrid polymer for high-performance lithium metal batteries. *J. Energy Chem.* **2022**, *76*, 631–638. [CrossRef]
39. Han, Y.H.; Jie, Y.L.; Huang, F.Y.; Chen, Y.W.; Lei, Z.W.; Zhang, G.Q.; Ren, X.D.; Qin, L.J.; Cao, R.G.; Jiao, S.H. Enabling Stable Lithium Metal Anode through Electrochemical Kinetics Manipulation. *Adv. Funct. Mater.* **2019**, *29*, 1904629. [CrossRef]
40. Zhou, B.X.; Bonakdarpour, A.; Stosevski, I.; Fang, B.Z.; Wilkinson, D.P. Modification of Cu current collectors for lithium metal batteries-A review. *Prog. Mater. Sci.* **2022**, *130*, 100996. [CrossRef]
41. Zhao, H.; Lei, D.N.; He, Y.B.; Yuan, Y.F.; Yun, Q.B.; Ni, B.; Lv, W.; Li, B.H.; Yang, Q.H.; Kang, F.Y.; et al. Compact 3D Copper with Uniform Porous Structure Derived by Electrochemical Dealloying as Dendrite-Free Lithium Metal Anode Current Collector. *Adv. Energy Mater.* **2018**, *8*, 1800266. [CrossRef]
42. Ke, X.; Liang, Y.H.; Ou, L.H.; Liu, H.D.; Chen, Y.M.; Wu, W.L.; Cheng, Y.F.; Guo, Z.P.; Lai, Y.Q.; Liu, P.; et al. Surface engineering of commercial Ni foams for stable Li metal anodes. *Energy Stor. Mater.* **2019**, *23*, 547–555. [CrossRef]
43. Fan, L.; Wei, S.Y.; Li, S.Y.; Li, Q.; Lu, Y.Y. Recent Progress of the Solid-State Electrolytes for High-Energy Metal-Based Batteries. *Adv. Energy Mater.* **2018**, *8*, 1702657. [CrossRef]
44. Yang, C.P.; Fu, K.; Zhang, Y.; Hitz, E.; Hu, L.B. Protected Lithium-Metal Anodes in Batteries: From Liquid to Solid. *Adv. Mater.* **2017**, *29*, 1701169. [CrossRef]
45. Zhou, D.; Shanmukaraj, D.; Tkacheva, A.; Armand, M.; Wang, G.X. Polymer Electrolytes for Lithium-Based Batteries: Advances and Prospects. *Chem* **2019**, *5*, 2326–2352. [CrossRef]
46. Chen, R.S.; Li, Q.H.; Yu, X.Q.; Chen, L.Q.; Li, H. Approaching Practically Accessible Solid-State Batteries: Stability Issues Related to Solid Electrolytes and Interfaces. *Chem. Rev.* **2020**, *120*, 6820–6877. [CrossRef]
47. Yu, Q.J.; Jiang, K.C.; Yu, C.L.; Chen, X.J.; Zhang, C.J.; Yao, Y.; Jiang, B.; Long, H.J. Recent progress of composite solid polymer electrolytes for all-solid-state lithium metal batteries. *Chin. Chem. Lett.* **2021**, *32*, 2659–2678. [CrossRef]
48. Zhao, B.L.; Ma, L.X.; Wu, K.; Cao, M.X.; Xu, M.G.; Zhang, X.X.; Liu, W.; Chen, J.T. Asymmetric double-layer composite electrolyte with enhanced ionic conductivity and interface stability for all-solid-state lithium metal batteries. *Chin. Chem. Lett.* **2021**, *32*, 125–131. [CrossRef]
49. Liu, H.; Cheng, X.B.; Huang, J.Q.; Yuan, H.; Lu, Y.; Yan, C.; Zhu, G.L.; Xu, R.; Zhao, C.Z.; Hou, L.P.; et al. Controlling Dendrite Growth in Solid-State Electrolytes. *ACS Energy Lett.* **2020**, *5*, 833–843. [CrossRef]
50. Sun, M.H.; Liu, T.F.; Yuan, Y.F.; Ling, M.; Xu, N.; Liu, Y.Y.; Yan, L.J.; Li, H.; Liu, C.Y.; Lu, Y.Y.; et al. Visualizing Lithium Dendrite Formation within Solid-State Electrolytes. *ACS Energy Lett.* **2021**, *6*, 451–458. [CrossRef]
51. Raj, V.; Venturi, V.; Kankanallu, V.R.; Kuiri, B.; Viswanathan, V.; Aetukuri, N.P.B. Direct correlation between void formation and lithium dendrite growth in solid-state electrolytes with interlayers. *Nat. Mater.* **2022**, *21*, 1050–1056. [CrossRef] [PubMed]
52. Yan, K.; Lu, Z.; Lee, H.-W.; Xiong, F.; Hsu, P.-C.; Li, Y.; Zhao, J.; Chu, S.; Cui, Y. Selective deposition and stable encapsulation of lithium through heterogeneous seeded growth. *Nat. Energy* **2016**, *1*, 16010. [CrossRef]
53. Pei, A.; Zheng, G.Y.; Shi, F.F.; Li, Y.Z.; Cui, Y. Nanoscale Nucleation and Growth of Electrodeposited Lithium Metal. *Nano Lett.* **2017**, *17*, 1132–1139. [CrossRef]
54. Jaeckle, M.; Gross, A. Microscopic properties of lithium, sodium, and magnesium battery anode materials related to possible dendrite growth. *J. Chem. Phys.* **2014**, *141*, 174710. [CrossRef]
55. Xiao, J. How lithium dendrites form in liquid batteries. *Science* **2019**, *366*, 426–427. [CrossRef] [PubMed]
56. Cheng, X.B.; Zhang, R.; Zhao, C.Z.; Zhang, Q. Toward Safe Lithium Metal Anode in Rechargeable Batteries: A Review. *Chem. Rev.* **2017**, *117*, 10403–10473. [CrossRef] [PubMed]
57. Fleury, V.; Chazalviel, J.N.; Rosso, M.; Sapoval, B. The role of the anions in the growth speed of fractal electrodeposits. *J. Electroanal. Chem. Interfacial Electrochem.* **1990**, *290*, 249–255. [CrossRef]
58. Chazalviel, J. Electrochemical aspects of the generation of ramified metallic electrodeposits. *Phys. Rev. A* **1990**, *42*, 7355–7367. [CrossRef]

59. Seong, I.W.; Hong, C.H.; Kim, B.K.; Yoon, W.Y. The effects of current density and amount of discharge on dendrite formation in the lithium powder anode electrode. *J. Power Sources* **2008**, *178*, 769–773. [CrossRef]

60. Monroe, C.; Newman, J. The impact of elastic deformation on deposition kinetics at lithium/polymer interfaces. *J. Electrochem. Soc.* **2005**, *152*, A396–A404. [CrossRef]

61. Ota, H.; Shima, K.; Ue, M.; Yamaki, J. Effect of vinylene carbonate as additive to electrolyte for lithium metal anode. *Electrochim. Acta* **2004**, *49*, 565–572. [CrossRef]

62. Fang, C.C.; Li, J.X.; Zhang, M.H.; Zhang, Y.H.; Yang, F.; Lee, J.Z.; Lee, M.H.; Alvarado, J.; Schroeder, M.A.; Yang, Y.Y.C.; et al. Quantifying inactive lithium in lithium metal batteries. *Nature* **2019**, *572*, 511–515. [CrossRef] [PubMed]

63. Xiang, Y.; Tao, M.; Zhong, G.; Liang, Z.; Zheng, G.; Huang, X.; Liu, X.; Jin, Y.; Xu, N.; Armand, M.; et al. Quantitatively analyzing the failure processes of rechargeable Li metal batteries. *Sci. Adv.* **2021**, *7*, eabj3423. [CrossRef] [PubMed]

64. Cheng, X.B.; Zhang, R.; Zhao, C.Z.; Wei, F.; Zhang, J.G.; Zhang, Q. A Review of Solid Electrolyte Interphases on Lithium Metal Anode. *Adv. Sci.* **2016**, *3*, 1500213. [CrossRef]

65. Xu, K. Electrolytes and Interphases in Li-Ion Batteries and Beyond. *Chem. Rev.* **2014**, *114*, 11503–11618. [CrossRef]

66. Cohen, Y.S.; Cohen, Y.; Aurbach, D. Micromorphological studies of lithium electrodes in alkyl carbonate solutions using in situ atomic force microscopy. *J. Phys. Chem. B* **2000**, *104*, 12282–12291. [CrossRef]

67. An, S.J.; Li, J.; Daniel, C.; Mohanty, D.; Nagpure, S.; Wood, D.L., III. The state of understanding of the lithium-ion-battery graphite solid electrolyte interphase (SEI) and its relationship to formation cycling. *Carbon* **2016**, *105*, 52–76. [CrossRef]

68. Aurbach, D.; Daroux, M.L.; Faguy, P.W.; Yeager, E. IDENTIFICATION OF SURFACE-FILMS FORMED ON LITHIUM IN PROPYLENE CARBONATE SOLUTIONS. *J. Electrochem. Soc.* **1987**, *134*, 1611–1620. [CrossRef]

69. Aurbach, D.; Zaban, A.; Ein-Eli, Y.; Weissman, I.; Chusid, O.; Markovsky, B.; Levi, M.; Levi, E.; Schechter, A.; Granot, E. Recent studies on the correlation between surface chemistry, morphology, three-dimensional structures and performance of Li and Li-C intercalation anodes in several important electrolyte systems. *J. Power Sources* **1997**, *68*, 91–98. [CrossRef]

70. Aurbach, D.; Zaban, A.; Gofer, Y.; Ely, Y.E.; Weissman, I.; Chusid, O.; Abramson, O. Recent studies of the lithium liquid electrolyte interface-electrochemical, morphological and spectral studies of a few important systems. *J. Power Sources* **1995**, *54*, 76–84. [CrossRef]

71. Aurbach, D. Review of selected electrode-solution interactions which determine the performance of Li and Li ion batteries. *J. Power Sources* **2000**, *89*, 206–218. [CrossRef]

72. Peled, E.; Ardel, G. Advanced Model for Solid Electrolyte Interphase Electrodes in Liquid and Polymer Electrolytes. *J. Electrochem. Soc.* **1997**, *144*, L208–L210. [CrossRef]

73. Li, Y.; Li, Y.; Pei, A.; Yan, K.; Sun, Y.; Wu, C.-L.; Joubert, L.-M.; Chin, R.; Koh, A.L.; Yu, Y.; et al. Atomic structure of sensitive battery materials and Interfaces revealed by cryo-electron microscopy. *Science* **2017**, *358*, 506–510. [CrossRef] [PubMed]

74. Ramasubramanian, A.; Yurkiv, V.; Foroozan, T.; Ragone, M.; Shahbazian-Yassar, R.; Mashayek, F.l. Lithium Diffusion Mechanism through Solid-Electrolyte Interphase in Rechargeable Lithium Batteries. *J. Phys. Chem. C* **2019**, *123*, 10237–10245. [CrossRef]

75. Zhang, X.Q.; Cheng, X.B.; Chen, X.; Yan, C.; Zhang, Q. Fluoroethylene Carbonate Additives to Render Uniform Li Deposits in Lithium Metal Batteries. *Adv. Funct. Mater.* **2017**, *27*, 1605989. [CrossRef]

76. Mogi, R.; Inaba, M.; Jeong, S.K.; Iriyama, Y.; Abe, T.; Ogumi, Z. Effects of some organic additives on lithium deposition in propylene carbonate. *J. Electrochem. Soc.* **2002**, *149*, A1578–A1583. [CrossRef]

77. Mukra, T.; Peled, E. Elucidation of the Losses in Cycling Lithium-Metal Anodes in Carbonate-Based Electrolytes. *J. Electrochem. Soc.* **2020**, *167*, 100520. [CrossRef]

78. Zhang, M.S.; Liu, R.J.; Wang, Z.K.; Xing, X.Y.; Liu, Y.G.; Deng, B.B.; Yang, T. Electrolyte additive maintains high performance for dendrite-free lithium metal anode. *Chin. Chem. Lett.* **2020**, *31*, 1217–1220. [CrossRef]

79. Cheng, X.B.; Zhao, M.Q.; Chen, C.; Pentecost, A.; Maleski, K.; Mathis, T.; Zhang, X.Q.; Zhang, Q.; Jiang, J.J.; Gogotsi, Y. Nanodiamonds suppress the growth of lithium dendrites. *Nat. Commun.* **2017**, *8*, 336. [CrossRef]

80. Hu, Y.; Chen, W.; Lei, T.; Jiao, Y.; Wang, H.; Wang, X.; Rao, G.; Wang, X.; Chen, B.; Xiong, J. Graphene Quantum Dots as the Nucleation Sites and Interfacial Regulator to Suppress Lithium Dendrites for High-Loading Lithium-Sulfur Battery. *Nano Energy* **2019**, *68*, 104373. [CrossRef]

81. Xu, S.; Zhao, T.; Ye, Y.; Yang, T.; Luo, R.; Li, L.; Wu, F.; Chen, R. A Designed Lithiophilic Carbon Channel on Separator to Regulate Lithium Deposition Behavior. *Small* **2022**, *18*, 2104390. [CrossRef] [PubMed]

82. Li, Z.; Peng, M.; Zhou, X.; Shin, K.; Tunmee, S.; Zhang, X.; Xie, C.; Saitoh, H.; Zheng, Y.; Zhou, Z.; et al. In Situ Chemical Lithiation Transforms Diamond-Like Carbon into an Ultrastrong Ion Conductor for Dendrite-Free Lithium-Metal Anodes. *Adv. Mater.* **2021**, *33*, 2100793. [CrossRef] [PubMed]

83. Wang, Z.H.; Wang, X.D.; Sun, W.; Sun, K.N. Dendrite-Free Lithium Metal Anodes in High Performance Lithium-Sulfur Batteries with Bifunctional Carbon Nanofiber Interlayers. *Electrochim. Acta* **2017**, *252*, 127–137. [CrossRef]

84. Liu, Y.; Liu, Q.; Xin, L.; Liu, Y.; Yang, F.; Stach, E.A.; Xie, J. Making Li-metal electrodes rechargeable by controlling the dendrite growth direction. *Nat. Energy* **2017**, *2*, 17083. [CrossRef]

85. Wondimkun, Z.T.; Beyene, T.T.; Weret, M.A.; Sahalie, N.A.; Huang, C.-J.; Thirumalraj, B.; Jote, B.A.; Wang, D.; Su, W.-N.; Wang, C.-H.; et al. Binder-free ultra-thin graphene oxide as an artificial solid electrolyte interphase for anode-free rechargeable lithium metal batteries. *J. Power Sources* **2020**, *450*, 227589. [CrossRef]

86. Liu, W.; Xia, Y.; Wang, W.; Wang, Y.; Jin, J.; Chen, Y.; Paek, E.; Mitlin, D. Pristine or Highly Defective? Understanding the Role of Graphene Structure for Stable Lithium Metal Plating. *Adv. Energy Mater.* **2019**, *9*, 1802918. [CrossRef]
87. Ma, Y.; Qi, P.; Ma, J.; Wei, L.; Zhao, L.; Cheng, J.; Su, Y.; Gu, Y.; Lian, Y.; Peng, Y.; et al. Wax-Transferred Hydrophobic CVD Graphene Enables Water-Resistant and Dendrite-Free Lithium Anode toward Long Cycle Li-Air Battery. *Adv. Sci.* **2021**, *8*, 2100488. [CrossRef] [PubMed]
88. Zhou, Y.; Zhang, X.; Ding, Y.; Zhang, L.; Yu, G. Reversible Deposition of Lithium Particles Enabled by Ultraconformal and Stretchable Graphene Film for Lithium Metal Batteries. *Adv. Mater.* **2020**, *32*, 2005763. [CrossRef]
89. Xie, K.; Wei, W.; Yuan, K.; Lu, W.; Guo, M.; Li, Z.; Song, Q.; Liu, X.; Wang, J.-G.; Shen, C. Toward Dendrite-Free Lithium Deposition via Structural and Interfacial Synergistic Effects of 3D Graphene@Ni Scaffold. *ACS Appl. Mater. Interfaces* **2016**, *8*, 26091–26097. [CrossRef]
90. Fan, L.; Sun, B.; Yan, K.; Xiong, P.; Guo, X.; Guo, Z.; Zhang, N.; Feng, Y.; Sun, K.; Wang, G. A Dual-Protective Artificial Interface for Stable Lithium Metal Anodes. *Adv. Energy Mater.* **2021**, *11*, 2102242. [CrossRef]
91. Zhu, J.; Li, P.; Chen, X.; Legut, D.; Fan, Y.; Zhang, R.; Lu, Y.; Cheng, X.; Zhang, Q. Rational design of graphitic-inorganic Bi-layer artificial SEI for stable lithium metal anode. *Energy Stor. Mater.* **2019**, *16*, 426–433. [CrossRef]
92. Tang, K.; Xiao, J.; Li, X.; Wang, D.; Long, M.; Chen, J.; Gao, H.; Chen, W.; Liu, C.; Liu, H. Advances of Carbon-Based Materials for Lithium Metal Anodes. *Front. Chem.* **2020**, *8*, 595972. [CrossRef] [PubMed]
93. Pomerantseva, E.; Bonaccorso, F.; Feng, X.; Cui, Y.; Gogotsi, Y. Energy storage: The future enabled by nanomaterials. *Science* **2019**, *366*, 969–980. [CrossRef] [PubMed]
94. Cao, W.; Chen, W.; Lu, M.; Zhang, C.; Tian, D.; Wang, L.; Yu, F. In situ generation of Li3N concentration gradient in 3D carbon-based lithium anodes towards highly-stable lithium metal batteries. *J. Energy Chem.* **2022**, *76*, 648–656. [CrossRef]
95. Liu, J.; Ma, H.; Wen, Z.; Li, H.; Yang, J.; Pei, N.; Zhang, P.; Zhao, J. Layered Ag-graphene films synthesized by Gamma ray irradiation for stable lithium metal anodes in carbonate-based electrolytes. *J. Energy Chem.* **2022**, *64*, 354–363. [CrossRef]
96. Chen, H.; Li, M.; Li, C.; Li, X.; Wu, Y.; Chen, X.; Wu, J.; Li, X.; Chen, Y. Electrospun carbon nanofibers for lithium metal anodes: Progress and perspectives. *Chin. Chem. Lett.* **2022**, *33*, 141–152. [CrossRef]
97. Zhang, Z.; Zhu, P.; Li, C.; Yu, J.; Cai, J.; Yang, Z. Needle-like cobalt phosphide arrays grown on carbon fiber cloth as a binder-free electrode with enhanced lithium storage performance. *Chin. Chem. Lett.* **2021**, *32*, 154–157. [CrossRef]
98. Liu, H.; Zhang, J.; Liu, Y.; Wei, Y.; Ren, S.; Pan, L.; Su, Y.; Xiao, J.; Fan, H.; Lin, Y.; et al. A flexible artificial solid-electrolyte interlayer supported by compactness-tailored carbon nanotube network for dendrite-free lithium metal anode. *J. Energy Chem.* **2022**, *69*, 421–427. [CrossRef]
99. Zhang, F.; Liu, X.Y.; Yang, M.H.; Cao, X.Q.; Huang, X.Y.; Tian, Y.; Zhang, F.; Li, H.X. Novel S-doped ordered mesoporous carbon nanospheres toward advanced lithium metal anodes. *Nano Energy* **2020**, *69*, 104443. [CrossRef]
100. Qi, J.; Li, Y.; Wei, G.; Li, J.; Sun, X.; Shen, J.; Han, W.; Wang, L. Nitrogen doped porous hollow carbon spheres for enhanced benzene removal. *Sep. Purif. Technol.* **2017**, *188*, 112–118. [CrossRef]
101. Jiang, H.; Zhou, Y.; Zhu, H.; Qin, F.; Han, Z.; Bai, M.; Yang, J.; Li, J.; Hong, B.; Lai, Y. Interconnected stacked hollow carbon spheres uniformly embedded with Ni2P nanoparticles as scalable host for practical Li metal anode. *Chem. Eng. J.* **2022**, *428*, 132648. [CrossRef]
102. Chen, W.; Hong, Y.; Zhao, Z.; Zhang, Y.; Pan, L.; Wan, J.; Nazir, M.; Wang, J.; He, H. Directing the deposition of lithium metal to the inner concave surface of graphitic carbon tubes to enable lithium-metal batteries. *J. Mater. Chem. A* **2021**, *9*, 16936–16942. [CrossRef]
103. Liu, L.; Yin, Y.-X.; Li, J.-Y.; Li, N.-W.; Zeng, X.-X.; Ye, H.; Guo, Y.-G.; Wan, L.-J. Free-Standing Hollow Carbon Fibers as High-Capacity Containers for Stable Lithium Metal Anodes. *Joule* **2017**, *1*, 563–575. [CrossRef]
104. Chen, C.; Guan, J.; Li, N.W.; Lu, Y.; Luan, D.; Zhang, C.H.; Cheng, G.; Yu, L.; Lou, X.W.D. Lotus-Root-Like Carbon Fibers Embedded with Ni-Co Nanoparticles for Dendrite-Free Lithium Metal Anodes. *Adv. Mater.* **2021**, *33*, 2100608. [CrossRef]
105. Wang, Z.Y.; Lu, Z.X.; Guo, W.; Luo, Q.; Yin, Y.H.; Liu, X.B.; Li, Y.S.; Xia, B.Y.; Wu, Z.P. A Dendrite-Free Lithium/Carbon Nanotube Hybrid for Lithium-Metal Batteries. *Adv. Mater.* **2021**, *33*, 2006702. [CrossRef]
106. Niu, C.J.; Pan, H.L.; Xu, W.; Xiao, J.; Zhang, J.G.; Luo, L.L.; Wang, C.M.; Mei, D.H.; Meng, J.S.; Wang, X.P.; et al. Self-smoothing anode for achieving high-energy lithium metal batteries under realistic conditions. *Nat. Nanotechnol.* **2019**, *14*, 594–601. [CrossRef]
107. Liu, K.; Li, Z.; Xie, W.; Li, J.; Rao, D.; Shao, M.; Zhang, B.; Wei, M. Oxygen-rich carbon nanotube networks for enhanced lithium metal anode. *Energy Stor. Mater.* **2018**, *15*, 308–314. [CrossRef]
108. Lin, D.; Liu, Y.; Liang, Z.; Lee, H.-W.; Sun, J.; Wang, H.; Yan, K.; Xie, J.; Cui, Y. Layered reduced graphene oxide with nanoscale interlayer gaps as a stable host for lithium metal anodes. *Nat. Nanotechnol.* **2016**, *11*, 626–632. [CrossRef]
109. Xu, Z.; Xu, L.; Xu, Z.; Deng, Z.; Wang, X. N, O-Codoped Carbon Nanosheet Array Enabling Stable Lithium Metal Anode. *Adv. Funct. Mater.* **2021**, *31*, 2102354. [CrossRef]
110. Wang, T.; Zhai, P.; Legut, D.; Wang, L.; Liu, X.; Li, B.; Dong, C.; Fan, Y.; Gong, Y.; Zhang, Q. S-Doped Graphene-Regional Nucleation Mechanism for Dendrite-Free Lithium Metal Anodes. *Adv. Energy Mater.* **2019**, *9*, 1804000. [CrossRef]
111. Li, Z.; Li, X.; Zhou, L.; Xiao, Z.; Zhou, S.; Zhang, X.; Li, L.; Zhi, L. A synergistic strategy for stable lithium metal anodes using 3D fluorine-doped graphene shuttle-implanted porous carbon networks. *Nano Energy* **2018**, *49*, 179–185. [CrossRef]
112. Fang, Y.; Zeng, Y.; Jin, Q.; Lu, X.F.; Luan, D.; Zhang, X.; Lou, X.W.D. Nitrogen-Doped Amorphous Zn-Carbon Multichannel Fibers for Stable Lithium Metal Anodes. *Angew. Chem. Int. Ed.* **2021**, *60*, 8515–8520. [CrossRef] [PubMed]

113. Yang, T.; Li, L.; Wu, F.; Chen, R. A Soft Lithiophilic Graphene Aerogel for Stable Lithium Metal Anode. *Adv. Funct. Mater.* **2020**, *30*, 2002013. [CrossRef]

114. Zhao, C.; Yu, C.; Li, S.; Guo, W.; Zhao, Y.; Dong, Q.; Lin, X.; Song, Z.; Tan, X.; Wang, C.; et al. Ultrahigh-Capacity and Long-Life Lithium-Metal Batteries Enabled by Engineering Carbon Nanofiber-Stabilized Graphene Aerogel Film Host. *Small* **2018**, *14*, 1803310. [CrossRef]

115. Li, Y.; Li, Y.; Zhang, L.; Tao, H.; Li, Q.; Zhang, J.; Yang, X. Lithiophilicity: The key to efficient lithium metal anodes for lithium batteries. *J. Energy Chem.* **2022**, *in press*. [CrossRef]

116. Zhang, F.; Liu, P.; Tian, Y.; Wu, J.; Wang, X.; Li, H.; Liu, X. Uniform lithium nucleation/deposition regulated by N/S co-doped carbon nanospheres towards ultra-stable lithium metal anodes. *J. Mater. Chem. A* **2022**, *10*, 1463–1472. [CrossRef]

117. Zhao, C.Y.; Yin, X.J.; Guo, Z.K.; Zhao, D.; Yang, G.Y.; Chen, A.S.; Fan, L.S.; Zhang, Y.; Zhang, N.Q. High lithiophilic nitrogen-doped carbon nanotube arrays prepared by in-situ catalyze for lithium metal anode. *Chin. Chem. Lett.* **2021**, *32*, 2254–2258. [CrossRef]

118. Chu, Y.T.; Xi, B.J.; Xiong, S.L. One-step construction of MoO2 uniform nanoparticles on graphene with enhanced lithium storage. *Chin. Chem. Lett.* **2021**, *32*, 1983–1987. [CrossRef]

119. Li, D.; Gao, Y.; Xie, C.; Zheng, Z. Au-coated carbon fabric as Janus current collector for dendrite-free flexible lithium metal anode and battery. *Appl. Phys. Rev.* **2022**, *9*, 011424. [CrossRef]

120. Tian, R.; Wan, S.L.; Guan, L.; Duan, H.N.; Guo, Y.P.; Li, H.; Liu, H.Z. Oriented growth of Li metal for stable Li/carbon composite negative electrode. *Electrochim. Acta* **2018**, *292*, 227–233. [CrossRef]

121. Gao, Y.; Wang, D.; Shin, Y.K.; Yan, Z.; Han, Z.; Wang, K.; Hossain, M.J.; Shen, S.; AlZahrani, A.; van Duin, A.C.T.; et al. Stable metal anodes enabled by a labile organic molecule bonded to a reduced graphene oxide aerogel. *Proc. Natl. Acad. Sci. USA* **2020**, *117*, 30135–30141. [CrossRef]

122. Martin, C.; Genovese, M.; Louli, A.J.; Weber, R.; Dahn, J.R. Cycling Lithium Metal on Graphite to Form Hybrid Lithium-Ion/Lithium Metal Cells. *Joule* **2020**, *4*, 1296–1310. [CrossRef]

123. Duan, J.; Zheng, Y.; Luo, W.; Wu, W.; Wang, T.; Xie, Y.; Li, S.; Li, J.; Huang, Y. Is graphite lithiophobic or lithiophilic? *Natl. Sci. Rev.* **2020**, *7*, 1208–1217. [CrossRef] [PubMed]

124. Yang, G.; Zhang, S.; Weng, S.; Li, X.; Wang, X.; Wang, Z.; Chen, L. Anionic Effect on Enhancing the Stability of a Solid Electrolyte Interphase Film for Lithium Deposition on Graphite. *Nano Lett.* **2021**, *21*, 5316–5323. [CrossRef] [PubMed]

125. Ye, H.; Xin, S.; Yin, Y.-X.; Li, J.-Y.; Guo, Y.-G.; Wan, L.-J. Stable Li Plating/Stripping Electrochemistry Realized by a Hybrid Li Reservoir in Spherical Carbon Granules with 3D Conducting Skeletons. *J. Am. Chem. Soc.* **2017**, *139*, 5916–5922. [CrossRef]

126. Sun, Y.; Zheng, G.; Seh, Z.W.; Liu, N.; Wang, S.; Sun, J.; Lee, H.R.; Cui, Y. Graphite-Encapsulated Li-Metal Hybrid Anodes for High-Capacity Li Batteries. *Chem* **2016**, *1*, 287–297. [CrossRef]

127. Zuo, T.T.; Wu, X.W.; Yang, C.P.; Yin, Y.X.; Ye, H.; Li, N.W.; Guo, Y.G. Graphitized Carbon Fibers as Multifunctional 3D Current Collectors for High Areal Capacity Li Anodes. *Adv. Mater.* **2017**, *29*, 1700389. [CrossRef]

128. Cai, W.; Yan, C.; Yao, Y.X.; Xu, L.; Chen, X.R.; Huang, J.Q.; Zhang, Q. The Boundary of Lithium Plating in Graphite Electrode for Safe Lithium-Ion Batteries. *Angew. Chem. Int. Ed.* **2021**, *60*, 13007–13012. [CrossRef]

129. Yang, G.; Zhang, S.; Tong, Y.; Li, X.; Wang, Z.; Chen, L. Minimizing carbon particle size to improve lithium deposition on natural graphite. *Carbon* **2019**, *155*, 9–15. [CrossRef]

130. Ho, A.S.; Parkinson, D.Y.; Finegan, D.P.; Trask, S.E.; Jansen, A.N.; Tong, W.; Balsara, N.P. 3D Detection of Lithiation and Lithium Plating in Graphite Anodes during Fast Charging. *ACS Nano* **2021**, *15*, 10480–10487. [CrossRef]

131. Luo, J.; Wu, C.E.; Su, L.Y.; Huang, S.S.; Fang, C.C.; Wu, Y.S.; Chou, J.; Wu, N.L. A proof-of-concept graphite anode with a lithium dendrite suppressing polymer coating. *J. Power Sources* **2018**, *406*, 63–69. [CrossRef]

132. Cui, C.; Yang, C.; Eidson, N.; Chen, J.; Han, F.; Chen, L.; Luo, C.; Wang, P.F.; Fan, X.; Wang, C. A Highly Reversible, Dendrite-Free Lithium Metal Anode Enabled by a Lithium-Fluoride-Enriched Interphase. *Adv. Mater.* **2020**, *32*, 1906427. [CrossRef] [PubMed]

133. Wang, S.; Liu, D.; Cai, X.; Zhang, L.; Liu, Y.; Qin, X.; Zhao, R.; Zeng, X.; Han, C.; Zhan, C.; et al. Promoting the reversibility of lithium ion/lithium metal hybrid graphite anode by regulating solid electrolyte interface. *Nano Energy* **2021**, *90*, 106510. [CrossRef]

134. Feng, W.; Dong, X.; Zhang, X.; Lai, Z.; Li, P.; Wang, C.; Wang, Y.; Xia, Y. Li/Garnet Interface Stabilization by Thermal-Decomposition Vapor Deposition of an Amorphous Carbon Layer. *Angew. Chem. Int. Ed.* **2020**, *59*, 5346–5349. [CrossRef] [PubMed]

135. Chen, L.; Zhang, J.; Tong, R.A.; Zhang, J.; Wang, H.; Shao, G.; Wang, C.A. Excellent Li/Garnet Interface Wettability Achieved by Porous Hard Carbon Layer for Solid State Li Metal Battery. *Small* **2022**, *18*, 2106142. [CrossRef] [PubMed]

136. Lee, Y.G.; Fujiki, S.; Jung, C.; Suzuki, N.; Yashiro, N.; Omoda, R.; Ko, D.S.; Shiratsuchi, T.; Sugimoto, T.; Ryu, S.; et al. High-energy long-cycling all-solid-state lithium metal batteries enabled by silver-carbon composite anodes. *Nat. Energy* **2020**, *5*, 299–308. [CrossRef]

137. Duan, J.; Wu, W.Y.; Nolan, A.M.; Wang, T.R.; Wen, J.Y.; Hu, C.C.; Mo, Y.F.; Luo, W.; Huang, Y.H. Lithium-Graphite Paste: An Interface Compatible Anode for Solid-State Batteries. *Adv. Mater.* **2019**, *31*, 1807243. [CrossRef]

138. Xing, X.; Li, Y.; Wang, S.; Liu, H.; Wu, Z.; Yu, S.; Holoubek, J.; Zhou, H.; Liu, P. YY Graphite-Based Lithium-Free 3D Hybrid Anodes for High Energy Density All-Solid-State Batteries. *ACS Energy Lett.* **2021**, *6*, 1831–1838. [CrossRef]

139. Bao, W.; Hu, Z.Y.; Wang, Y.Y.; Jiang, J.H.; Huo, S.K.; Fan, W.Z.; Chen, W.J.; Jing, X.; Long, X.Y.; Zhang, Y.F. Poly(ionic liquid)-functionalized graphene oxide towards ambient temperature operation of all-solid-state PEO-based polymer electrolyte lithium metal batteries. *Chem. Eng. J.* **2022**, *437*, 135420. [CrossRef]
140. Nicotera, I.; Simari, C.; Agostini, M.; Enotiadis, A.; Brutti, S. A Novel Li+-Nafion-Sulfonated Graphene Oxide Membrane as Single Lithium-Ion Conducting Polymer Electrolyte for Lithium Batteries. *J. Phys. Chem. C* **2019**, *123*, 27406–27416. [CrossRef]
141. Li, Z.Y.; Liu, F.; Chen, S.S.; Zhai, F.; Li, Y.; Feng, Y.Y.; Feng, W. Single Li ion conducting solid-state polymer electrolytes based on carbon quantum dots for Li-metal batteries. *Nano Energy* **2021**, *82*, 105698. [CrossRef]

batteries

Article

Unveil Overcharge Performances of Activated Carbon Cathode in Various Li-Ion Electrolytes

Xianzhong Sun [1,2,3,*], Yabin An [1,2,3], Xiong Zhang [1,2,3], Kai Wang [1,2,3], Changzhou Yuan [4,*], Xiaohu Zhang [1,2], Chen Li [1,2,3], Yanan Xu [1,2] and Yanwei Ma [1,2,3,*]

[1] Institute of Electrical Engineering, Chinese Academy of Sciences, Beijing 100190, China
[2] Institute of Electrical Engineering and Advanced Electromagnetic Drive Technology (AEDT), Qilu Zhongke, Jinan 250013, China
[3] School of Engineering Sciences, University of Chinese Academy of Sciences, Beijing 100049, China
[4] School of Materials Science & Engineering, University of Jinan, Jinan 250022, China
* Correspondence: xzsun@mail.iee.ac.cn (X.S.); mse_yuancz@ujn.edu.cn (C.Y.); ywma@mail.iee.ac.cn (Y.M.)

Abstract: Typically, the practical lithium-ion capacitor (LIC) is composed of a capacitive cathode (activated carbon, AC) and a battery-type anode (graphite, soft carbon, hard carbon). There is a risk of the LIC cell overcharging to an unsafe voltage under electrical abuse conditions. Since the anode potential is usually quite low during the charging process and can be controlled by adjusting the amount of anode materials, the overcharge performances of LIC full-cell mainly depend on the AC cathode. Thus, it is necessary to independently investigate the overcharge behaviors of the AC cathode in nonaqueous Li-ion electrolytes without the interference of the anode electrode. In this work, the stable upper potential limits of the AC electrode in three types of lithium-ion electrolytes were determined to be 4.0−4.1 V via the energy efficiency method. Then, the AC//Li half-cells were charged to 5.0 V and 10.0 V, respectively, to investigate the overcharge behaviors. For the half-cells with propylene carbonate (PC)-based electrolytes, the voltage increased sharply to 10.0 V with a vertical straight line at the end of the overcharging process, indicating that the deposits of electrolyte decomposition had separated the AC electrode surface from the electrolytes, forming a self-protective passivation film with a dielectric capacitor behavior. The dense and compact passivation film is significant in separating the AC electrode surface from the electrolytes and preventing LIC cells from volume expansion and explosion risks under electrical abuse and overcharging conditions.

Keywords: lithium-ion capacitors; energy efficiency; electrolyte; overcharge; gas evolution; passivation film

Citation: Sun, X.; An, Y.; Zhang, X.; Wang, K.; Yuan, C.; Zhang, X.; Li, C.; Xu, Y.; Ma, Y. Unveil Overcharge Performances of Activated Carbon Cathode in Various Li-Ion Electrolytes. *Batteries* **2023**, *9*, 11. https://doi.org/10.3390/batteries9010011

Academic Editor: Pascal Venet

Received: 3 November 2022
Revised: 11 December 2022
Accepted: 21 December 2022
Published: 24 December 2022

1. Introduction

The lithium-ion capacitor (LIC) is a type of novel asymmetric supercapacitor that combines the lithium-ion intercalated electrode of Li-ion batteries (LIBs) with the electrical double-layer electrode of supercapacitors [1–6]. In 2002, Kanebo Ltd. and Fuji Heavy Industry developed practical LICs employing porous carbon cathodes, pre-lithiated polyacenic semiconductor (PAS) anodes, and electrolytes using a laminated structure. After that, other companies, e.g., the JM Energy corporation (Musashi Energy Solutions Co., Yamanashi, Japan) and TAIYO YUDEN company (Tokyo, Japan), also pushed the commercialization of LICs based on various technical routes [7]. After pre-lithiation treatments [8,9], the anode potential reduces, and the stable working voltage of the practical LIC cell increases to 3.8 V [10,11].

However, sometimes LIC cells can be charged to a voltage higher than the rated voltage, also known as overcharge, under electrical abuse conditions such as the malfunction of charging equipment and/or the inappropriate design of the capacitor management system (CMS) [12–15]. Oca et al. [12] examined the overcharge process of commercial 1100 F LIC pouch and 3300 F LIC prismatic cells. When the LIC cell was charged to the high voltage of 4.6 V, the LIC cell revealed an exothermic reaction, accompanied by gaseous products and a remarkable swelling of the pouch cell. For the prismatic LIC cell, the endothermic process was observed for a voltage lower than 4.4 V, and with the overcharge period continuing, the cell temperature increased gradually. After the LIC cell voltage reached 5.0 V, the cell voltage dropped, and the cell temperature could be as high as 76 °C. Due to a large amount of gas products and the high pressure in the metallic case, the vent of the laminate cell opened [12]. Bolufawi et al. [13] studied the overcharging performance of a 200-F LIC cell manufactured by General Capacitor LLC. The LIC cell was charged from 4.0 V to the maximum voltage of 6.7 V, and then the voltage dropped. Accordingly, the cell temperature increased from the ambient temperature to about 37 °C, along with swelling and gassing.

Considering that the anode potential of the LIC full-cell is usually confined to a relatively small range (e.g., 0.1–0.5 V vs. Li^+/Li), and the anode potential can even be ca. 0 V vs. Li^+/Li in a fully charged state [16–18], it is presumed that the upper potential of the cathode can exceed the oxidative decomposition limits of electrolyte in the case of overcharging. Therefore, the overcharge behaviors of LICs mainly depend on the activated carbon (AC) cathode [19,20]. Thus, it is necessary to independently investigate the overcharge behaviors of AC cathode in nonaqueous Li-ion electrolytes without the interference of the anode electrode. In this work, the stable upper potential limits of AC electrodes in various electrolytes were determined through the energy efficiency (EE) method, and the AC//Li half-cells were overcharged to 5.0 V to assess the overcharge performances of the AC electrode. Furthermore, the AC//Li half-cells were charged as high as 10.0 V, and a substantial self-protective passivation effect was observed.

2. Materials and Methods

The AC electrodes were provided by the Tianjin Planno Energy Technology company. Three common nonaqueous electrolytes were adopted in this study. 1 M lithium bis(fluorosulfonyl)imide (LiFSI) solution in propylene carbonate (PC), denoted as PCE; 1 M lithium hexafluorophosphate ($LiPF_6$) dissolved in the mixture solvent of ethylene carbonate (EC), diethyl carbonate (DEC), and dimethyl carbonate (DMC) in a 1:1:1 volume ratio, denoted as EDD; and 1 M LiFSI dissolved in the solvent mixture with EC, PC and DEC in a volume ratio of 3:1:4, denoted as EPD.

The AC electrodes were cut into a suitable shape with a width of 35 mm and a length of 40 mm, cold pressed under 6 MPa, and then dried in a vacuum oven for more than 8 h at 80 °C. In this work, the pouch cell was employed, which consisted of one single-faced AC electrode and one Li foil counter electrode. The AC electrode was separated with a Li electrode by an FPC3018 composite cellulose separator composed of cellulose and polyester (Mitsubishi Paper Mills, Ltd., Tokyo, Japan). In an argon-filled controlled-atmosphere glove box (MBRAUN, Shanghai, China, H_2O < 0.1 ppm, O_2 < 0.1 ppm). After the Li-ion electrolyte solution was injected, the aluminum-plastic package was hermetically sealed.

The AC//Li half-cells were galvanostatic charged and discharged using a Neware battery charger (CT-4008T-5V10mA-164, Shenzhen, China) and overcharged to 10.0 V using a VMP3 electrochemical station (BioLogic, Seyssinet-Pariset, France). The surface morphologies of AC electrodes after overcharging process were observed by a CIQTEK scanning electron microscope (SEM3100, Hefei, China) coupled with energy-dispersive X-ray spectroscopy analysis (Element EDS, EDAX, Pleasanton, CA, USA). The released gaseous products at high voltage were analyzed on a Thermo Scientific TRACE 1300 series gas chromatograph (Shanghai, China).

3. Results

3.1. Stable Upper Potential Limits of AC Electrode

Energy efficiency (EE) variation can be adopted as a useful tool to determine the stable upper and lower potential limits of AC electrodes in aqueous or nonaqueous electrolytes [16,21–23]. According to the galvanostatic charging and discharging profiles of the AC//Li half-cell, EE is defined as the ratio of the discharge energy to the charge energy, which considers the impacts of the capacity loss and the voltage drop during the charging and discharging processes. In addition, the Coulombic efficiency (CE) is termed as the ratio of the discharged capacity to the charged capacity, and the voltage efficiency (VE) is termed as the ratio between the discharging plateau voltage and the charging plateau voltage of the AC//Li half-cell. When the subtle redox reaction (charge transfer reaction) occurs at the electrode/electrolyte interface on the AC electrode, the electrolyte decomposition reactions can result in the deposition of insoluble products and the formation of passivation film (cathode electrolyte interphase, CEI) on the surface or in the accessible pores of the AC electrode. Here, EE is much more sensitive to irreversible redox reactions than CE due to the cumulative effect of VE and CE [23]. Therefore, the sudden drop in EE can be used as a basis for judgment of the stable upper potential limit of the AC electrode. The change trends of CE and VE can be used as references.

The AC//Li half-cells with different Li-ion electrolytes were galvanostatic charged and discharged from the potential of zero charge (abbreviated as *pzc*, about 3.08 V) [24] to various cut-off voltages at 2 mA. The vertex voltage increased from 3.4 V to 4.5 V with a 0.1 V-step, as shown in Figure 1a. The EE, CE, and VE values as functions of vertex voltage are shown in Figure 1b. According to the EE criterion to determine the cut-off voltage, the stable upper potential limits of AC electrodes in EDD, EPD, and PCE electrolytes can be determined to be 4.0 V, 4.0 V, and 4.1 V, respectively, which are consistent with the experimental results reported in our previous publication [16].

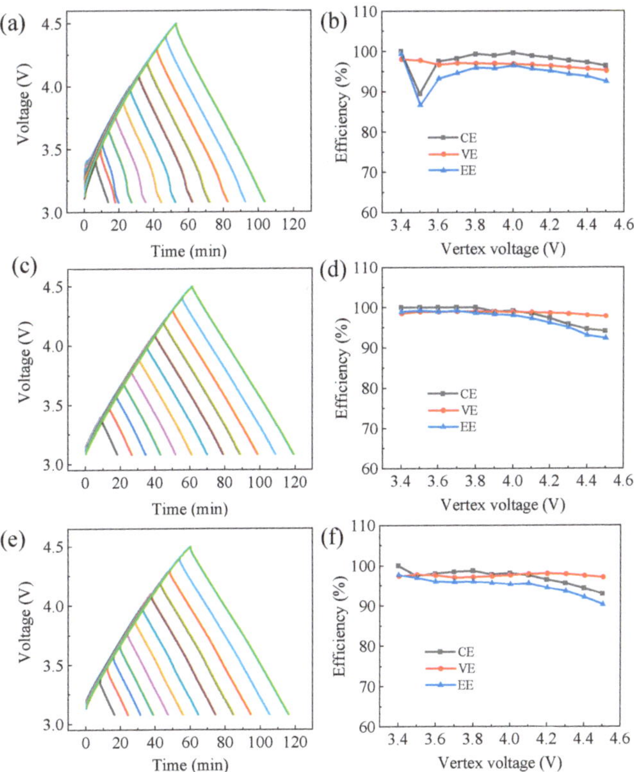

Figure 1. Galvanostatic charging and discharging curves of AC//Li half-cells upon the voltage ranges from *pzc* to various vertex voltages in different Li-ion electrolytes: (**a**) EDD, (**c**) EPD, and (**e**) PCE; the corresponding EE, CE, and VE values as functions of the vertex voltage: (**b**) EDD, (**d**) EPD, and (**f**) PCE.

3.2. Overcharge to 5.0 V

Considering that the high-voltage limits of LICs mainly depend on the oxidative decomposition of Li-ion electrolytes on the AC surface, the overcharging behaviors of AC//Li half-cells have been investigated. The AC//Li half-cells using EPD, PCE, and EDD electrolytes were first overcharged to 5.0 V at a current of 0.5 mA. The voltage-versus-time curves for the AC//Li half-cells are plotted and shown in Figure 2a. The voltage changed with time almost linearly below 4.2 V for all the half-cells. When the half-cells were charged to higher voltages of over 4.2 V, the upward trend became slow (knee point). Then the voltage dropped at several voltage points, ascribed to the decomposition of electrolyte components [25]. As shown in Figure 2b, the mutation points are 4.59 V, 4.69 V, and 4.94 V for the EPD half-cell, 4.81 V, 4.89 V, and 4.95 V for the PCE half-cell, and 4.67 V and 4.94 V for the EDD half-cell, representing the cathodic peaks of electrolyte oxidation reaction on the interface between electrolyte and AC.

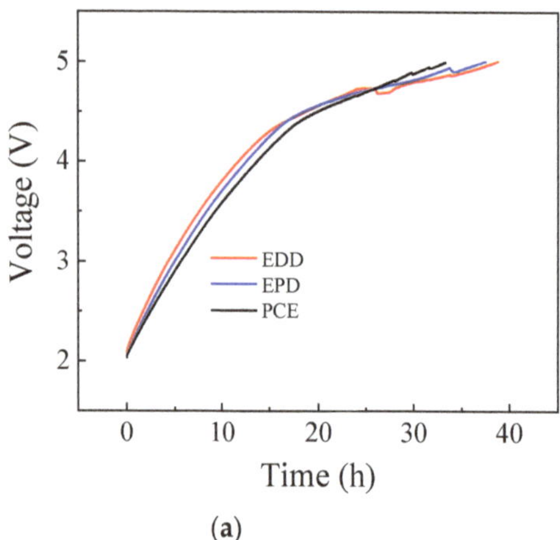
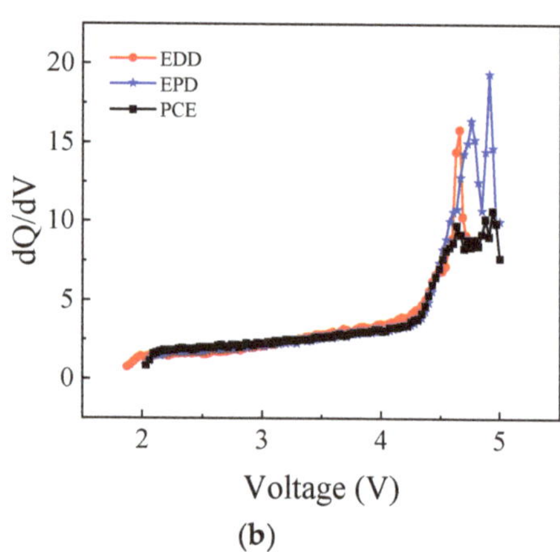

(a)   **(b)**

Figure 2. Overcharged to 5.0 V of AC//Li half-cells using EDD, EPD, and PCE electrolytes, respectively: (**a**) Voltage versus time curves, (**b**) dQ/dV versus voltage curves.

3.3. Overcharge to 10.0 V

Furthermore, the half-cells were set to charge to 10.0 V at a current of 25 mA using the VMP3 electrochemical station. The voltage versus time profiles for AC electrodes in EDD, EPD, and EDD electrolytes are depicted in Figure 3a. For instance, when the voltage of the EPD half-cell reached 5 V, it decreased rapidly with the increase of charge time and charged capacity. The half-cell voltage dropped to 4.4 V after being charged to 21.6 times the rated capacity (about 4.4 h). It stayed nearly unchanged for the subsequent period, during which 28 times of rated capacity was additionally charged (5.6 h). Then, the half-cell voltage increased to 6.8 V linearly after being charged an additional 22.5 times the rated capacity (4.5 h). Finally, the half-cell voltage sharply increased to 10.0 V with a vertical straight line after being charged for a total of 14.4 h, which indicated that the deposits of electrolyte decomposition had separated the surface of the AC cathode and electrolyte, revealing a dielectric capacitor behavior. The PCE AC//Li half-cell revealed similar behaviors. The voltage rose to about 5 V voltage and dropped to 4.6 V, and then reached the fully polarized state with less dwell time. During these processes, there was only a very slight gas evolution. On the contrary, the voltage of the EDD half-cell only rose to about 8 V, accompanying continuous electrolyte decomposition and huge volume expansion. The photos of the half-cells with EPD, PCE, and EDD electrolytes after the overcharging process are respectively illustrated in Figure 3b–d. For all three half-cells, a less noticeable temperature rise was observed just by touching them. After the overcharging processes, the voltage of half-cells rapidly dropped to 4.2–4.3 V upon rest period, which can be attributed to the large leakage current at a voltage higher than the stable voltage range [26].

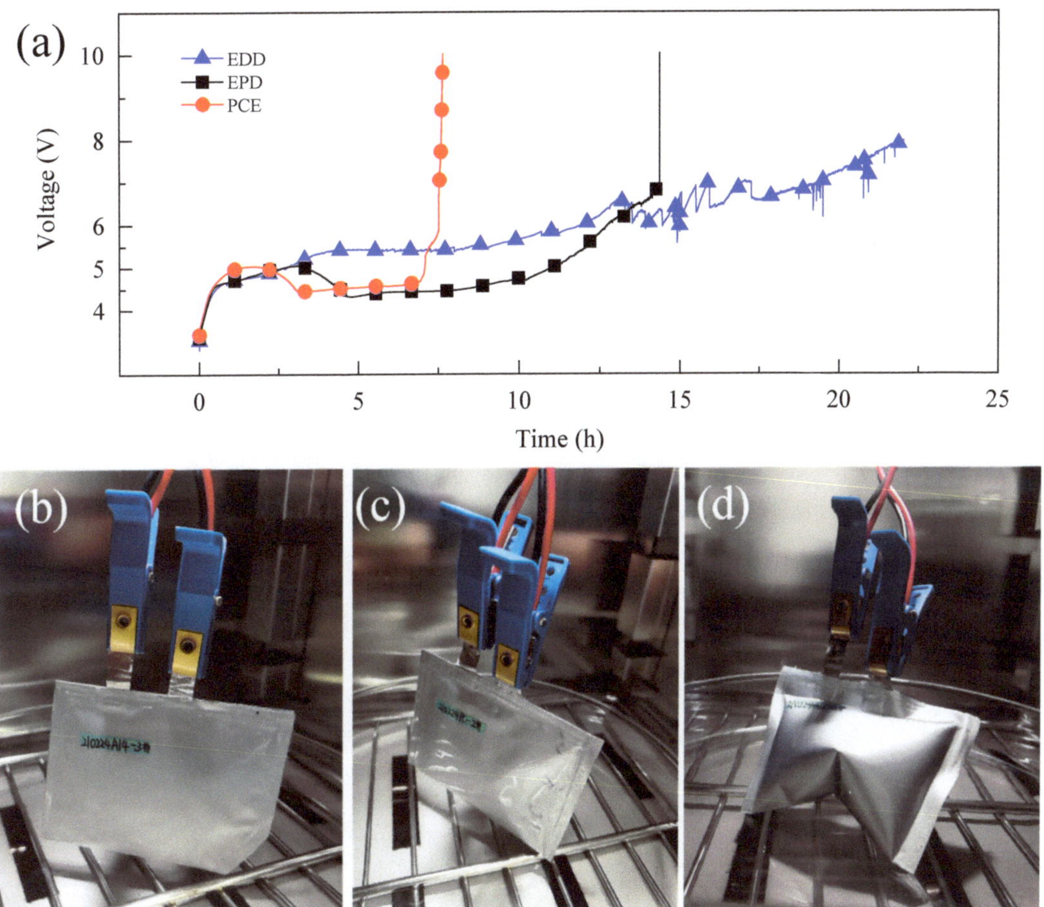

Figure 3. Overcharged to 10.0 V of AC//Li half-cells: (**a**) voltage versus time profiles, and photos of the half-cells after overcharging process with different electrolytes: (**b**) EPD, (**c**) PCE and (**d**) EDD.

3.4. Post-Mortem Analysis of Half-Cells

Since the AC//Li half-cell with PCE electrolyte can be overcharged to the high voltage of 10.0 V, it is assumed that the deposits of electrolyte decomposition had separated the AC electrode surface and electrolyte through the substantial passivation film. After the overcharging process, three half-cells were disassembled in the glove box, and the AC electrodes were thoroughly washed with DMC solvent and immersed in DMC overnight. Then the electrodes were dried in an argon atmosphere. The SEM morphologies and the compositions of the AC electrode surface were characterized. As reported in our previous publication [16], after the AC electrode was charged to 4.5 V in the PCE electrolyte, a thin film layer was formed on the electrode surface, which made the edges of the AC and carbon black particles no longer clear. The AC electrode has been covered with thicker electrolyte decomposition deposits after being cycled 2000 times between *pzc* and 4.5 V. By comparing the SEM images of the fresh AC electrode (Figure 4a) and overcharged AC electrode (Figure 4b), it can be found that a very thick passivation film covers the overcharged AC electrode surface, which consists of two phases, i.e., sphere-like small particles (zone 1), and the amorphous deposits (zone 2). The EDS spectra illustrated that the contents of O and S elements are high in zone 1, as shown in Figure 5a. On the contrary, zone 2 has a large content of F and S elements, and it contains fewer O elements, as shown in Figure 5b. From the cross-sectional view of the SEM images in Figure 4c, it can be observed that the thickness of the passivation film is about 1 μm. Similarly, the thick passivation films formed on the AC electrode surface, which was obtained from the AC//Li half-cell with EPD electrolyte after overcharging, as shown in Figure 4c.

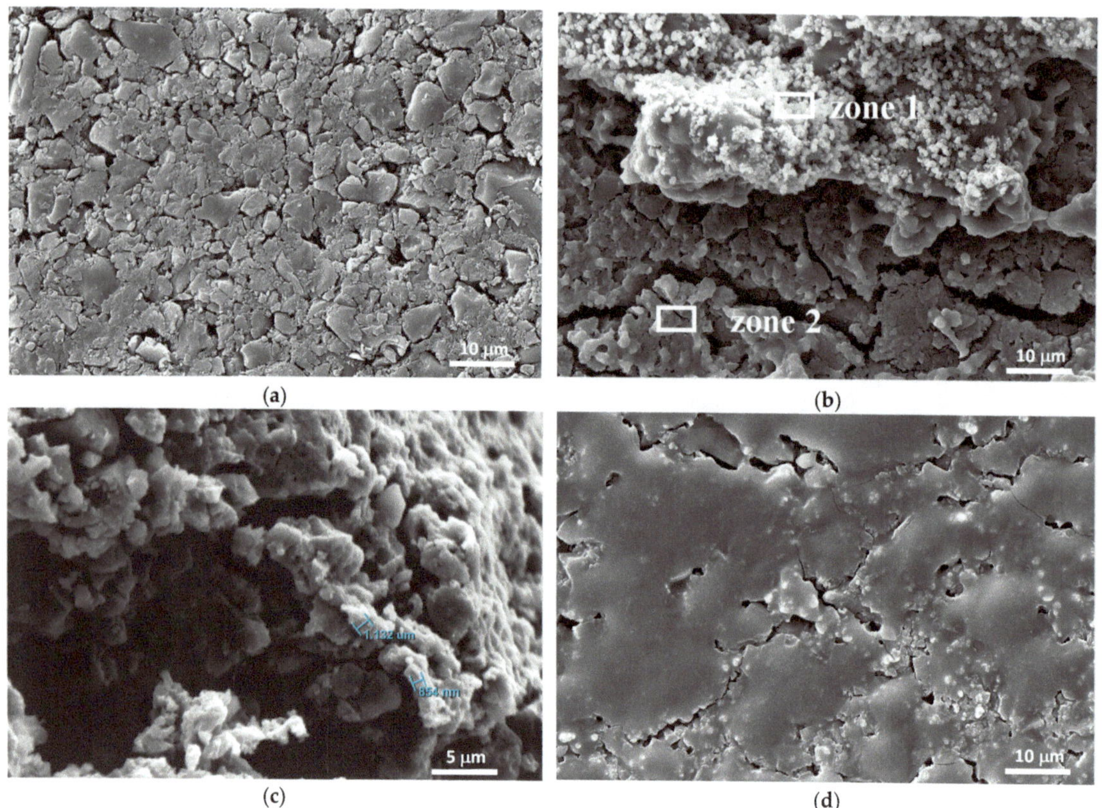

Figure 4. SEM images of (**a**) fresh AC electrode and the AC electrodes from the AC//Li half-cell charged to the high voltage of 10.0 V with different electrolytes; (**b**,**c**) PCE; (**d**) EPD.

For the overcharged half-cell with EDD electrolytes, the gaseous products were collected and analyzed through a gas chromatograph, as shown in Figure 6a,b. It was identified that the emitted gas mainly consisted of H_2, CO_2, CH_4, and CO, and the relative volume ratio is 46.0%, 45.3%, 3.6%, and 5.1%, respectively. Other organic alkane gas is hardly found. The relative volume ratio between N_2 and O_2 is 3.0, which is near the composition of air (78%:21%), and the two gas components are mainly ascribed to the mixed air during sampling. The gas evolution of CO_2 is attributed to the oxidative decomposition of solvent on the AC electrode surface. CO, CH_4, and H_2 gas are mainly ascribed to the reductive decomposition of solvent and trace amount of H_2O/HF on metallic Li electrode surface [27–29]. Since the linear solvents, e.g., dimethyl carbonate (DMC), are more prone to degradation than those of cyclic solvents of ethylene carbonate (EC) and propylene carbonate (PC) [30], the decomposition of DMC is taken as an example, and the possible reactions are as follows [31,32]:

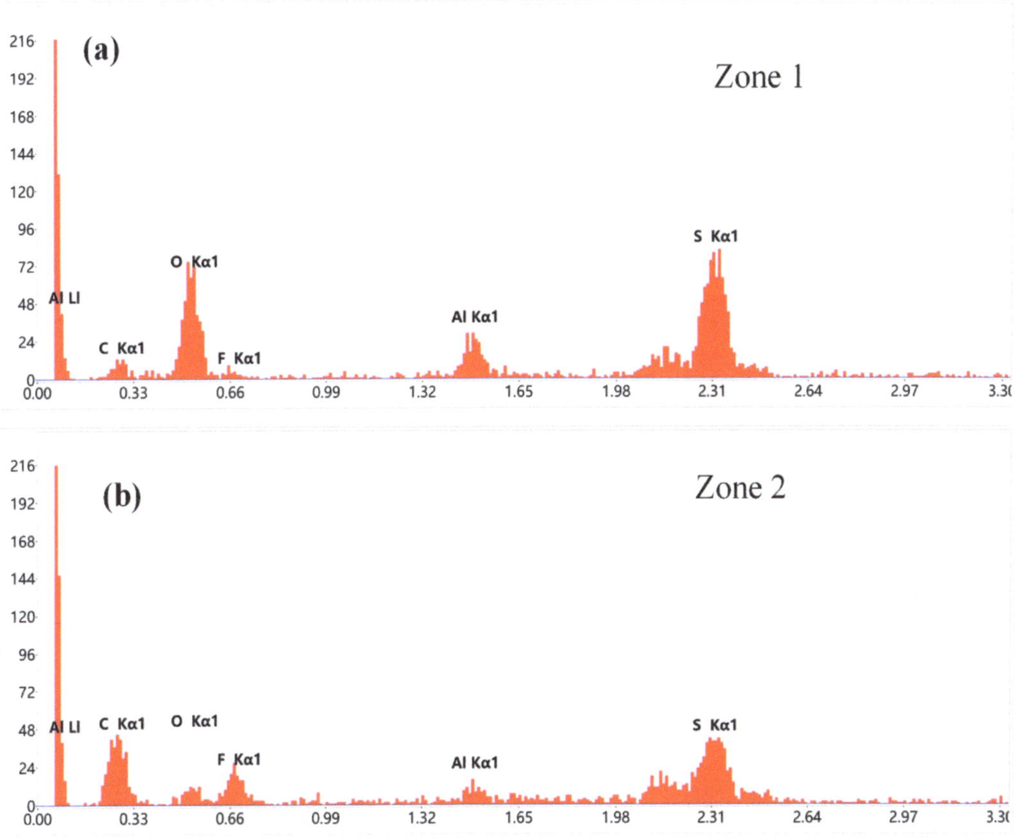

Figure 5. EDS spectra of AC electrodes obtained from the AC//Li half-cell with PCE electrolyte after overcharging to the high voltage of 10.0 V: (**a**) zone 1; (**b**) zone 2.

When the EDD half-cell was disassembled, the remaining electrolyte had turned from the clear color of fresh electrolyte to brownish black, and the Li electrode was covered with light black deposits. The SEM image indicates that the morphology of deposits on the AC electrode surface for the EDD sample is different from the abovementioned samples, as shown in Figure 7a,b. The deposits show a sheet-like shape, which tends to exfoliate from the AC electrode surface due to the low adhesion. In some areas, there are fewer deposits on the AC electrode surface. Moreover, the deposited layer seems to be not dense enough to completely separate the electrolyte and AC electrodes. The EDS results illustrated that the deposits have a large content of Al, O, and F elements, as shown in Figure 7c. The composition of these light black deposits may consist of AlF_3, Li_2CO_3, ROLi, and $ROCO_2Li$.

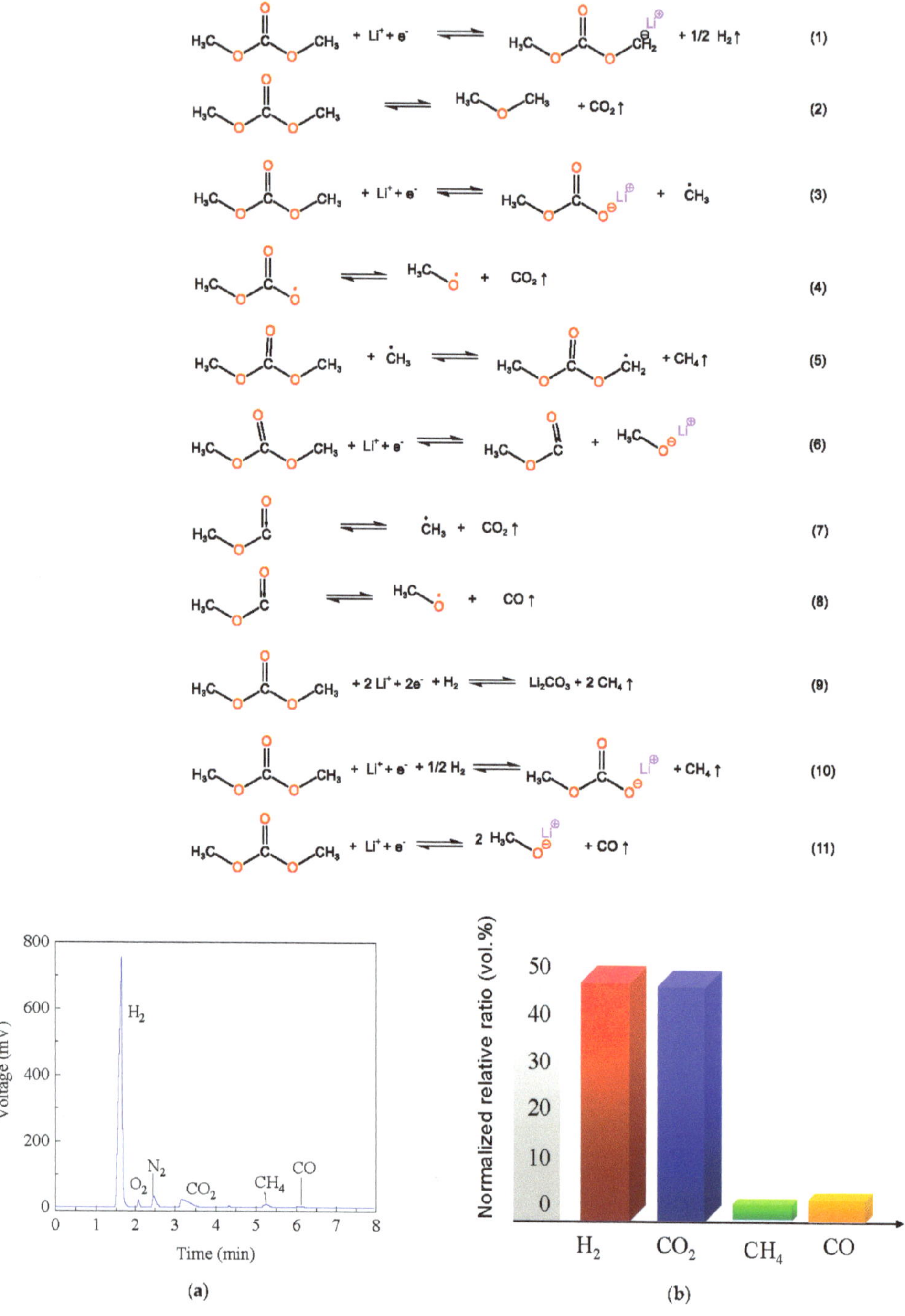

Figure 6. Identification and quantification of emitted gases: (**a**) gas chromatograph spectrum; (**b**) normalized relative volume ratio of the released gas products.

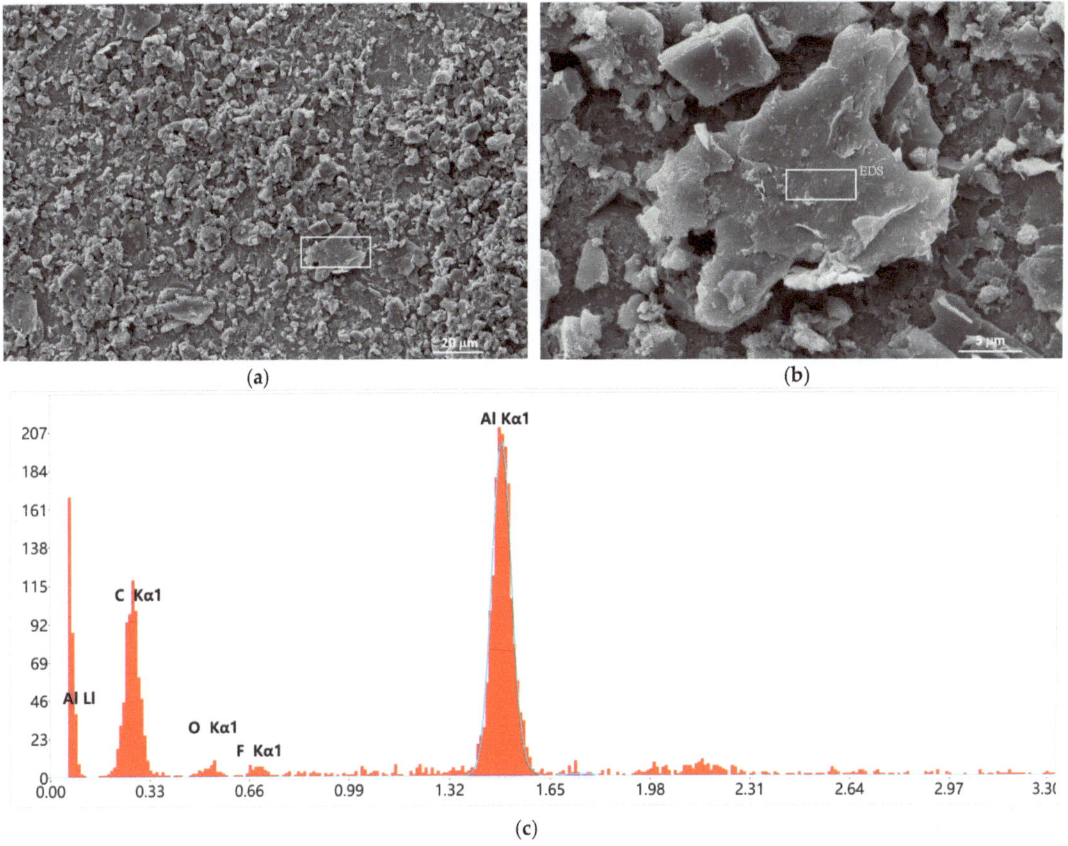

Figure 7. (**a,b**) SEM image of AC electrode after overcharging, and (**c**) EDS analysis.

4. Discussion

From the above results, it can be found that the self-protective passivation film can be formed on an AC electrode surface by EPD and PCE electrolytes. The mechanism of passivation is still not very clear, as we need more advanced interface characterizations to distinguish the contributions of lithium salt and organic solvent in this system, e.g., transmission electron microscopy (TEM), X-ray photoelectron spectroscopy (XPS), Fourier transform infrared spectroscopy (FT-IR), and cryo-electron microscopy (cryo-EM). A possible explanation is that the passivation effects are due to the synergistic effect of both lithium salt and solvent. It is noticed that the passivation film is rich in S elements, especially the sphere-like small particles. This phase is thought to involve the inorganic compounds containing Li, S, F, and O elements, which are derived from LiFSI salt and solvents. The composition of the passivation layer is presumed to be composed of LiF, Li_2CO_3, Li_2SO_3, ROLi, $ROCO_2Li$, and RSO_2Li. The dense and compact passivation film, called the solid electrolyte interphase (SEI) or cathodic electrolyte interface (CEI), is of great significance in separating the AC electrode surface from the electrolytes upon overcharging period. This self-protective passivation effect is desirable, and its formation process can be illustrated in Figure 8. Under electrical abuse and overcharging conditions, the superior overcharge endurance of EPD and PCE systems can prevent LIC cells from volume expansion, explosion, and catching fire risks. In further work, the composition and the formation mechanism of the passivation layer will be investigated in more depth. Accordingly, the electrolyte composition is expected to be optimized to bring about the self-protective passivation effect in a more reversible way.

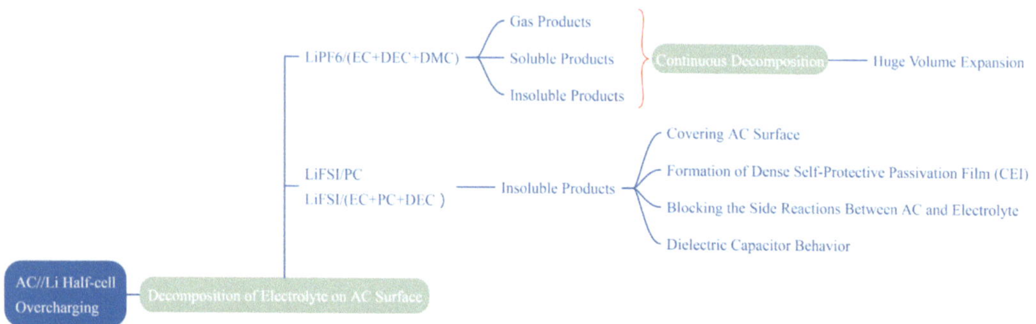

Figure 8. Schematic diagram of the decomposition of electrolyte on AC surface upon overcharging.

5. Conclusions

In this work, the stable upper potential limits of AC electrodes in three types of Li-ion electrolytes were evaluated and determined to be 4.0−4.1 V through the energy efficiency approach. Then, the AC//Li half-cells were charged to 5.0 V and 10.0 V to study the decomposition of electrolyte components. It was noticed that the self-protective passivation film could form on the AC electrode surface in EPD and PCE electrolytes, mainly ascribed to the insoluble decomposition products of LiFSI salt and organic solvent. The composition of the passivation layer is presumed to be composed of LiF, Li_2CO_3, Li_2SO_3, ROLi, $ROCO_2Li$, and RSO_2Li, and the dense and compact passivation film is greatly significant in separating the AC electrode surface from the electrolytes and restraining further electrolyte decomposition. This self-protective passivation effect is beneficial for preventing LIC cells from volume expansion and explosion risks under electrical abuse and overcharging conditions.

Author Contributions: X.S.: Conceptualization, Investigation, Writing—original draft, Writing—review & editing. Y.A.: Investigation. X.Z. (Xiong Zhang): Formal analysis, Writing—review & editing, Funding acquisition, Resources. K.W.: Funding acquisition, Formal analysis, Writing—review & editing. C.Y.: Funding acquisition, Formal analysis, Writing—review & editing. X.Z. (Xiaohu Zhang): Investigation, Validation. C.L.: Funding acquisition, Resources. Y.X.: Funding acquisition, Resources. Y.M.: Conceptualization, Methodology, Resources, Supervision, Funding acquisition. All authors have read and agreed to the published version of the manuscript.

Funding: This work was financially supported by the National Natural Science Foundation of China (Nos. 52077207, 51907193, 52171211), Youth Innovation Promotion Association, CAS (No. Y2021052), Natural Science Foundation of Shandong Province (No. ZR2021QE012), and Taishan Scholars (No. ts201712050).

Institutional Review Board Statement: Not applicable.

Informed Consent Statement: Not applicable.

Data Availability Statement: Not applicable.

Acknowledgments: Linbin Geng is thanked for her support of this work.

Conflicts of Interest: The authors declare no conflict of interest.

References

1. Simon, P.; Gogotsi, Y. Perspectives for electrochemical capacitors and related devices. *Nat. Mater* **2020**, *19*, 1151–1163. [PubMed]
2. Feng, X.; Shi, X.; Ning, J.; Wang, D.; Zhang, J.; Hao, Y.; Wu, Z.-S. Recent advances in micro-supercapacitors for AC line-filtering performance: From fundamental models to emerging applications. *eScience* **2021**, *1*, 124–140. [CrossRef]

3. Karimi, D.; Behi, H.; Van Mierlo, J.; Berecibar, M. A Comprehensive Review of Lithium-Ion Capacitor Technology: Theory, Development, Modeling, Thermal Management Systems, and Applications. *Molecules* **2022**, *27*, 3119. [CrossRef]
4. Eguchi, T.; Sugawara, R.; Abe, Y.; Tomioka, M.; Kumagai, S. Impact of full prelithiation of Si-based anodes on the rate and cycle performance of Li-ion capacitors. *Batteries* **2022**, *8*, 49. [CrossRef]
5. Han, F.; Qian, O.; Meng, G.; Lin, D.; Chen, G.; Zhang, S.; Pan, Q.; Zhang, X.; Zhu, X.; Wei, B. Structurally integrated 3D carbon tube grid-based high-performance filter capacitor. *Science* **2022**, *377*, 1004.
6. Ding, J.; Hu, W.; Paek, E.; Mitlin, D. Review of hybrid ion capacitors: From aqueous to lithium to sodium. *Chem. Rev.* **2018**, *118*, 6457–6498. [CrossRef]
7. Han, P.X.; Xu, G.J.; Han, X.Q.; Zhao, J.W.; Zhou, X.H.; Cui, G.L. Lithium ion capacitors in organic electrolyte system: Scientific problems, material development, and key technologies. *Adv. Energy Mater.* **2018**, *8*, 1801243. [CrossRef]
8. Arnaiz, M.; Canal-Rodríguez, M.; Carriazo, D.; Villaverde, A.; Ajuria, J. Enabling versatile, custom-made lithium-ion capacitor prototypes: Benefits and drawbacks of using hard carbon instead of graphite. *Electrochim. Acta* **2023**, *437*, 141456. [CrossRef]
9. Arnaiz, M.; Ajuria, J. Pre-lithiation strategies for lithium ion capacitors: Past, present, and future. *Batter. Supercaps* **2021**, *4*, 733–748. [CrossRef]
10. Yuan, J.; Qin, N.; Lu, Y.; Jin, L.; Zheng, J.; Zheng, J.P. The effect of electrolyte additives on the rate performance of hard carbon anode at low temperature for lithium-ion capacitor. *Chin. Chem. Lett.* **2022**, *33*, 3889–3893. [CrossRef]
11. Jin, L.M.; Yuan, J.M.; Shellikeri, A.; Naderi, R.; Qin, N.; Lu, Y.Y.; Fan, R.L.; Wu, Q.; Zheng, J.S.; Zhang, C.M.; et al. An overview on design parameters of practical lithium-ion capacitors. *Batter. Supercaps* **2021**, *4*, 749. [CrossRef]
12. Oca, L.; Guillet, N.; Tessard, R.; Iraola, U. Lithium-ion capacitor safety assessment under electrical abuse tests based on ultrasound characterization and cell opening. *J. Energy Storage* **2019**, *23*, 29. [CrossRef]
13. Bolufawi, O.; Shellikeri, A.; Zheng, J.P. Lithium-ion capacitor safety testing for commercial application. *Batteries* **2019**, *5*, 74. [CrossRef]
14. Soltani, M.; Beheshti, S.H. A comprehensive review of lithium ion capacitor: Development, modelling, thermal management and applications. *J. Energy Storage* **2021**, *34*, 102019. [CrossRef]
15. Wang, Z.Y.; Jiang, L.H.; Liang, C.; Zhao, C.P.; Wei, Z.S.; Wang, Q.S. Effects of 3-fluoroanisol as an electrolyte additive on enhancing the overcharge endurance and thermal stability of lithium-ion batteries. *J. Electrochem. Soc.* **2020**, *167*, 130517. [CrossRef]
16. Sun, X.; Zhang, X.; Wang, K.; An, Y.; Zhang, X.; Li, C.; Ma, Y. Determination strategy of stable electrochemical operating voltage window for practical lithium-ion capacitors. *Electrochim. Acta* **2022**, *428*, 140972. [CrossRef]
17. Liang, J.; Wang, D.W. Design Rationale and Device Configuration of Lithium-Ion Capacitors. *Adv. Energy Mater* **2022**, *12*, 2200920. [CrossRef]
18. Jin, L.; Guo, X.; Shen, C.; Qin, N.; Zheng, J.; Wu, Q.; Zhang, C.; Zheng, J.P. A universal matching approach for high power-density and high cycling-stability lithium ion capacitor. *J. Power Sources* **2019**, *441*, 227211. [CrossRef]
19. El Ghossein, N.; Sari, A.; Venet, P. Development of a capacitance versus voltage model for lithium-ion capacitors. *Batteries* **2020**, *6*, 54. [CrossRef]
20. Zheng, W.; Li, Z.; Han, G.; Zhao, Q.; Lu, G.; Hu, X.; Sun, J.; Wang, R.; Xu, C. Nitrogen-doped activated porous carbon for 4.5 V lithium-ion capacitor with high energy and power density. *J. Energy Storage* **2022**, *47*, 103675. [CrossRef]
21. Santos, M.C.G.; Silva, G.G.; Santamaria, R.; Ortega, P.F.R.; Lavall, R.L. Discussion on operational voltage and efficiencies of ionic-liquid-based electrochemical capacitors. *J. Phys. Chem. C* **2019**, *123*, 8541. [CrossRef]
22. Bahdanchyk, M.; Hashempour, M.; Vicenzo, A. Evaluation of the operating potential window of electrochemical capacitors. *Electrochim. Acta* **2020**, *332*, 135503. [CrossRef]
23. Cougnon, C. Exploring the interdependence between the coulombic, voltage and energy efficiencies. *Electrochem. Commun.* **2020**, *120*, 106832. [CrossRef]
24. Levi, M.; Levy, N.; Sigalov, S.; Salitra, G.; Aurbach, D.; Maier, J. Electrochemical quartz crystal microbalance (EQCM) studies of ions and solvents insertion into highly porous activated carbons. *J. Am. Chem. Soc.* **2010**, *132*, 13220. [CrossRef] [PubMed]
25. Qin, N.; Jin, L.; Lu, Y.; Wu, Q.; Zheng, J.; Zhang, C.; Chen, Z.; Zheng, J.P. Over-potential tailored thin and dense lithium carbonate growth in solid electrolyte interphase for advanced lithium ion batteries. *Adv. Energy Mater.* **2022**, *12*, 2103402. [CrossRef]
26. Sun, X.; An, Y.; Geng, L.; Zhang, X.; Wang, K.; Yin, J.; Huo, Q.; Wei, T.; Zhang, X.; Ma, Y. Leakage current and self-discharge in lithium-ion capacitor. *J. Electroanal. Chem.* **2019**, *850*, 113386. [CrossRef]
27. Ishimoto, S.; Asakawa, Y.; Shinya, M.; Naoi, K. Degradation responses of activated-carbon-based EDLCs for higher voltage operation and their factors. *J. Electrochem. Soc.* **2009**, *156*, A563. [CrossRef]
28. An, Z.; Xu, X.; Fan, L.; Yang, C.; Xu, J. Investigation of Electrochemical Performance and Gas Swelling Behavior on Li$_4$Ti$_5$O$_{12}$/Activated Carbon Lithium-Ion Capacitor with Acetonitrile-Based and Ester-Based Electrolytes. *Electronics* **2021**, *10*, 2623. [CrossRef]
29. Cao, W.; Li, Q.; Yu, X.; Li, H. Controlling Li deposition below the interface. *eScience* **2022**, *2*, 47. [CrossRef]
30. Fernandes, Y.; Bry, A.; de Persis, S. Thermal degradation analyses of carbonate solvents used in Li-ion batteries. *J. Power Sources* **2019**, *414*, 250. [CrossRef]

31. Wang, Q.; Mao, B.; Stoliarov, S.I.; Sun, J. A review of lithium ion battery failure mechanisms and fire prevention strategies. *Prog. Energy Combust. Sci.* **2019**, *73*, 95. [CrossRef]
32. Fernandes, Y.; Bry, A.; de Persis, S. Identification and quantification of gases emitted during abuse tests by overcharge of a commercial Li-ion battery. *J. Power Sources.* **2018**, *389*, 106. [CrossRef]

Article

Efficient Battery Models for Performance Studies-Lithium Ion and Nickel Metal Hydride Battery

Umapathi Krishnamoorthy [1,*], Parimala Gandhi Ayyavu [2], Hitesh Panchal [3], Dayana Shanmugam [1], Sukanya Balasubramani [1], Ali Jawad Al-rubaie [4], Ameer Al-khaykan [5], Ankit D. Oza [6], Sagram Hembrom [7], Tvarit Patel [8], Petrica Vizureanu [9,10] and Diana-Petronela Burduhos-Nergis [10,*]

[1] Department of Electrical and Electronics Engineering, M. Kumarasamy College of Engineering, Karur 639113, India

[2] Department of Electronics and Communication Engineering, Kalaignar Karunanidhi Institute of Technology, Coimbatore 641402, India

[3] Department of Mechanical Engineering, Government Engineering College, Patan 384265, India

[4] Department of Medical Instrumentation Engineering Techniques, Al-Mustaqbal University College, Hillah 51001, Iraq

[5] Department of Air Conditions and Refrigeration Techniques, Al-Mustaqbal University College, Hillah 51001, Iraq

[6] Department of Computer Sciences and Engineering, Institute of Advanced Research, The University for Innovation, Gandhinagar 382426, India

[7] Department of Metallurgical Engineering, B.I.T. Sindri, Dhanbad 828123, India

[8] Department of Biological Sciences and Biotechnology, Institute of Advanced Research, The University for Innovation, Gandhinagar 382426, India

[9] Faculty of Materials Science and Engineering, "Gheorghe Asachi" Technical University, 700050 Iasi, Romania

[10] Technical Sciences Academy of Romania, Dacia Blvd 26, 030167 Bucharest, Romania

* Correspondence: umapathi.uit@gmail.com (U.K.); diana.burduhos@tuiasi.ro (D.-P.B.-N.)

Citation: Krishnamoorthy, U.; Gandhi Ayyavu, P.; Panchal, H.; Shanmugam, D.; Balasubramani, S.; Al-rubaie, A.J.; Al-khaykan, A.; Oza, A.D.; Hembrom, S.; Patel, T.; et al. Efficient Battery Models for Performance Studies-Lithium Ion and Nickel Metal Hydride Battery. *Batteries* **2023**, *9*, 52. https://doi.org/10.3390/batteries9010052

Academic Editors: Junsheng Zheng and Carlos Ziebert

Received: 7 October 2022
Revised: 10 December 2022
Accepted: 9 January 2023
Published: 12 January 2023

Abstract: Apart from being emission-free, electric vehicles enjoy benefits such as low maintenance and operating costs, noise-free, easy to drive, and the convenience of charging at home. All these benefits are directly dependent on the performance of the battery used in the vehicle. In this paper, one-dimensional modeling of Li-ion and NiMH batteries was developed, and their performances were studied. The performance characteristics of the batteries, such as the charging and discharging characteristics, the constituent losses of over-potential voltage, and the electrolyte concentration profile at various stages of charge and discharge cycles, were also studied. It is found that the electrolyte concentration profiles of Li-ion batteries show a drooping behavior at the start of the discharge cycle and a rising behavior at the end of discharge because of the concentration polarization due to the low diffusion coefficient. The electrolyte concentration profiles of NiMH batteries show rising behavior throughout the discharge cycle without any deviations. The reason behind this even behavior throughout the discharge cycle is attributed to the reduced concentration polarization due to electrolyte transport limitations. It is found that the losses associated with the NiMH battery are larger and almost constant throughout the battery's operation. Whereas for the Li-ion batteries, the losses are less variable. The electrolyte concentration directly affects the overpotential losses incurred during the charging and discharging phases.

Keywords: electric vehicles (EV); batteries; Li-ion; NiMH; concentration profile; losses

1. Introduction

Enlightened to create a green and safe environment for our future generations, humans have started exploring ways to stop global warming. One-fifth of the total carbon dioxide emissions contributing to global warming is from the transportation division. In addition, this reveals to us the importance of moving toward greener transportation systems. With this said, countries have started encouraging the research, development, and utilization of electric vehicles by funding researchers, reducing taxes and giving subsidies for electric vehicles.

A varied range of research are being done on electric vehicles from the performance metrics of electric drives used, charging station utilization, Electromagnetic Interferences, testing of EVs., etc and a review of them is done with the objective of listing the recent research areas in electric vehicle design. A finite element analysis is carried out on a permanent magnet synchronous motor at different operating speeds in order to evaluate its performance metrics in [1]. Optimized charge scheduling is a means of reducing traffic jams at charging stations and with this objective an effective charge scheduling algorithm is developed say Grey Sail Fish Optimization based on grey wolf and sail fish optimizers in [2]. An emulator framework is proposed based on arduino UNO for testing of E-bicycles in [3]. A switched reluctance motor design for high torque is proposed in [4] based on the concept of slotting the stator's periphery. The various electromagnetic interferences that intrude an electric vehicle performance are studied and an LC filter design explained in [5]. A simple electric vehicle model is created and analyzed in [6] whereas a fuzzy logic based energy management technique is proposed in [7] for hybrid vehicles. A simulation model of the electric drive system is carried out in [8] with the aim of improving the speed torque performance of a BLDC motor drive system. A neuro-fuzzy based ANFIS controller is proposed with the aim of improving the power quality of a PV powered system in [9]. A deep learning based driver assistance and detection system is proposed in [10].

Most portable electronic devices, from smartphones to laptops to huge electric vehicles, use Li-ion batteries, preferably due to their many benefits. The primary advantages of using Li-ion batteries include their higher energy density, lower self-discharge rates, low maintenance, higher cell voltage, constant load characteristics, and no priming required. The major drawback of LIBs is that they are not robust and require a protection circuit to keep the charging and discharging levels within safe limits and a cell temperature monitoring circuit to prevent temperature extremes. Another drawback is the aging of the battery with the number of charging and discharging cycles, and lithium-ion batteries are 40% more expensive than nickel-cadmium batteries. Li-ion batteries are still blossoming, and many performance improvements are needed, creating much room for research and development.

NiMH batteries exhibit a lower energy density than Li-ion batteries but can tolerate overcharging and discharging, and environmental compatibility and safety, making them suitable for portable power tools and hybrid electric vehicles. NiMH batteries are less expensive than Li-ion batteries but are less durable than the others. A Li-ion battery interacts with fabric properties to determine how depleted and charged it is. A developer can examine the effects of various design factors such as separator length, electrode, charge duration, and an initial electrolyte salt.

Modeling results for Li-ion batteries are made, and it is found that there is a good agreement between simulation and experimental results [11]. Neglecting the diffusion of nickel oxide active material, a NiMH battery model is explained in [12]. It is also discovered to be sensitive to the material's kinetic parameters. Comprehensive modeling considering the kinetic parameters, Ohm's law, and charge balances are explained in [13]. A 3D model of a Li-ion battery is created and used to investigate thermal behaviors. It is found that overcharging of the battery leads to a massive rise in temperature and a sharp gradient within the battery [14]. Ref. [15] introduced a 2D model of the NiMH battery to study temperature variations. Results in [16] indicate that a nominally uniform temperature profile can be achieved when the thermal conductivity increases. A planar electrode approximation is used in modeling a NiMH cell, accounting for active species, electrochemical kinetics, and ohmic effects. The predictions show results on par with the experimental data presented in [17].

A fast-rechargeable NiMH battery model is presented in [18]. Online simulation is used to model the high-capacity battery cell for plug-in HEV research [19]. The various electrode materials used to improve battery performance are summarized in [20,21]. A thin-film lithium-ion battery design is presented in [22]. A model that helps find the operational parameters for the battery is presented in [23]. A review of the effects of lithium deposition in Li-cells is discussed in [24]. Efficient techniques for the characterization and detection of battery degradation are explained in the references [25,26]. The effect of rapid battery charging on the age of the battery is described in [27]. A detailed review of the literature reveals that modeling Li-ion and NiMH batteries will be necessary for research. Thus, this paper addresses the modeling of both batteries and highlights their various characteristics.

As the battery model based on the Adaptive Battery State Estimator (ABSE) is inaccurate because load dependencies are eliminated, a new, improved ABSE-based model is proposed in [28]. It estimates a battery's state of charge and available power. A performance-enhanced reduced electrochemical model along with a dual non-linear filter to obtain the online parameters is suggested in [29] for the estimation of the state of charge (SoC) and state of health (SoH). In [30], an electrolyte-enhanced composite single-particle model of lithium-ion batteries is proposed, and the performance of the same is found to be better than the standard single-particle model.

The non-linear behavior of the lithium batteries is accounted for by introducing the Weiner structure into the traditional equivalent circuit model, and a 1.5% improvement in SOC concentration is attained [31]. When modeling a battery, it is well known that reduced-order models are faster and data-driven models are more accurate; thus, in [32], a two-level model of a lithium battery is created by combining the two. In addition, it was demonstrated that combining the two produced better results than operating separately.SOH estimation using a 2RC model is proposed in [33] and has shown to be effective with lower root mean square error when compared to the 1RC model. A convolutional neural network that learns the temporal and conditional dynamics of the lithium-ion battery is used to study the degradation trends of the battery [34] and such a model is found to predict future conditions accurately. Ref. [35] conducts a study of the features and modeling of lithium-ion batteries. A data model is described to predict the battery characteristics in [36] effectively. A generic model that can be used for evaluating any battery cell is modeled and verified in [37]. A finite element analysis is used to predict the thermal performance capability of a NiMH battery, and this model is found to help simulate charging and discharging cycles [38]. A non-chemical, partial non-linear model is developed for lead-acid batteries and found to predict SOC accurately. This model described in [39] can also be applied to other battery types by making proper modifications to parameter values. In [40], a simple model independent of self-discharge, temperature, and the number of cycles of discharge is developed for NiMH and Li-ion batteries, and it is found that the model build data is less than 0.4% in error from the validated experimental data. In [41], battery models were developed based on Thevenin's equivalent circuit. It is found that the third-order model matches the experimental data very well compared to the other lower-order models.

A mechanical analogy of an electrochemical battery is modeled in [42], and it was found to produce near-experimental results for the time-domain response of the battery. An ANN-based prediction model based on a dataset developed from Thevenin's equivalent model is described in [43] and found to work with an average error of 4%. The basics of various parameters and characteristics to be studied for effective modeling of NiMH batteries are explained in [44]. A temperature analysis model based on the finite element method is detailed in [45]. It compared the temperature variation with the geometry of the cell. A thermal model of a battery with embedded and distributed temperature sensors is proposed to measure the real-time distributed measurement of lithium-ion battery internal and surface temperature profiles [46]. An algorithmic framework based on a deep reinforcement learning-based optimizer is introduced for the fast charging of the lithium-ion battery [47].

2. Lithium-Ion Battery Model

The negative porous electrode domain dimensions of the Battery are 320 mm, the polymer electrolyte domain dimensions are 52 mm, and the positive porous electrode domain dimensions are 183 mm. This model is built by having an electrolyte of 2 M LiPF6 salt with LiyMn$_2$O$_4$ for the anode and a carbon-based cathode. The parameters of Material balance and Ionic Charge balance are modeled using a polymer matrix and a 1:1 electrolyte of plasticized Ethylene Carbonate/Di-Methyl Carbonate. The total electrolyte volume is the composition of the electrolyte in a liquid state that is available as a polymer.

The cell's voltage while in a conducting state is calculated using ionic charge balance and Ohm's law. The charge transfer reactions result in high and low voltages. As the electrodes are porous, while modeling their conductivities, porosity and tortuosity are considered and are given by, σ_s^{eff}

$$\sigma_s^{eff} = \sigma_s \varepsilon^\gamma \tag{1}$$

γ is the Bruggeman co-efficient set to 3.3 when the diffusivity and electrolyte conductivity are treated similarly. Spherical coordinates describe the diffusion equation. In addition, Butler-Volmer equations introduce high and low terms.

2.1. Boundary Conditions

After generating the current density waveform with 0 V applied to the negative electrode and a given potential at the positive electrode, it was discovered that the current density discharged, went to zero for a brief period, and then entered the charging stage. The boundaries are considered insulating for both material and ionic charge balance.

2.2. Material Properties

As said before, an electrolyte of 2 M LiPF6 salt with LiyMn$_2$O$_4$ for the anode and carbon-based cathode. Moreover, the electrolyte of plasticized Ethylene Carbonate/Di-Methyl Carbonate in the ratio of 1:2 is used. Significant variations in the electrolyte conductivity and the equilibrium potential are noticed during the charging and discharging phases of the battery's characteristics. Figure 1 shows the interpolation of the parameter ionic conductivity with the increase in electrolyte concentration whereas Figures 2 and 3 depicts the potential variations at the electrodes.

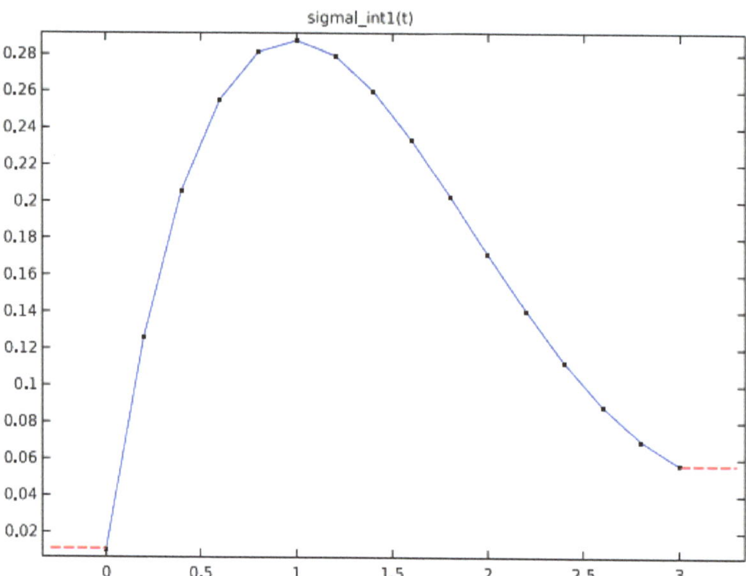

Figure 1. The Interpolation of ionic conductivity with electrolyte concentration.

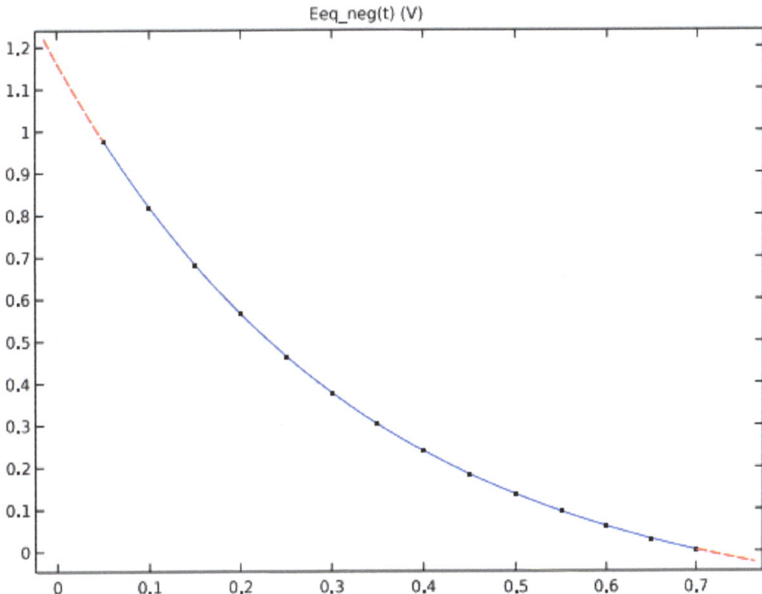

Figure 2. Potential of the two electrodes at Equilibrium based on (SOC).

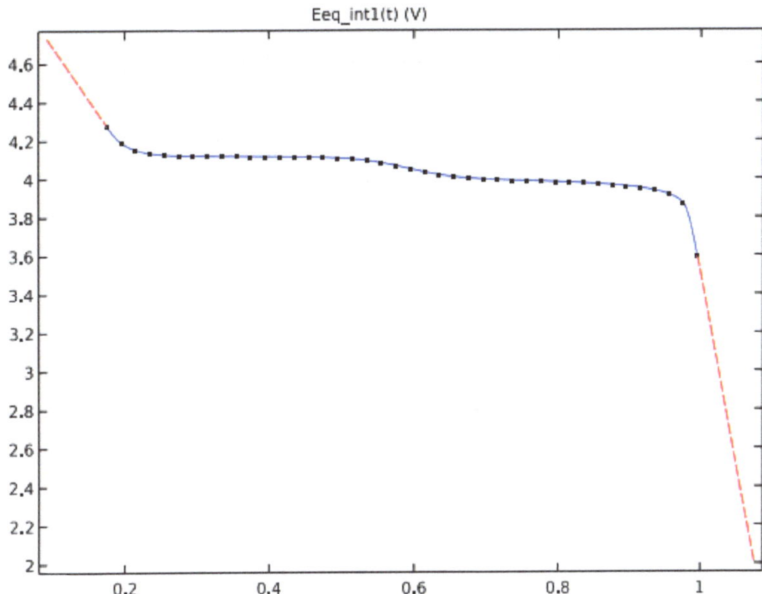

Figure 3. Potential Variation in the electrode at equilibrium.

2.3. Discharge Characteristics

The battery capacity is studied at different discharge rates by studying the behavior of the battery at various current densities. When the cell voltage drops below 3 volts, it is considered to be the end of discharge. It is found that the full theoretical discharge occurs for 1 C, and the corresponding current density is 17.6 A/m^2.

Initial SoC values of 0.17 and 0.56 are assigned for the positive and negative electrodes that contribute approximately 4.22 V at full charge, which is nothing but the open circuit cell voltage. Figure 4 shows that the maximum discharge occurs for 1 C, which is 1.75 A/m^2. A charge density grows, discharge ends quickly, and for 4 C, the battery delivers over 50% before 3 V, the end of discharge.

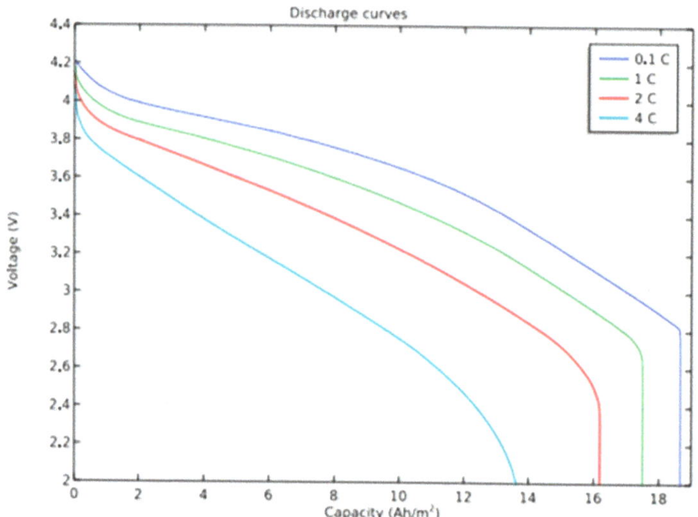

Figure 4. Discharge Curves for different rates of discharge.

2.4. Charging and Discharging Cycles

Figure 5 below shows the charge-discharge cycles of the model. The graph shows that the cycle has the 2000 s of discharge, 300 s of open circuit, and then 2000 s of charging at a relatively small current density before becoming an open circuit. At zero current, the ohmic loss is around 100 mV, and the concentration increases over a potential of 50 mV. The constituent losses due to overpotential at the start and end of discharging can be visualized for further analysis. These variations can be brought into the same plot by varying the bias values at the start and end of the discharge plot. The reaction over potential is shown without bias to obtain the beginning of discharge (at 148 mV) and ending of discharge (558 mV). This way, all the plots will be within the same range of potential.

Figure 6 is the plot that shows the details of the constituents of the over-potential losses during the start and end of the discharge cycle, along with the reaction over-potential. The charging and discharging curves for different electrolyte concentration profiles can be used to investigate the cause of the steep voltage decrease.

Figure 7 depicts the electrolyte concentration profile (*y*-axis) across the cell's width (*x*-axis) at various stages of the discharging and charging cycles for the Li-ion battery model. The electrolyte occupies a distance of 100 μm to 150 μm across the width of the cell, and the electrolyte concentration across this distance remains almost constant. For simulation, a discharge of 2000 s at nominal current density, 300 at open circuit, and 2000 of charging at nominal current density is applied before the circuit is finally open-circuited. Thus, the interval from the 2000 s to the 2400 s depicts the losses due to ohmic resistance and concentration overpotential when a cell is in open circuit condition. Then it transitions to the charging stage. These two losses are reflected as a drooping electrolyte concentration profile across the cell width, whereas the charging cycle starts at 3000 s, increasing the electrolyte concentration. Hence, at 3000 s, the concentration variation is reversed. The reason for this concentration variation is the concentration polarization induced by a low diffusion coefficient. These variations are studied to find the reason for the variation in over-potential losses during the discharging and charging periods, as depicted in Figure 6. For lithium-ion batteries, over-potential losses are greater at the end of the discharge than at the beginning.

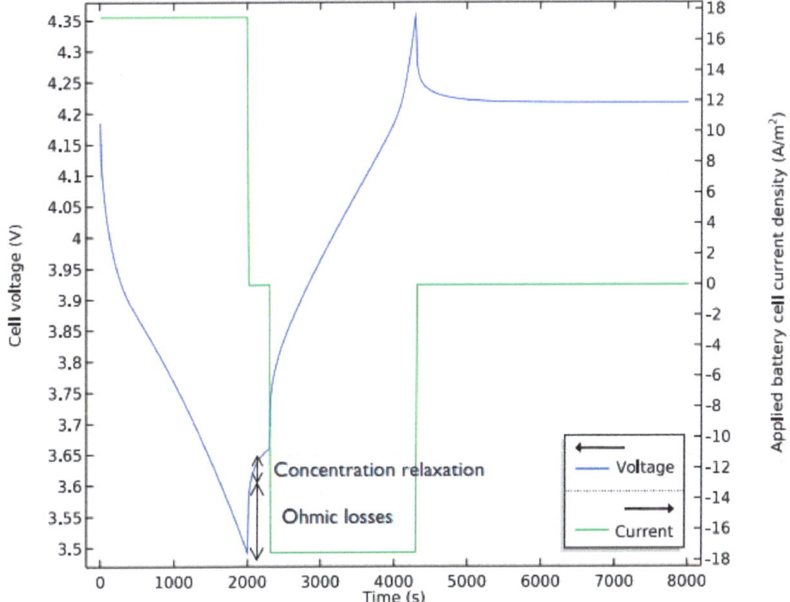

Figure 5. V&I curve over the discharging and charging time.

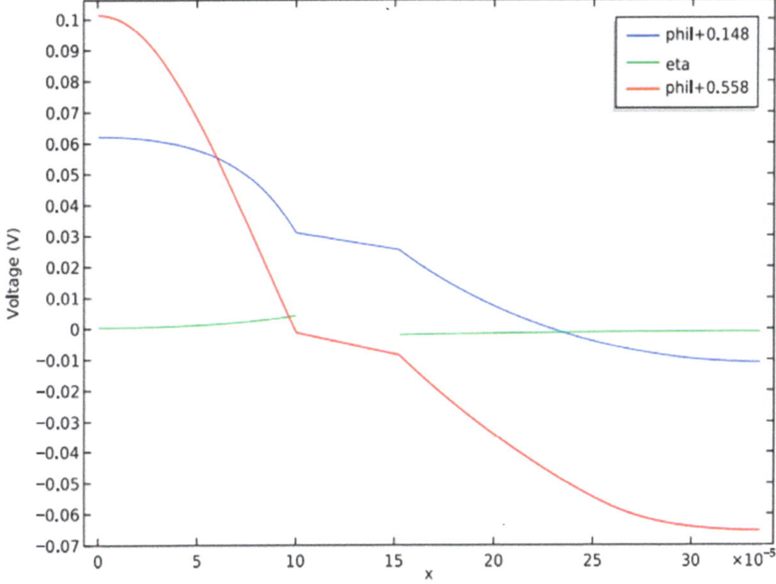

Figure 6. Constituent losses of over potential loss in Battery.

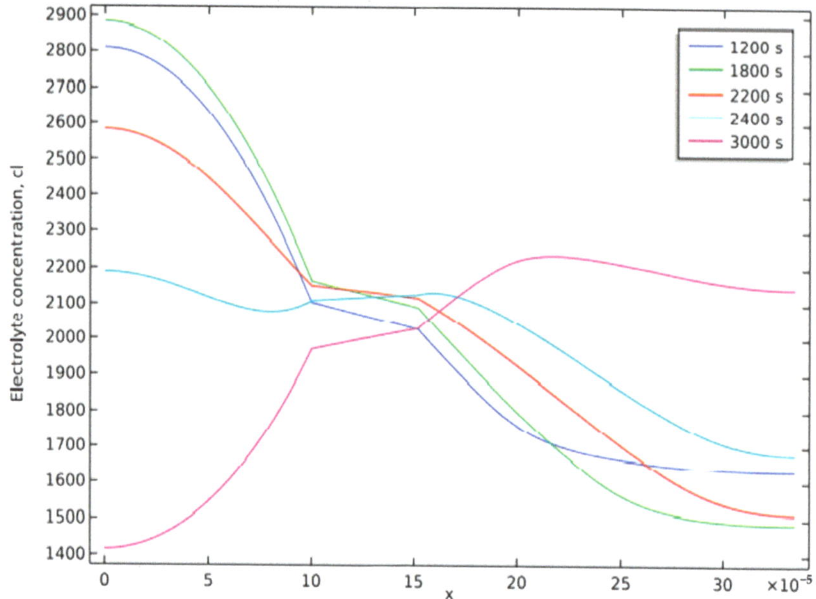

Figure 7. Concentration profile at various stages.

3. Nickel Metal Hydride Battery Model

The battery domain dimensions are 350 μm metal hydride for the porous cathode, 250 μm electrolyte separator, and 443 μm nickel oxide for the porous anode. This model is built by having an electrolyte of KOH diluted in water to a 30% (wt) solution. The parameters of material balance and ionic charge balance are modeled for a 1:1 binary electrolyte. The electrode reactions involve both an ion (OH^-) and a solvent (H_2O).

The cell's voltage while in a conducting state is calculated using ionic charge balance and Ohm's law. The charge transfer reactions result in high and low voltages. Because the electrodes are porous, porosity and tortuosity are taken into account when modeling their conductivities and are given by σ_s^{eff} (same as Equation (1)). the Bruggeman coefficient is set to 1.5, corresponding to spherical particles. Spherical coordinates describe the diffusion equation. In addition, the Butler-Volmer equations define the charge density in the electrodes by introducing high and low terms.

3.1. Boundary Conditions

The negative electrode is set to 0 V, the potential at the positive electrode is specified, and then the current density waveform is generated. It is found that the current density discharges, then becomes zero for a short duration, and finally enters the charging stage. The boundaries are considered insulating for both material and ionic charge balance.

3.2. Material Properties

A metal hydride ($LaNi_5Hx$)-based negative electrode, a nickel oxide ($NiOHOHy$) for the positive electrode, and an electrolyte of KOH, diluted in water to 30%,wt., are used. Figures 8 and 9 depicts the potential variations at negative and positive electrodes at equilibrium based on State of Charge.

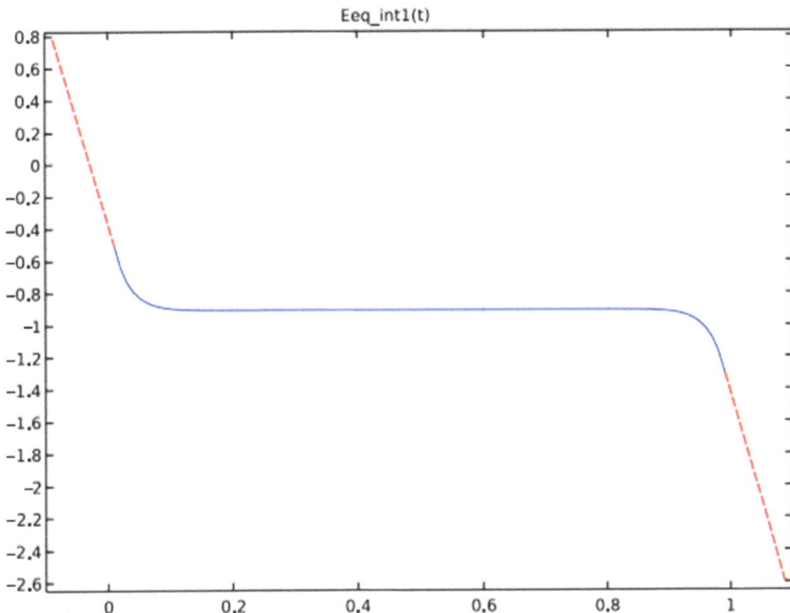

Figure 8. Potential of the two electrodes at Equilibrium based on (SOC).

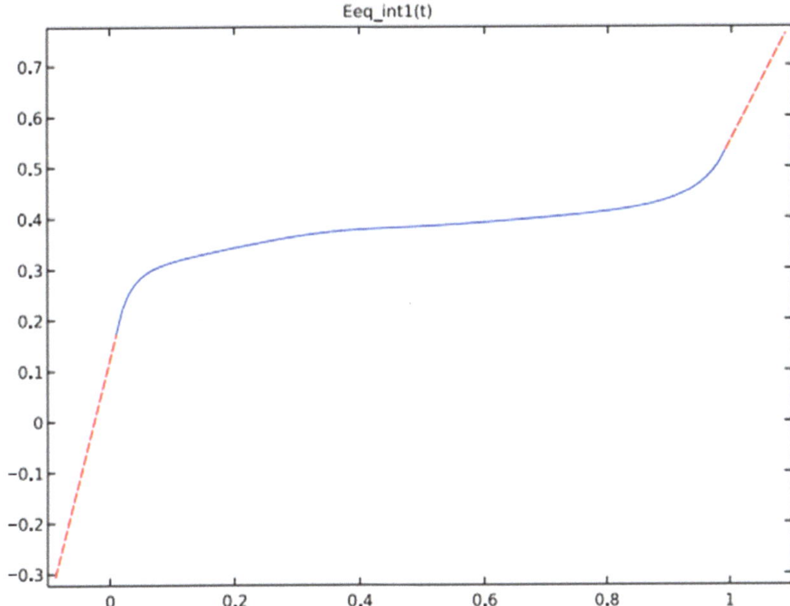

Figure 9. Equilibrium voltage of electrode materials.

Significant variations in electrolyte conductivity and the potential at equilibrium are noticed during the charging and discharging phases of the battery's life.

3.3. Discharge Characteristics

Initially, the battery is assumed to be fully charged, and the discharges at two different current densities are simulated to study its characteristics. When the cell voltage drops below 0.99 V, it is considered the end of discharge. It is found that the full theoretical discharge occurs at 1 C, and the corresponding current density is 430 A/m^2.

Figure 10 shows that the maximum discharge capacity is obtained for a current density of 43 A/m^2 at 0.1 C. At 1 C, the battery reaches 90% of its theoretical capacity before it reaches 1 V. As there is no side reaction in the model, the voltage is higher than normal as discharge begins. Similar to the case with Li-ion batteries, the contribution of other losses to total overpotential can be studied by introducing a bias of 0.91 V. Figure 11 shows other losses contributing to the over-potential loss at 1 C discharge.

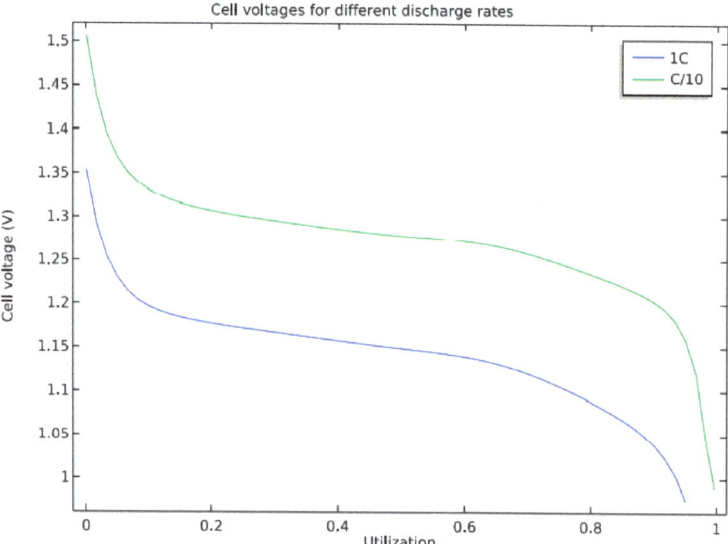

Figure 10. Discharge Curves for a different rate of discharge.

The causes of the steep voltage decrease can be studied by investigating the charging and discharging cycles at different time intervals. As electrolyte transport has limitations, the cell experiences little polarization due to concentration variations at different intervals. The causes of the steep voltage decrease can be studied by investigating the charging and discharging cycles at different time intervals. As electrolyte transport has limitations, the cell experiences little polarization due to concentration variations at different intervals.

Figure 12 depicts the NiMH battery model's electrolyte concentration profile (*y*-axis) across the cell's width (*x*-axis) at various discharging and charging cycle stages. The electrolyte occupies a distance of 350 μm to 600 μm across the cell's width, and the electrolyte concentration across this distance remains almost constant. For simulation, similar to Li-ion battery modeling, a discharge of 2000 s at nominal current density, 300 s at open circuit, and 2000 s of charging at nominal current density are applied before the battery is finally open-circuited. Because of electrolyte transport limitations, there is only little polarization; hence, gradients are pretty low across the discharging and charging cycles. This results in a more or less constant local charge transfer current density, but it is a considerable over-voltage loss in the battery.

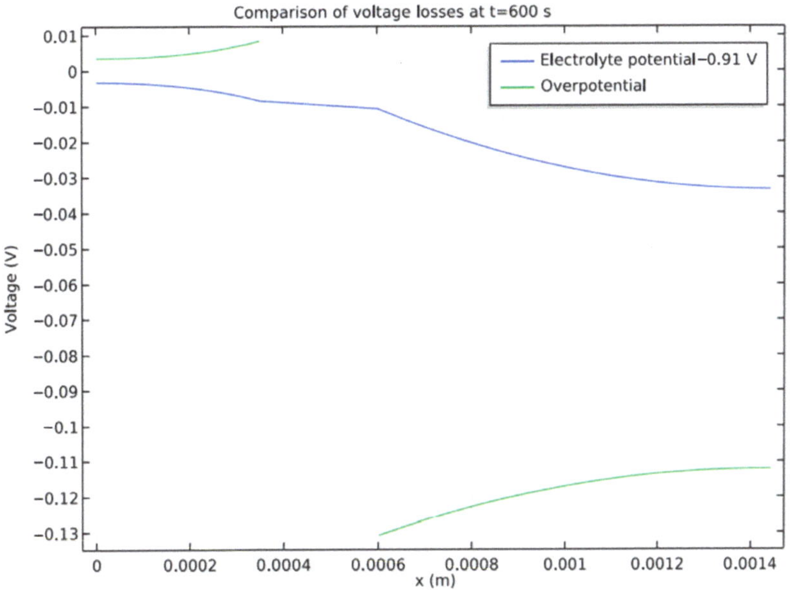

Figure 11. Constituent losses of over potential loss in Battery.

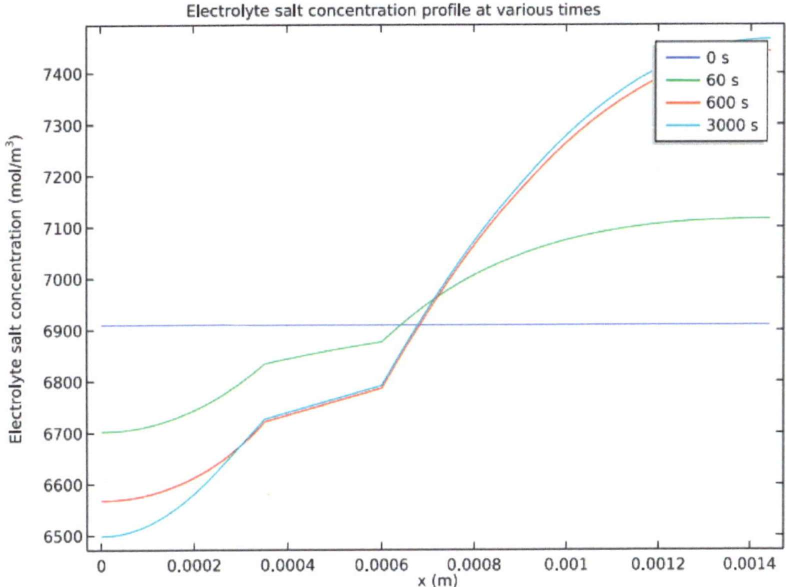

Figure 12. Concentration profile at various stages.

4. Conclusions

This paper analyzes the behavior of the Li-ion and NiMH battery 1D models under isothermal operating conditions. The advantage of using battery models is their ability to predict the cell's current, voltage distribution, and electrolyte concentration under operating conditions. The uniqueness of this work is that it compares the electrolyte concentration variation across the battery's discharging, open circuit, and charging phases, i.e., from 0 to 2000 s, 2000 s to 2300 s, 2300 s to 3000 s, respectively, for a 1D model of Li-ion and

NiMH batteries. The performance characteristics of the batteries, such as the charging and discharging characteristics, the constituent losses of over-potential voltage, and the electrolyte concentration profile at various stages of charge and discharge cycles, were also studied. This work helps analyze and understand the variation in over-potential losses associated with the different stages of battery operation. Voltage and current changes during operation are analyzed, as well as the effect of electrolyte concentration on the losses. These models help to understand the battery performance under different operating conditions for electric vehicles. The results prove that the electrolyte concentration directly affects the losses incurred during the charging and discharging of batteries. This model can be integrated with a machine-learning algorithm for prediction in the future to improve its performance.

Author Contributions: U.K.–Conceptualization; writing—original draft preparation; P.G.A.— methodology; writing—original draft preparation; A.J.A.-r.—formal analysis; data curation; methodology; H.P.—writing—original draft preparation; formal analysis; data curation; D.S.— methodology; data curation; S.B.—investigation, methodology; A.A.-k.—validation, investigation; A.D.O.—project administration; formal analysis; S.H.- formal analysis; data curation; methodology; T.P.—investigation, methodology; P.V.—writing—review and editing; funding acquisition; D.-P.B.-N.—writing—review and editing; funding acquisition. All authors have read and agreed to the published version of the manuscript.

Funding: This work was supported by the Gheorghe Asachi Technical University of Iaşi—TUIASI-Romania, Scientific Research Funds, FCSU-2022.

Institutional Review Board Statement: Not applicable.

Informed Consent Statement: Not applicable.

Data Availability Statement: Not applicable.

Conflicts of Interest: The authors declare that they have no conflict of interest.

References

1. Sheela, A.; Suresh, M.; Shankar, V.G.; Panchal, H.; Priya, V.; Atshaya, M.; Sadasivuni, K.K.; Dharaskar, S. FEA based analysis and design of PMSM for electric vehicle applications using magnet software. *Int. J. Ambient. Energy* **2022**, *43*, 2742–2747. [CrossRef]
2. Rajamoorthy, R.; Arunachalam, G.; Kasinathan, P.; Devendiran, R.; Ahmadi, P.; Pandiyan, S.; Muthusamy, S.; Panchal, H.; Kazem, H.A.; Sharma, P. A novel intelligent transport system charging scheduling for electric vehicles using Grey Wolf Optimizer and Sail Fish Optimization algorithms. *Energy Sources Part A Recovery Util. Environ. Eff.* **2022**, *44*, 3555–3575. [CrossRef]
3. Ashokkumar, R.; Suresh, M.; Sharmila, B.; Panchal, H.; Gokul, C.; Udhayanatchi, K.V.; Sadasivuni, K.K.; Israr, M. A novel method for Arduino based electric vehicle emulator. *Int. J. Ambient. Energy* **2022**, *43*, 4299–4304. [CrossRef]
4. Patel, M.A.; Asad, K.; Patel, Z.; Tiwari, M.; Prajapati, P.; Panchal, H.; Suresh, M.; Sangno, R.; Israr, M. Design and optimisation of slotted stator tooth switched reluctance motor for torque enhancement for electric vehicle applications. *Int. J. Ambient. Energy* **2022**, *43*, 4283–4288. [CrossRef]
5. Karthik, M.; Usha, S.; Venkateswaran, K.; Panchal, H.; Suresh, M.; Priya, V.; Hinduja, K.K. Evaluation of electromagnetic intrusion in brushless DC motor drive for electric vehicle applications with manifestation of mitigating the electromagnetic interference. *Int. J. Ambient. Energy* **2020**. [CrossRef]
6. Sharmila, B.; Srinivasan, K.; Devasena, D.; Suresh, M.; Panchal, H.; Ashokkumar, R.; Meenakumari, R.; Kumar Sadasivuni, K.; Shah, R.R. Modelling and performance analysis of electric vehicle. *Int. J. Ambient. Energy* **2022**, *43*, 5034–5040. [CrossRef]
7. Anbazhagan, G.; Jayakumar, S.; Muthusamy, S.; Sundararajan, S.C.; Panchal, H.; Sadasivuni, K.K. An effective energy management strategy in hybrid electric vehicles using Taguchi based approach for improved performance. *Energy Sources Part A Recovery Util. Environ. Eff.* **2022**, *44*, 3418–3435. [CrossRef]
8. Balan, G.; Sundararajan, S.C.; Arumugam, S.; Manoharan, M.; Muthukrishnan, K.; Muthusamy, S.; Panchal, H.; Varadharaj, K.; Sadasivuni, K.K.; Jayakumar, S. Performance analysis and enhancement of brain emotion-based intelligent controller and its impact on PMBLDC motor drive for electric vehicle applications. *Energy Sources Part A Recovery Util. Environ. Eff.* **2022**, *44*, 5640–5664. [CrossRef]
9. Cholamuthu, P.; Irusappan, B.; Paramasivam, S.K.; Ramu, S.K.; Muthusamy, S.; Panchal, H.; Nuvvula, R.S.S.; Kumar, P.P.; Khan, B. A Grid-Connected Solar PV/Wind Turbine Based Hybrid Energy System Using ANFIS Controller for Hybrid Series Active Power Filter to Improve the Power Quality. *Int. Trans. Electr. Energy Syst.* **2022**, *2022*, 9374638. [CrossRef]

10. Balan, G.; Arumugam, S.; Muthusamy, S.; Panchal, H.; Kotb, H.; Bajaj, M.; Ghoneim, S.S.M.; Kitmo, U. An Improved Deep Learning-Based Technique for Driver Detection and Driver Assistance in Electric Vehicles with Better Performance. *Int. Trans. Electr. Energy Syst.* **2022**, *2022*, 854817. [CrossRef]

11. Doyle, M.; Newman, J.; Gozdz, A.S.; Schmutz, C.N.; Tarascon, J.M. Comparison of Modeling Predictions with Experimental Data from Plastic Lithium Ion Cells. *J. Electrochem. Soc.* **1996**, *143*, 1890–1903. [CrossRef]

12. Paxton, B.; Newman, J. Modeling of Nickel/Metal Hydride Batteries. *J. Electrochem. Soc.* **1997**, *144*, 3818–3831. [CrossRef]

13. Albertus, P.; Christensen, J.; Newman, J. Modeling Side Reactions and Non-isothermal Effects in Nickel Metal-Hydride Batteries. *J. Electrochem. Soc.* **2008**, *155*, A48–A60. [CrossRef]

14. Jahantigh, N.; Afsharia, E. Thermal analysis of nickel/metal (Ni/MH) hydride battery during charge cycle. In Proceedings of the 3rd IASME/WSEAS international conference on Energy & Environment (EE'08), Cambridge, UK, 23–25 February 2008; World Scientific and Engineering Academy and Society (WSEAS): Stevens Point, WI, USA, 2008; pp. 73–78.

15. Shi, J.; Wu, F.; Chen, S.; Zhang, C. Thermal analysis of rapid charging nickel/metal hydride batteries. *J. Power Sources* **2006**, *157*, 592–599. [CrossRef]

16. Araki, T.; Nakayama, M.; Fukuda, K.; Onda, K. Thermal Behavior of Small Ni/MH Battery during Rapid Charge and Discharge Cycles. *J. Electrochem. Soc.* **2005**, *152*, A1128–A1135. [CrossRef]

17. Wu, B.; Mohammed, M.; Brigham, D.; Elder, R.; White, R.E. Anon–isothermal model of a nickel/metal hydridecell. *J. Power Sources* **2001**, *101*, 149–157. [CrossRef]

18. Xiao, G.Y. Fast Charging Nickel-Metal Hydride Traction Batteries. *J. Electrochem. Soc.* **2007**, *171*, A265–A272.

19. Shin, D.-H.; Jeong, J.-B.; Kim, T.-H.; Kim, H.-J. Modeling of Lithium Battery Cells for Plug-In Hybrid Vehicles. *J. Power Electron.* **2013**, *13*, 429–436. [CrossRef]

20. Hayner, C.M.; Zhao, X.; Kung, H.H. Materials for Rechargeable Lithium-Ion Batteries. *Annu. Rev. Chem. Biomol. Eng.* **2012**, *3*, 445–471. [CrossRef]

21. Ali, E. Low Voltage Anode Materials for Lithium–Ion Batteries. *Energy Storage Mater.* **2017**, *7*, 157–180.

22. Hu, L.; Wu, H.; La Mantia, F.; Yang, Y.; Cui, Y. Thin, Flexible Secondary Li-Ion Paper Batteries. *ACSNANO* **2010**, *4*, 5843–5848. [CrossRef]

23. Krieger, E.M.; Arnold, C.B. Effects of undercharge and internal loss on the rate dependence of battery charge storage efficiency. *J. Power Sources* **2012**, *210*, 286–291. [CrossRef]

24. Waldmann, T.; Hogg, B.-I.; Wohlfahrt-Mehrens, M. Li plating as unwanted side reaction in commercial Li-ion cells—A review. *J. Power Sources* **2018**, *384*, 107–124. [CrossRef]

25. Smith, A.J.; Svens, P.; Varini, M.; Lindbergh, G.; Lindström, R.W. Expanded in Situ Aging Indicators for Lithium-Ion Batteries with a Blended NMC-LMO Electrode Cycled at Sub-Ambient Temperature. *J. Electrochem. Soc.* **2021**, *168*, 110530. [CrossRef]

26. Epding, B.; Rumberg, B.; Jahnke, H.; Stradtmann, I.; Kwade, A. Investigation of significant capacity recovery effects due to long rest periods during high current cyclic aging tests in automotive lithium ion cells and their influence on lifetime. *J. Energy Storage* **2019**, *22*, 249–256. [CrossRef]

27. Mussa, A.S.; Liivat, A.; Marzano, F.; Klett, M.; Philippe, B.; Tengstedt, C.; Lindbergh, G.; Edström, K.; Lindström, R.W.; Svens, P. Fast-charging effects on ageing for energy-optimized automotive LiNi1/3Mn1/3Co1/3O2/graphite prismatic lithium-ion cells. *J. Power Sources* **2019**, *422*, 175–184. [CrossRef]

28. Zhang, W.; Wang, L.; Wang, L.; Liao, C.; Zhang, Y. Joint State-of-Charge and State-of-Available-Power Estimation Based on the Online Parameter Identification of Lithium-Ion Battery Model. *IEEE Trans. Ind. Electron.* **2021**, *69*, 3677–3688. [CrossRef]

29. Gao, Y.; Liu, K.; Zhu, C.; Zhang, X.; Zhang, D. Co-Estimation of State-of-Charge and State-of- Health for Lithium-Ion Batteries Using an Enhanced Electrochemical Model. *IEEE Trans. Ind. Electron.* **2021**, *69*, 2684–2696. [CrossRef]

30. Gopalakrishnan, K.; Offer, G.J. A Composite Single Particle Lithium-Ion Battery Model Through System Identification. *IEEE Trans. Control Syst. Technol.* **2021**, *30*, 1–13. [CrossRef]

31. Naseri, F.; Schaltz, E.; Stroe, D.-I.; Gismero, A.; Farjah, E. An Enhanced Equivalent Circuit Model with Real-Time Parameter Identification for Battery State-of-Charge Estimation. *IEEE Trans. Ind. Electron.* **2021**, *69*, 3743–3751. [CrossRef]

32. Shui, Z.Y.; Li, X.H.; Feng, Y.; Wang, B.C.; Wang, Y. Combining Reduced-Order Model with Data-Driven Model for Parameter Estimation of Lithium-Ion Battery. *IEEE Trans. Ind. Electronics* **2023**, *70*, 1521–1531. [CrossRef]

33. Amir, S.; Gulzar, M.; Tarar, M.O.; Naqvi, I.H.; Zaffar, N.A.; Pecht, M.G. Dynamic Equivalent Circuit Model to Estimate State-of-Health of Lithium-Ion Batteries. *IEEE Access* **2022**, *10*, 18279–18288. [CrossRef]

34. Wang, J.; Xiang, Y. Fast Modeling of the Capacity Degradation of Lithium-Ion Batteries via a Conditional Temporal Convolutional Encoder–Decoder. *IEEE Trans. Transp. Electrif.* **2022**, *8*, 1695–1709. [CrossRef]

35. Vermeer, W.; Mouli, G.R.C.; Bauer, P. A Comprehensive Review on the Characteristics and Modeling of Lithium-Ion Battery Aging. *IEEE Trans. Transp. Electrif.* **2022**, *8*, 2205–2232. [CrossRef]

36. Ni, Z.; Yang, Y. A Combined Data-Model Method for State-of-Charge Estimation of Lithium-Ion Batteries. *IEEE Trans. Instrum. Meas.* **2022**, *71*, 2503611. [CrossRef]

37. Cao, Y.; Kroeze, R.C.; Krein, P.T. Multi-timescale Parametric Electrical Battery Model for Use in Dynamic Electric Vehicle Simulations. *IEEE Trans. Transp. Electrif.* **2016**, *2*, 432–442. [CrossRef]

38. Renhart, W.; Magele, C.; Schweighofer, B. FEM-Based Thermal Analysis of NiMH Batteries for Hybrid Vehicles. *IEEE Trans. Magn.* **2008**, *44*, 802–805. [CrossRef]

39. Agarwal, V.; Uthaichana, K.; DeCarlo, R.A.; Tsoukalas, L.H. Development and Validation of a Battery Model Useful for Discharging and Charging Power Control and Lifetime Estimation. *IEEE Trans. Energy Convers.* **2010**, *25*, 821–835. [CrossRef]
40. Chen, M.; Rincon-Mora, G.A. Accurate electrical battery model capable of predicting runtime and I-V performance. *IEEE Trans. Energy Convers.* **2006**, *21*, 504–511. [CrossRef]
41. Hu, T.; Zanchi, B.; Zhao, J. Simple Analytical Method for Determining Parameters of Discharging Batteries. *IEEE Trans. Energy Convers.* **2011**, *26*, 787–798. [CrossRef]
42. Alsharif, K.I.; Pesch, A.H.; Borra, V.; Li, F.X.; Cortes, P.; Macdonald, E.; Choo, K. A Novel Modal Representation of Battery Dynamics. *IEEE Access* **2022**, *10*, 16793–16806. [CrossRef]
43. Piao, C.; Yang, X.; Teng, C.; Yang, H. An improved model based on artificial neural networks and Thevenin model for nickel metal hydride power battery. In Proceedings of the 2010 International Conference on Optics, Photonics and Energy Engineering (OPEE), Wuhan, China, 10–11 May 2010; pp. 115–118. [CrossRef]
44. Tarabay, J.; Karami, N. Nickel Metal Hydride battery: Structure, chemical reaction, and circuit model. In Proceedings of the 2015 Third International Conference on Technological Advances in Electrical, Electronics and Computer Engineering (TAEECE), Beirut, Lebanon, 29 April –1 May 2015; pp. 22–26. [CrossRef]
45. Bharathan, D.; Pesaran, A.; Vlahinos, A.; Kim, G.-H. Improving battery design with electro-thermal modeling. In Proceedings of the 2005 IEEE Vehicle Power and Propulsion Conference, Chicago, IL, USA, 7 September 2005; p. 8. [CrossRef]
46. Wei, Z.; Hu, J.; He, H.; Yu, Y.; Marco, J. Embedded Distributed Temperature Sensing Enabled Multistate Joint Observation of Smart Lithium-Ion Battery. *IEEE Trans. Ind. Electron.* **2023**, *70*, 555–565. [CrossRef]
47. Wei, Z.; Quan, Z.; Wu, J.; Li, Y.; Pou, J.; Zhong, H. Deep Deterministic Policy Gradient-DRL Enabled Multiphysics-Constrained Fast Charging of Lithium-Ion Battery. *IEEE Trans. Ind. Electron.* **2022**, *69*, 2588–2598. [CrossRef]

Article

Carbon Nano-Onion-Encapsulated Ni Nanoparticles for High-Performance Lithium-Ion Capacitors

Xiaohu Zhang [1,2,†], Keliang Zhang [2,†], Weike Zhang [3], Xiong Zhang [1,2,4], Lei Wang [1,4], Yabin An [1,2,4], Xianzhong Sun [1,2,4], Chen Li [1,2,4], Kai Wang [1,2,4] and Yanwei Ma [1,2,4,*]

[1] Institute of Electrical Engineering, Chinese Academy of Sciences, Beijing 100190, China
[2] Institute of Electrical Engineering and Advanced Electromagnetic Drive Technology, Qilu Zhongke, Jinan 250013, China
[3] College of Environmental Science and Engineering, Taiyuan University of Technology, Taiyuan 030600, China
[4] School of Engineering Sciences, University of Chinese Academy of Sciences, Beijing 100049, China
* Correspondence: ywma@mail.iee.ac.cn
† These authors contributed equally to this work.

Abstract: Lithium-ion capacitors (LICs) feature a high-power density, long-term cycling stability, and good energy storage performance, and so, LICs will be widely applied in new energy, new infrastructure, intelligent manufacturing. and other fields. To further enhance the comprehensive performance of LICs, the exploration of new material systems has become a focus of research. Carbon nano-onions (CNOs) are promising candidates in the field of energy storage due to the properties of their outstanding electrical conductivity, large external surface area, and nanoscopic dimensions. Herein, the structure, composition, and electrochemical properties of carbon nano-onion-encapsulated Ni nanoparticles (Ni@CNOs) have been characterized first in the present study. The initial discharge and charge capacities of Ni@CNOs as anodes (in half-cells (vs. Li)) were 869 and 481 mAh g^{-1} at 0.1 A g^{-1}, respectively. Even at a current density of 10 A g^{-1}, the reversible specific capacity remained at 111 mAh g^{-1}. Ni@CNOs were used as anode materials to assemble LICs (full pouch cells (vs. activated carbon)), which exhibited compelling electrochemical performance and cycle stability after optimizing the mass ratio of the positive and negative electrodes. The energy density of the LICs reached 140.1 Wh kg^{-1} at 280.2 W kg^{-1} and even maintained 76.6 Wh kg^{-1} at 27.36 kW kg^{-1}. The LICs also demonstrated excellent cycling stability with a 94.09% capacitance retention over 40,000 cycles. Thus, this work provides an effective solution for the ultra-rapid fabrication of Ni-cored carbon nano-onion materials to achieve high-performance LICs.

Keywords: lithium-ion capacitors; carbon nano-onion; high energy density; high power density; long cycle life

Citation: Zhang, X.; Zhang, K.; Zhang, W.; Zhang, X.; Wang, L.; An, Y.; Sun, X.; Li, C.; Wang, K.; Ma, Y. Carbon Nano-Onion-Encapsulated Ni Nanoparticles for High-Performance Lithium-Ion Capacitors. *Batteries* **2023**, *9*, 102. https://doi.org/10.3390/batteries9020102

Academic Editors: Carolina Rosero-Navarro and Carlos Ziebert

Received: 13 December 2022
Revised: 20 January 2023
Accepted: 27 January 2023
Published: 2 February 2023

1. Introduction

Developing new energy storage technologies with high performance, low cost, and high security have been the goal of researchers. In recent years, people's awareness of environmental protection has been intensifying and the scale of energy storage for new energy vehicles has been expanding, making more individuals develop an interest in environmental protection and energy efficiency. Various energy storage technologies have been widely applied in our lives, such as electric vehicles, electric forklifts, and electric heavy-duty trucks [1–3]. Among the various new electrochemistry energy storage technologies, electrochemical energy storage devices are the most widely used in everyday life, and relevant research is constantly being conducted to further enhance the overall performance of products. LIBs possess a high energy density of about 80~300 Wh kg^{-1}, with a limited power density of less than 1 kW kg^{-1}, a narrow temperature difference range (−20~65 °C), and a short cycle life (~5000 cycles) [4]. In contrast, EDLCs are characterized by their fast charge and discharge speed (high power density, ≥10 kW kg^{-1}), wide temperature range ($-40\sim65$ °C), long cycle life ($\geq$500,000 cycles), and high security, but poor energy density ($\leq$10 Wh kg^{-1}) [5,6]. However, neither is suitable in some mechanical fields such as machine instantaneous stop and standby power supply. Therefore, developing an energy storage device with high energy, high output, and a long cycle life is of paramount importance in the future.

Lithium-ion capacitors (LICs), as hybrid electrochemical capacitors, typically use activated carbon and graphite as their positive and negative terminals [7–9]. LICs make up for the defect between LIBs and EDLCs, which promote the energy density by several times and increase the working voltage compared to EDLCs [10]. As a result, LICs have a wide range of broad application prospects, especially in the fields of new energy, new infrastructure, and intelligence [11–13]. Despite the many advantages of LICs, however, recent studies have suggested that the non-Faradaic reaction limits the specific capacity of LICs in cathodes, while the sluggish Li$^+$ intercalation kinetics of the anode limit the intercalation/deintercalation of the device [14,15]. Therefore, it is key to improve this mismatch phenomenon of the capacity and kinetics of cathodes and anodes which represent the characters of the battery type and the double layer. To address this issue, the development of electrode materials has accelerated in recent years. The capacitive-type cathode materials of LICs mainly include carbonaceous materials, such as activated carbon (AC) [16], templated carbons [17], graphene [18,19], and graphene oxide (GO) [20]. The battery-type anode materials mainly include carbonaceous materials [21], transition metal oxide [22], Mxene [23], SiOx [24], Li$_4$Ti$_5$O$_{12}$ [25], and polyanion and carbon composites [26,27].

Among the various carbon forms, carbon nano-onions (CNOs) are therefore promising candidates for LICs applications, because CNOs exhibit controlled surface qualities, high electrical conductivity (similar to carbon black), and nanoscopic dimensions (typically 2 to 600 nm) [28–30]. CNOs are zero-dimensional carbon nanoparticles with a cage within a cage structure. They consist of smaller fullerenes nested within larger fullerenes and multiple closed shells wrapped around each other and they resemble an onion in structure. CNOs have been extensively investigated as electrodes for ultrafast charge/discharge devices for energy storage. Permana et al. [31,32] studied onion-like carbon (OCL) prepared using different synthetic routes towards LICs. Nanodiamond-derived OLCs can effectively improve the electrochemical performance of LICs. LICs with OLCs reached a maximum energy density (243 Wh kg^{-1}), a high power density (20,149 W kg^{-1}), and a stable capacity retention of 78% after 1000 cycles. Aref et al. [33] fabricated symmetric LICs by using a pre-lithiated coalesced carbon onion, whose energy density was 67 Wh kg^{-1} and the power density was 14.5 kW kg^{-1}. As anode materials of LICs, CNOs improve the energy density of LICs. However, the synergistic enhancement of energy density, power density, and cycle life is an important research direction for achieving CNOs' application in LICs.

Carbon nano-onion-encapsulated Ni nanoparticles (Ni@CNOs) were modified and prepared to improve their electrochemical performance as anodes in LICs. Ni@CNOs is a kind of composite material with a special core/shell structure and embedded with metal Ni. In this manuscript, fistly, the structure, composition, and electrochemical properties of Ni@CNOs were characterized. Ni@CNOs were adopted as anode materials to assemble LICs and test their performance. The electrochemical performance of Ni@CNOs as anodes was evaluated between 0.01 and 3 V vs. Li/Li$^+$. LICs exhibited excellent electrochemical performance and cycle stability after optimizing the mass ratio of positive and negative electrodes. The LICs achieved an energy density of 140.1 Wh kg^{-1} when the power density was 280.2 W kg^{-1} and even maintained 76.6 Wh kg^{-1} when the power density was 27.36 kW kg^{-1}. At the same time, LICs demonstrated excellent cycling stability and reliability with a capacitance retention of 94.09% over 40,000 cycles. Thus, Ni@CNOs are highly attractive anode materials for practical LICs.

2. Materials and Methods

2.1. Materials

The Ni@CNOs were prepared by a chemical vapor deposition (CVD) procedure. The carbon source was CH_4 which was gradually deposited on the nickel catalyst at 800 °C until the Ni catalyst particles were all coated with graphite to form Ni@CNOs. The CVD reaction took place in a horizontal atmosphere furnace with quartz tubes as the reactors. The conversion reaction of CH_4 was completed according to the following formula:

$$CH_4 \rightarrow 2H_2 + C \text{ (carbon nano-onions)} \tag{1}$$

Catalyst preparation: Firstly, nickel acetate and ferric nitrate were weighed according to a certain proportion, added into deionized water, and fully stirred until dissolved in a constant temperature water bath at 60 °C in a magnetic agitator. Secondly, the configured citric acid solution was slowly added with a certain concentration to continue stirring for 20 min, and the mixed solution was transferred to the muffle furnace for calcination for 5 h. Finally, the product was taken out and fully ground to obtain the Ni catalyst. It should be noted that the catalyst was in the NiOx state at this time, and it needed to be reduced to stimulate the catalytic activity before the methane cracking reaction. The Ni catalyst consisted of a large number of particles with a diameter of about 100 nm, and the morphology was mostly circular or elliptical.

Ni@CNO preparation: Firstly, a certain amount of prepared catalyst was uniformly dispersed and paved in a quartz boat and placed into the middle of the furnace. After sealing the pipes on both sides, nitrogen was injected into the air pipe to remove the air and act as a protective gas. Secondly, a heating rate of 10 °C/min was set to increase the temperature in the furnace to 800 °C, and the reducing gas H_2 (N_2: H_2 as a carrier gas) was introduced and kept for 5 h. Finally, after the catalyst was activated, the temperature was raised to a certain temperature at the rate of 5 °C/min, and CH_4 was injected for a constant temperature reaction for a period of time. In the pyrolysis reaction, the Ni catalyst was formed as a closed CNO coating. At the end of the reaction, when the reactor was brought down to room temperature naturally, the Ni@CNOs were obtained.

Since methane cracking is an irreversible endothermic reaction, temperature is one of the decisive factors that determines whether it can be carried out and whether the reaction is complete. In order to explore the influence of temperature on the morphology and structure of CNOs, the reaction temperatures were set as 700 °C, 800 °C, and 900 °C. The productivity of the Ni@CNOs increased with the increase inof temperature, and the maximum productivity was achieved at 900 °C, as shown Table S1. However, there were a large number of carbon nanotubes and chain Ni@CNOs at 900 °C. In order to improve the purity of the product, 800 °C was used for pyrolysis in the subsequent tests. The experimental results show that the longer the time, the higher the utilization rate of the catalyst, the more uniform the size of the Ni@CNOs, and the more spherical the morphology. Considering the actual production and the deactivation of the catalyst, the reaction time cannot be extended indefinitely, so the reaction time was set as 5 h. The different flow conditions show that the higher the flow rate, the higher the productivity, as shown Table S2. However, increasing the gas flow rate in the actual production will increase production costs. In order to control the morphology, particle size, and costs, a 300 sccm flow rate was adopted as the optimal condition in this manuscript.

The other materials used in this work were obtained from different suppliers. The AC [activated carbon (YP-80)] was purchased from Kuraray chemical co, Japan. The conductive carbon additive was obtained from TIMCAL Graphite and Carbon, and in a dynamic vacuum, PVDF (HSV900, provided by Kynar) was dried at 80 °C for 12 h. The N-methyl pyrrolidone (NMP) solvent was purchased from Macklin. The electrolyte solution containing 1.2 M $LiPF_6$ in EC/DMC/DMC (1:1:1 by volume) was purchased from Dongguan Shanshan battery material Co., Ltd., and a cellulose paper separator (TF40-30) was provided by Nippon Kodashi Corporation.

2.2. Cell Preparation

For the Ni@CNOs electrode, the active material (Ni@CNOs), carbon black, and PVDF were mixed in the weight ratio of 8:1:1. Then, the slurry was coated on 9 μm-thick Cu foil and dried in an oven at 80 °C for 24 h. The performance of the Ni@CNOs was measured using 2032-type coin cells with circular electrodes (1.1 cm diameter and mass loading of 1~2 mg cm^{-2}). A half-cell was assembled into the 2032type coin cell, and the 1 M LiPF$_6$ was added into the EC/DEC/DMC (1:1:1 by volume) electrolyte. The coin cells were fabricated in an inert atmosphere, which was in an argon-filled glove box (MBRAUN, H$_2$O < 0.1 ppm, O$_2$ < 0.1 ppm).

2.3. LIC Pouch Cells' Preparation

Preparation of the cathode electrodes: The AC (activated carbon), carbon black, and PVDF were mixed in the weight ratio of 8:1:1. Then, the slurry was coated on 16 μm-thick Al foil and dried in an oven at 110 °C for 12 h.

Preparation of anode electrodes: The active material (Ni@CNOs), carbon black, and PVDF were mixed in the weight ratio of 8:1:1. Then, the slurry was coated on 9 μm-thick Cu foil and dried in an oven at 80 °C for 12 h.

For this assembly, two electrodes were punched from the laminates, of which the area was 14 cm^2 (dimension 35 × 40 mm) with an additional coated tab area (wiped off with alcohol later). Two porous electrodes were prepared by the drilling equipment immediately. A cellulose paper separator was used between these porous electrodes. The lithium foil was fixed near the anode as a reference electrode/auxiliary electrode. The cathode and anode electrodes were welded by aluminum tabs and nickel tabs, respectively. In addition, then, in order to completely remove the trace water in the electrode, the pouch cell was dried at 120 °C for 24 h in a vacuum. Subsequently, a metal lithium auxiliary electrode was added onto the side close to the negative electrode, an electrolyte was injected, and it was sealed to obtain flexible packaging LIC pouch cells.

Pre-lithiation process: With the anode as the working electrode and the lithium foil as the auxiliary electrode, discharge to 0.01 V at 0.1 A g^{-1} took place, which meant that the lithium ions were electrochemically driven to migrate from the lithium electrode to the anode, and finally, the pre-lithiation process was completed.

2.4. Material Characterization

X-ray diffraction patterns of the Ni@CNOs samples were identified by powder X-ray diffraction (XRD, Bruker D8 ADVANCE). The patterns were scanned between 10 and 80 at a scanning speed of 5 (°) min^{-1}. The morphology and microstructure of the Ni@CNOs powders were characterized by using a scanning electron microscope (SEM, Hitachi S4800) and LabRam HR-800 (Horiba Jobin Yvon). The HRTEM images were analyzed using a high-resolution transmission electron microscope (JEOL JSM-2010). The nitrogen adsorption and desorption measurements were carried out on a Micromeritics ASAP 2020 HD analyzer. The specific surface area of the sample was measured and analyzed by the Brunauer–Emmett–Teller (BET) theory and the pore size distributions were acquired using nitrogen adsorption–desorption isotherms on the surface area to analyze and examine the surface area and the porous structure of the materials. Raman spectroscopic analysis was performed with a confocal Raman system (LabRam HR-800, Horiba Jobin Yvon) with an excitation wavelength of 532 nm. The elemental analysis of the samples was conducted through X-ray photoelectron spectroscopy (XPS, ESCALAB 250Xi, Thermo Scientific with Al Kα X-ray).

2.5. Electrochemical Measurements

The galvanostatic charge–discharge (GCD) and cyclic performance were measured through the battery testing system (NEWARE, Shenzhen, China). The energy density (*E*: Wh kg^{-1}), power density (*P*: W kg^{-1}), specific capacity (*C*: mAh g^{-1}), and specific capacitance (F: F g^{-1}) were calculated according to the following Equations (2), (3), (4), and (5), respectively:

$$E = \frac{I \times \int_{t_1}^{t_2} V dt}{m \times 3.6} \tag{2}$$

$$P = \frac{3600 \times E}{\Delta t} \tag{3}$$

$$F = \frac{3.6 \times E}{0.5 \times (V_{max}^2 - V_{min}^2)} \tag{4}$$

$$C = \frac{I \times \Delta t}{m \times 3.6} \tag{5}$$

where I is the current in A; it is also used to specify the discharging current. Δt is the discharge time in s, m is the total mass of the cathode and anode, and V_{max} and V_{min} are the maximum and minimum voltage, respectively. The discharge accumulated electric energy represented in a watt-hour notation is calculated by dividing W by 3600. The specific test method's malicious reference standard is BS EN 62813:2015 "Lithium-ion capacitors for use in electric and electronic equipment—Test methods for electrical characteristics".

Electrochemical impedance spectroscopy (EIS) was conducted by applying a small amplitude of 10 mV under open circuit conditions in the frequency range of 0.01 Hz–100 kHz. In order to further study the electrochemical properties of Ni@CNOs, by using Bio-Logic VMP3, CV was performed at different scan rates of 0.1–2 mV s^{-1}.

3. Results and Discussion

3.1. Morphology and Structure Analysis of Ni@CNOs

To investigate the composition and structure of Ni@CNOs, X-ray diffraction (XRD) and Raman spectroscopy were performed. As shown in Figure 1a, the XRD of Ni@CNOs exhibits the presence of high-intensity peaks centered at 2θ = 26.4° and 43.7°, which correspond to the (002) plane of graphite and the (111) plane of single-phase Ni [34]. In addition, the absence of a NiO peak suggests that the Ni nucleus is not oxidized, and it can be speculated that the inside of the Ni nucleus is embedded in the graphite shell layer. The Ni nucleus is not carbonized or oxidized, which also shows that Ni@CNOs have a nuclear shell structure. Compared with the Ni (111) plane, the intensity of the graphite (002) plane is not very strong, indicating that the carbon shells are at a relatively medium graphite degree. In addition, the graphitization extent (G) was determined by average d$_{002}$ spacing using the following equation [35]:

$$G = \frac{3.440 - d_{002}}{3.440 - 3.354} \tag{6}$$

where the graphitization extent (G) is the degree of graphitization (%), 3.440 is the interlayer distance at the border to non-graphitic carbon, 3.354 is the interlayer distance of perfectly stacked graphite, and d(002) is the interlayer spacing derived from the XRD (nm).

It can be seen that the degree of graphitization is calculated to be 58.7%, which also indicates the medium degree of graphitization for Ni@CNOs.

Figure 1b shows two strong Raman peaks of Ni@CNOs, corresponding to the disordered structure carbon (i.e., the D-band, 1340 cm^{-1}) and graphite mode (i.e., the G-band, 1590 cm^{-1}) with a ratio of I_D/I_G = 0.84, respectively. The structure of the sp^2 hybridized carbon can be described by the G-band, and the intensity ratio between the G- toD-bands (i.e., I_G/I_D) can determine the extent of the graphite mode in carbon materials [36,37]. In addition, the presence of more porous disordered structures contributes to enhancing the diffusion of lithium-ion. The higher intensity of the G-band and the lower value of the ID/IG also confirm the medium degree of graphitization for Ni@CNOs. Figure S1 shows the XPS survey spectrum of Ni@CNOs, in which it can be seen that only C and O elements were detected on the material surface, and there was no Ni element. This is due to the thick carbon layer; the X-ray cannot penetrate the sample, so the Ni element cannot be detected in the XPS. In addition, different valence states of the C element in the composites can be seen in Figure 2e. The binding energies at 284.7, 285.9, and 288.1 eV are consistent with the Sp2 hybrid carbon (C-C), C-0 bond, and C=O bond, respectively.

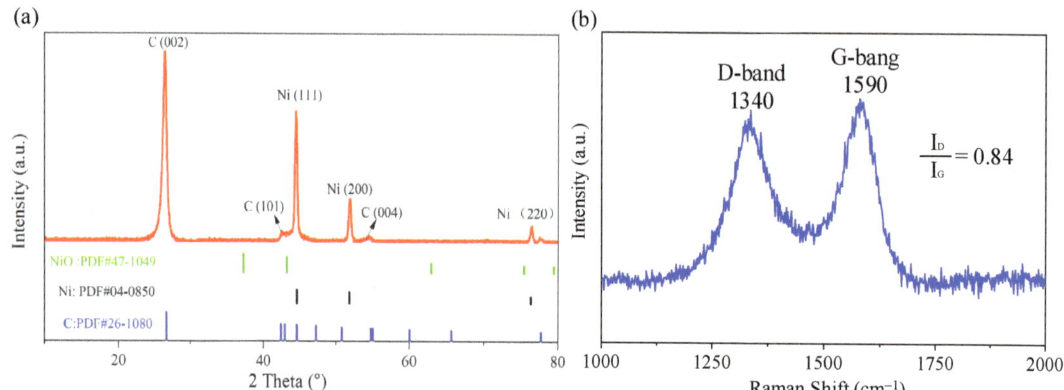

Figure 1. Structural information on Ni@CNOs: (**a**) XRD pattern and (**b**) Raman spectra.

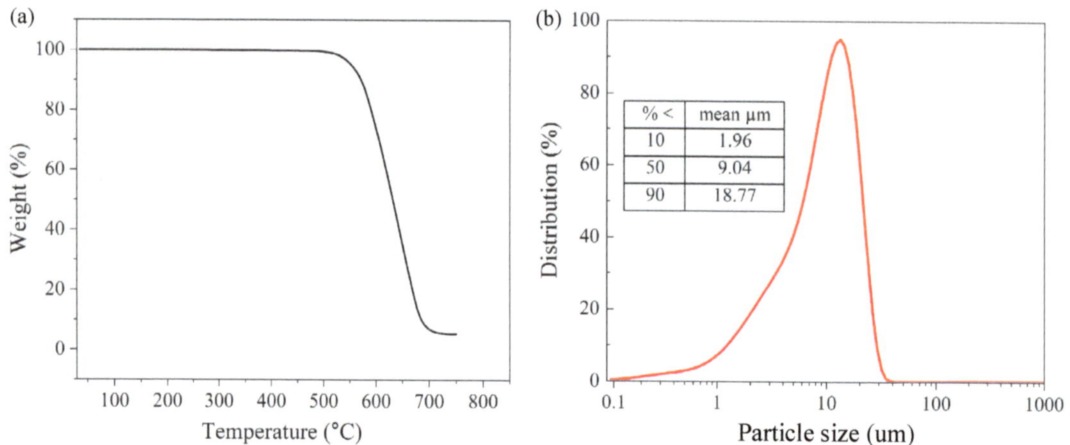

Figure 2. (**a**) TGA image of Ni@CNOs and (**b**) particle size distribution of Ni@CNOs.

The thermal gravimetric analyzer (TGA) for the Ni@CNOs is shown in Figure 2a, where we recorded the relationship between the sample mass and the change in temperature. The Figure 2a samples have an obvious weight-loss range at T > 540 °C with a main weight loss of 90%. The remaining 10% of the residue should be NiO produced by the reaction of the nickel core and oxygen in the air. The results of the TGA analysis indirectly confirmed the prepared core–shell structure of the Ni@CNOs, which is consistent with the results obtained by SEM and TEM. Figure 2b shows the particle size distribution of Ni@CNOs. The average particle size of the samples is 9.04 um, which is bigger than the size from the observed SEM image; this test result is caused by the agglomeration of nanoparticles.

Figure 3a,b demonstrates the spherical morphologies of Ni@CNO powders in the SEM image, which are rich in irregularly shaped nanoparticles with a wide size distribution. Meanwhile, these spherical nanoparticles are agglomerated together to form carbon onion clusters due to intermolecular and electrostatic forces. The TEM image of the Ni@CNOs reveals a mostly shell/core structure, consisting of a crystalline core and a coating onion-like shell, as shown in Figure 3c,d. The diameter distribution of the nickel core is 30 to 150 nm and the thickness of the carbon shell is 40 to 80 nm. As can be seen from the HR-TEM image, the carbon structure is concentric with some degree of structural disorder, indicating multiple layers of fullerene-like carbon, as shown in Figure 3e. Ten to twenty graphitic layers are observed in the sample, with an average value of 0.34 nm for the interplanar distances (D-spacing) between the graphitic layers, corresponding to the (002) lattice plane of the graphite layers. Moreover, the selected area electron diffraction patterns (SAED) suggest the different diffraction spots of Ni@CNOs, including the graphitic layers (002) lattice plane and nickel (111) and (200) and (220) lattice plane, which is consistent with the results of the XRD patterns, as shown in Figure 3f. Apparently, nickel is a neutral metal atom in an onion-like shell, which means it is in a zero valence state. With these fullerene-like layers, the possibility for the infiltration of lithium-ion electrolytes greatly shortens the ion diffusion pathway in the Ni@CNOs anode. Carbon onion anode electronic conductivity may also be affected by the coalesced microstructure. Obviously, Ni nanoparticles are encapsulated in the onion-shaped graphite shell. In order to determine the content and distribution of elements on the surface of the CNO composite electrode material, we have carried out a spectroscopic analysis for the CNOs. The EDS pattern (Figure 3g–i) and the energy spectra (Figure S3) of the Ni@CNOs reveals the existence of nickel and carbon. Among them, the content of the C element (89.8 wt. %) is dominant, and the content of the Ni element (9.9 wt. %) is less, which corresponds to the peak value of Ni in the XRD test in Figure 3d, and their contents are consistent with the TGA test results.

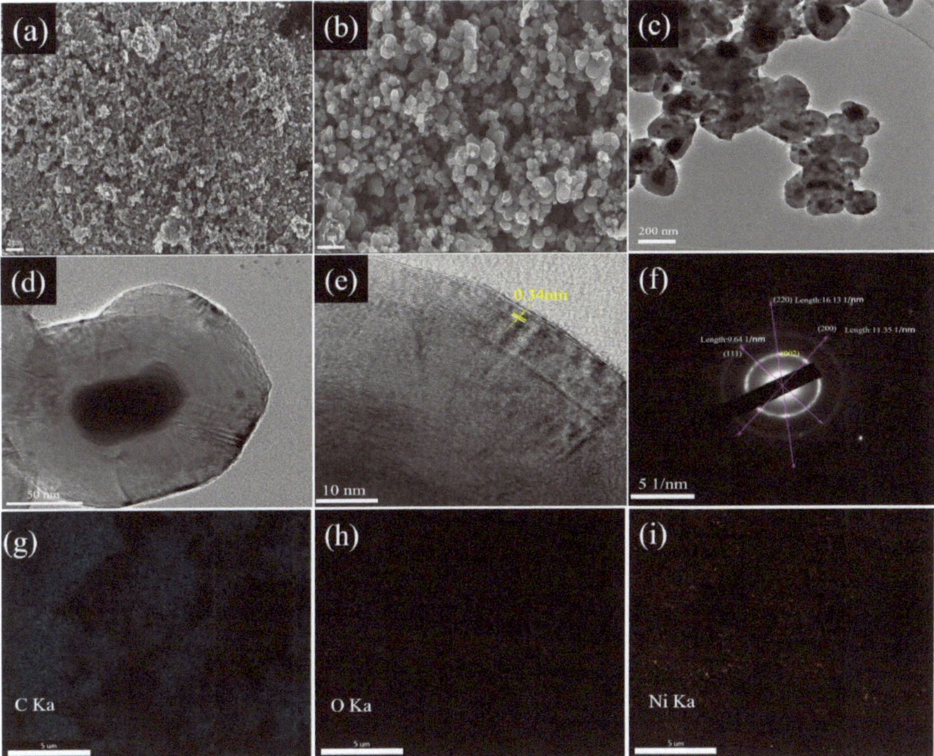

Figure 3. SEM (**a**,**b**) images of Ni@CNOs. TEM (**c**,**d**) and HR-TEM (**e**) image of Ni@CNOs and (**f**) the corresponding SAED pattern. (**g**–**i**) SEM-EDS images.

Such structures provide good electrochemical stability and also increase the electrical conductivity of electrode materials. By comparing several common carbon materials on the market, as shown in Figure 4, Ni@CNOs show an electrical conductivity of 25.84 S cm^{-1} at 10 Mpa, which is higher than that of other carbon materials, such as commercial hard carbon (14.93 S cm^{-1}), soft carbon (20.22 S cm^{-1}), commercial graphite (18.65 S cm^{-1}), and AC (2.56 S cm^{-1}).

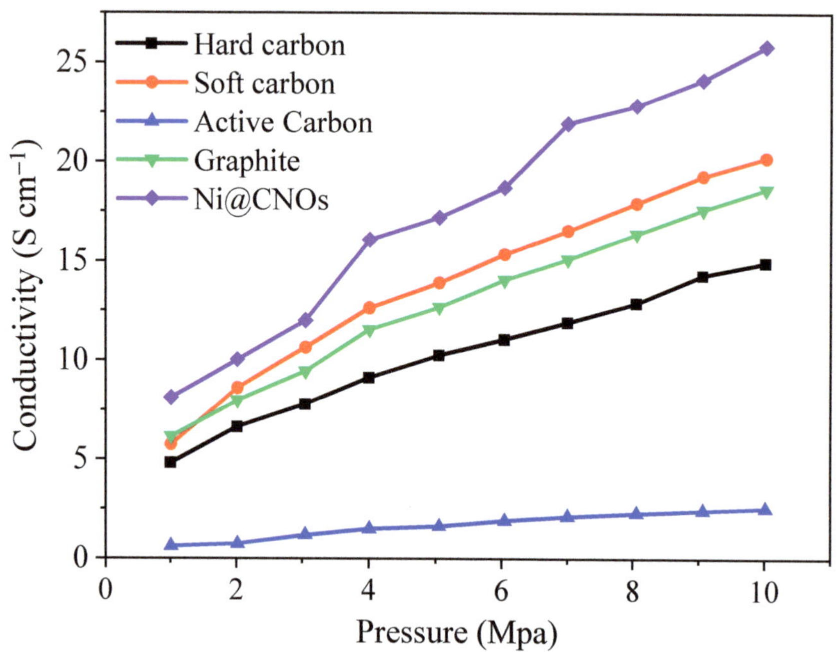

Figure 4. Ni@CNOs and other carbon materials' conductivity comparison.

As shown in Figure 5a, the BET specific surface area (surface area determined by the Brunauer–Emmett–Teller method) for Ni@CNOs is about 22.3 m^2 g^{-1}. The isotherm shape reveals an IV-type isotherm shape with a distinct hysteresis loop, which indicates that Ni@CNOs have porous structures. Furthermore, the pore size distribution calculated by the BJH method indicates that there is a pore size mode around 1.8 nm in Ni@CNOs, as shown in Figure 5b. Therefore, as anode electrodes for application in LICs, it is expected that Ni@CNOs can form a wide range of electrochemical reaction interfaces, while additional ion diffusion channels will reduce the interface resistance and enhance diffusion dynamics.

3.2. Electrochemical Performance of Ni@CNOs

Further investigations into the electrochemical performances of Ni@CNOs were evaluated within the potential window of 0.01–3.0 V vs. Li/Li$^+$. Figure 6a presents the cyclic voltammetry (CV) profile of the Ni@CNOs electrode at a scan rate of 0.1 mV s^{-1}. A pair of obvious redox peaks were observed in the range of 0.001–0.3 V and a pair of weak broad peaks were observed in 0.8–1.3 V. The obvious redox peaks are attributed to lithium intercalation/de-intercalation and the weak broad peaks are caused by edge adsorption and surface adsorption [38]. In addition, an additional peak located at around 0.6 V is attributed to the formation of SEI. There is no redox pair of nickel in the CV curve, so Ni plays a role in improving the conductivity of materials in Ni@CNOs [39,40]. The energy consumed in the SEI film process leads to an initial irreversible capacity. According to Figure 6b, the initial discharge and charge capacities were 869 and 481 mAh g^{-1} at 0.1 A g^{-1}, respectively. From the result, the initial coulomb efficiency of the Ni@CNOs electrode is about 55.35%, which can be attributed to the solid electrolyte interface (SEI) film. The subsequent charging and discharging process was stabilized with a capacity of 480 mAh g^{-1} and a coulomb efficiency of almost 100%.

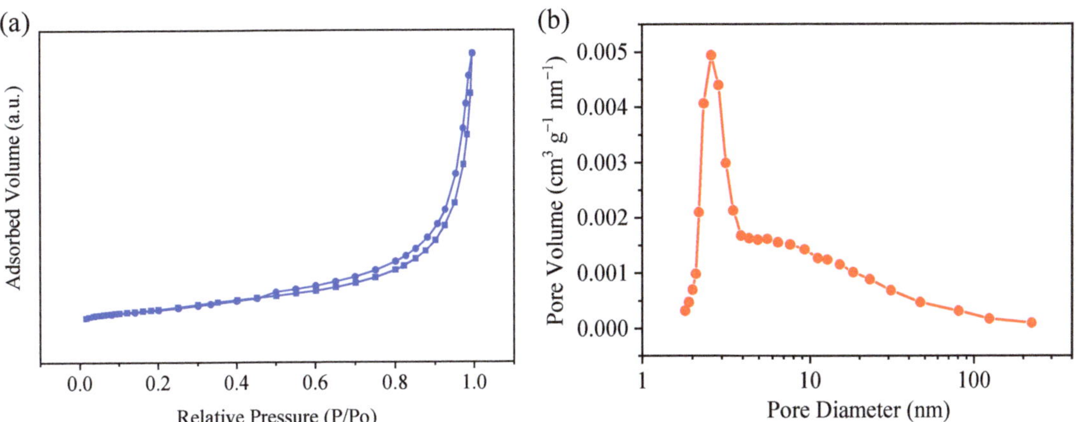

Figure 5. Structural information of the Ni@CNOs. (**a**) N2 adsorption/desorption isotherms and (**b**) pore size distribution on the BJH model.

Figure 6c depicts the CV curves of Ni@CNOs at scan rates of 0.1 to 2.0 mV s^{-1}. It can be seen that the sharp reduction peak appeared at the potentials of 0.75 V during the first cycle, which is generally ascribed to the formation of solid electrolyte interface (SEI) film on the Ni@CNOs surface. After the second circle, the peak becomes stable; it is gleamingly obvious that there are a pair of obvious redox peaks in the range of 0.001–0.3 V and a pair of weak broad peaks in 0.8–1.3 V. The obvious redox peaks are attributed to the lithium intercalation/de-intercalation and the weak broad peaks are caused by edge adsorption and surface adsorption. As shown in Figure 6d, the reversible specific capacity of the Ni@CNOs anode at 0.1 A g^{-1} is 480 mAh g^{-1}. Moreover, when the current density was increased to 2 A g^{-1}, the reversible specific capacity of the Ni@CNOs anode remained at 242 mAh g^{-1} and especially when the current density was increased to 10 A g^{-1}, the reversible specific capacity of the Ni@CNOs anode remained at 111 mAh g^{-1}. The reversible specific capacity reached its initial value when the C-rate returned to 0.1 A g^{-1} after high-speed measurements, implying stable electrochemical performance of the Ni@CNOs anode. Thus, in general, because of the additional interface area and pore structure of the graphene in the Ni@CNOs composite, reaction kinetics and ion transfer can be greatly improved. In addition, the presence of metal Ni increases the electronic conductivity of the Ni@CNOs anode. In addition, the cycling stability of the Ni@CNOs at a high current density has also been investigated. As shown in Figure 6e, the Ni@CNOs exhibit an outstanding capacity retention of 96.5% at 1 A g^{-1} and a coulombic efficiency approaching 100% after 1000 cycles in the half-cells. As a result of incomplete activation of the anode, the specific capacity first decreased and then gradually increased [41]. After about 1000 cycles, the device operated stably, which can be owed to the highly stable SEI film formed on the Ni@CNO surface.

3.3. Performance of LIC Pouch Cells Based on an Ni@CNO Anode and AC (Activated Carbon) Cathode

To deal with the low coulomb efficiency of the Ni@CNO anode materials in the first cycle and to reduce the influence of the electrolyte ion concentration on device performance during the formation of SEI film, the anode Ni@CNOs material was pre-embedded with lithium and assembled into a complete battery with AC cathode material. The pre-lithiation process of the anode is typically utilized, and therefore the working voltage window can be enlarged to boost the energy density [42,43]. According to the conservation of electric quantity, the maximum capacity that can be released by the energy storage device is mainly determined by the cathode material. However, at the same time, the surface load of the pole plate cannot expand infinitely. In the actual operation process, excessively thick electrode plates are prone to powder dropping, which does not adhere well to the collector and also affects the conductivity of the electrode plate. Therefore, in order to maximize the material capacity, the energy density and power density were balanced. Then, the LIC pouch cells were assembled with three mass ratios of 1:1, 1:2, and 1:3 (anode vs. cathode) and the electrochemical performance was measured to determine the optimal ratio of the anode and cathode.

Galvanostatic cycling of AC//Ni@CNOs LIC pouch cell systems at current densities of 0.1 A g^{-1} to 10 A g^{-1} was conducted to evaluate their rate performance. As shown in Figure 7a, LIC pouch cells with a mass ratio of 2:1 provided highly specific capacities of 78.48, 75.42, 70.38, 66.96, 61.92, and 55.26 F g^{-1} at 0.1, 0.2, 1, 2, 5, and 10 A g^{-1}, respectively, within a potential range of 2.0–4.2 V, which is much better than the other two LIC pouch cells. Based on the above experimental results, the optimized cathode/anode mass ratio of AC [activated carbon (YP-80)]//Ni@CNOs LIC pouch cells is 2:1. Figure 7b shows the CV curves of the LIC pouch cells at various scan rates from 2 to 20 mV s^{-1} in the voltage range of 2.0–4.2 V. As expected, the LIC pouch cells display a rectangular shape, and no apparent shape distortion was observed at a scan rate of 20 mV s^{-1}, indicating good capacitance behavior. Figure 7c shows the linear and symmetric voltage profiles of the AC [activated carbon (YP-80)]//Ni@CNOs (the mass ratio of 2:1) LIC pouch cells within the voltage range of 2.0–4.2 V, which clearly demonstrates its excellent capacitive behavior. This result is again consistent with the CV measurements. Figure 7d reveals a comparison of the rate capability of AC//Ni@CNOs LIC pouch cells and pure AC//SC LIC pouch cells in the voltage range of 2.0–4.2 V, with the rate performance of the AC//Ni@CNOs LIC pouch cells being much better than that of the AC//SC LIC pouch cells at current densities from 0.1 to 10 A g^{-1}. This excellent electrochemical performance can be ascribed to the special structure of Ni@CNOs. For the EIS analysis, EIS was employed to further explore the electrochemical behavior of the LIC pouch cells. Figure 6e displays the Nyquist plot for the LIC pouch cells device. In the corresponding equivalent circuit model, R_s represents the e ohmic internal resistance, Rct is the charge transfer resistance, and C_{dl} denotes the double-layer capacitance and capacity of the surface layer. The ohmic internal resistance and charge transfer internal resistance of the capacitor are very small, which is attributed to the coalescence and interparticle connectivity of the Ni@CNOs. According to the Nyquist plot, the Rct value of SC is greater than Ni@CNOs. This result reveals that the reaction resistance is decreased by the additional interface area provided by the Ni@CNOs network in the composite. Figure 7f shows the long-term high rate cycling of the AC (activated carbon (YP-80))//Ni@CNOs LIC pouch cells cell at a rate of 2 A g^{-1} in the voltage range of 2.2–3.8 V. After a further 40,000 cycles, the LIC pouch cells maintain good cycle ability, with 94.09% capacity retention and a coulombic efficiency close to 100% at full cycling. The result clearly demonstrates the superior rate capability of the AC [activated carbon (YP-80)]//Ni@CNOs system, which indicates that the Ni@CNOs anode remains stable with cyclic stability against cycling, without li$^+$ consumption and particle pulverization.

The porous structure of Ni@CNOs increases the electrode reaction interface, thus providing more favorable ion diffusion channels and a wider range of electrochemical reaction interfaces. Ultimately this reduces the interface reaction impedance and internal resistance caused by concentration polarization. Specifically, the presence of Ni in CNOs increases the electronic conductivity of the electrode and improves the multiplier performance of the device.

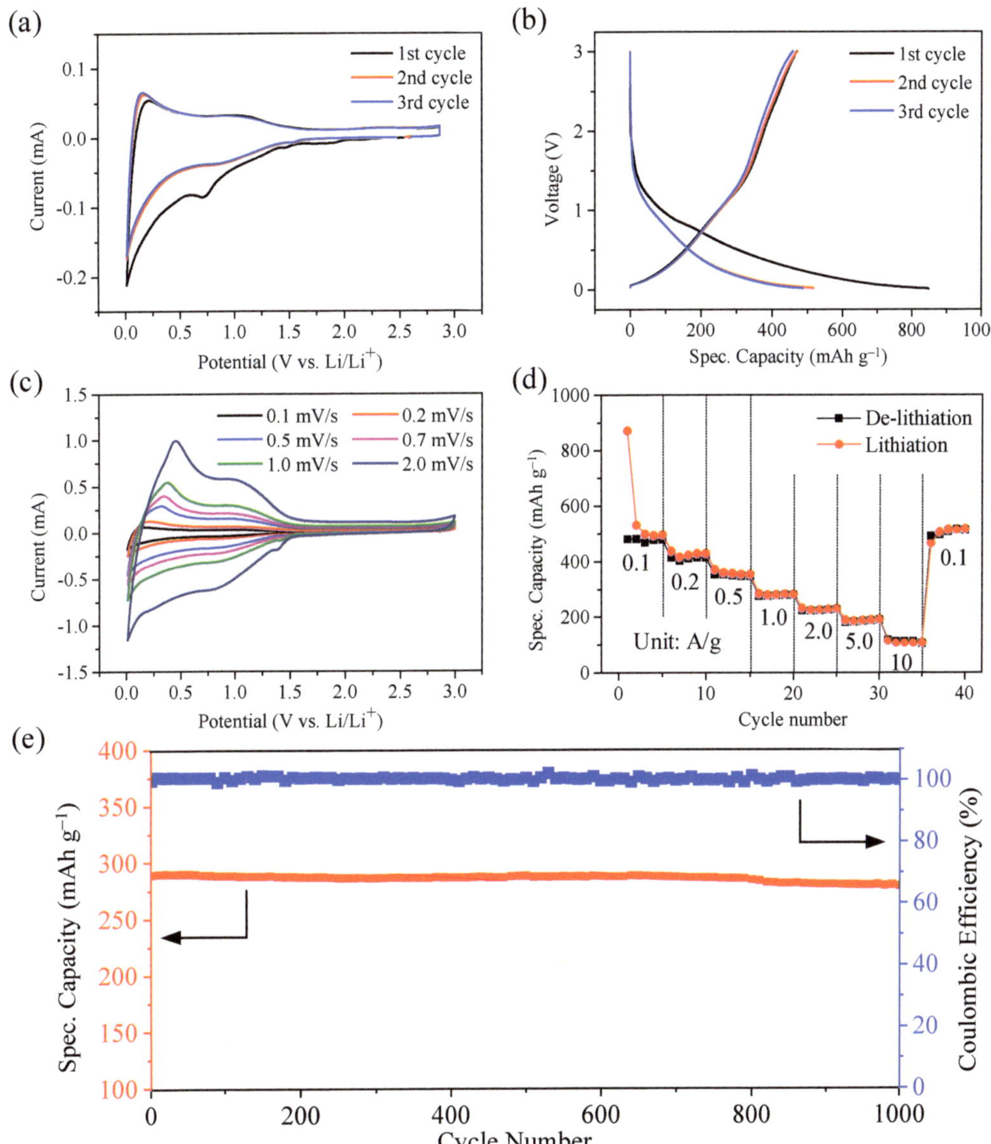

Figure 6. Electrochemical properties of the Ni@CNOs anode. (**a**) CV curves of the Ni@CNOs anode, (**b**) initial three charge/discharge cycles of the Ni@CNOs anode, (**c**) CV curves at scan rates of 0.1 to 10 mV s^{-1}, (**d**) specific capacities of the Ni@CNOs anode at different current densities, and (**e**) the cycling performance and its corresponding coulombic efficiency at a current density of 1 A g^{-1}.

Figure 8 shows the gravimetric Ragone plot of the AC [activated carbon (YP-80)]//Ni@CNOs and the LIC pouch cells reported in the reference. As expected, the AC//Ni@CNOsLIC pouch cells reach an energy density of up to 140.1 Wh Kg^{-1} at 275 W kg^{-1}. Even at an ultra-high power density of 27 kW kg^{-1}, the electrode can deliver an energy density of 76.6 Wh kg^{-1}. Compared to several representative LIC pouch cells systems, our AC [activated carbon (YP-80)]//Ni@CNOsLIC pouch cells exhibit superior energy and power performance, such as nitrogen-enriched mesoporous carbon nanospheres/graphene (N-

GMCS)//pre-lithiated microcrystalline graphite (PLMG) [44], URGO (urea (H2N-CO-NH2) RGO)//AC (commercial products) [45], HC (commercial products)//AC (commercial products) [46], and HC (commercial products)//AC (commercial products) [47], with more detailed information provided in Table 1. From the above discussions, this study offers a promising platform for the Ni@CNOs anodes as prospective electrode materials for electrochemical energy storage applications.

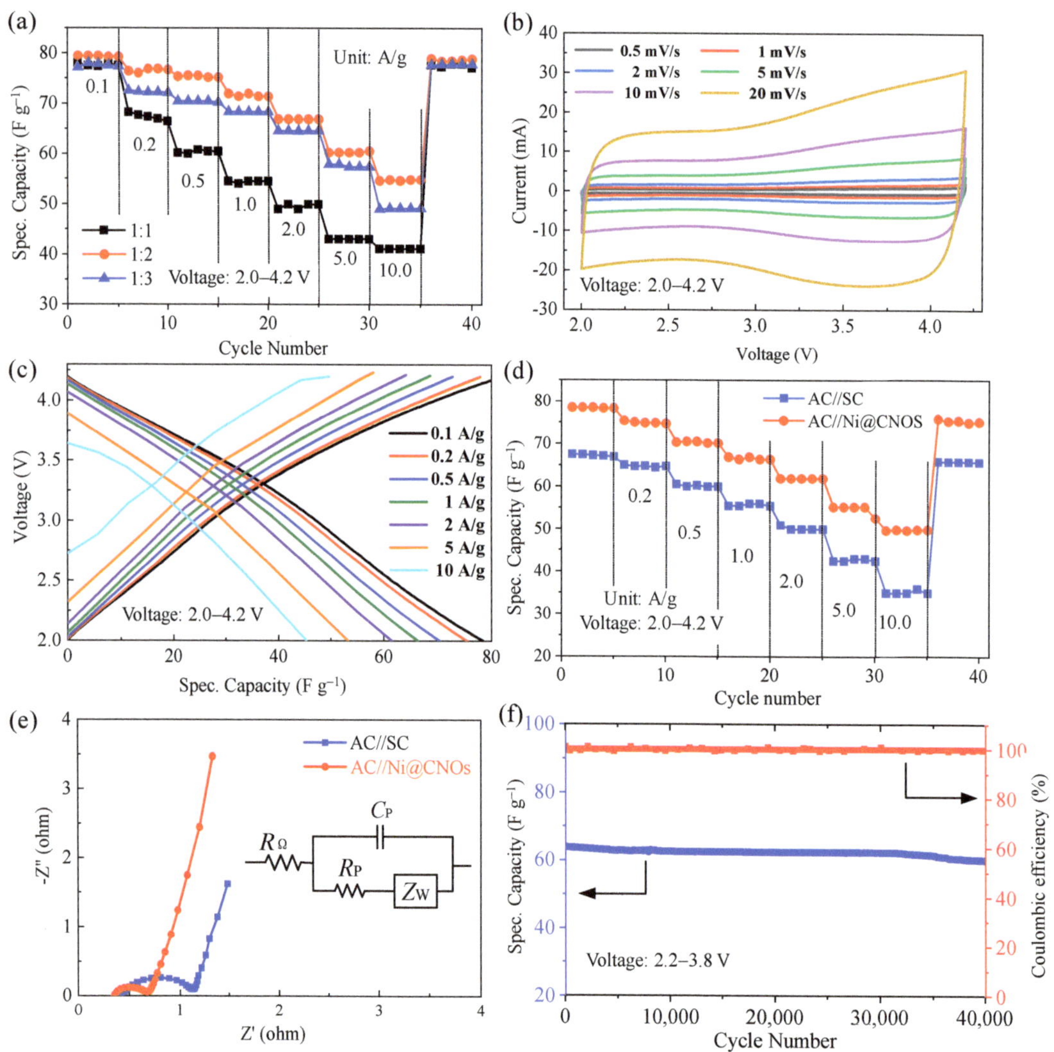

Figure 7. Electrochemical properties of the LIC pouch cells. (**a**) Rate performance of three LIC pouch cells with various mass ratios (anode vs. cathode) (**b**) CV curves of the LIC pouch cells with different scan rates in the voltage range of 2.0–4.2 V, (**c**) charge and discharge profiles of LIC from 0.1 to10 A g^{-1} in the voltage range of 2.0–4.2 V, (**d**) comparison of rate capability of the AC//Ni@CNOs LIC pouch cells and the pure AC//SC LIC pouch cells, (**e**) Nyquist plot of the LIC, and (**f**) long-term high-rate cycling performance at 2 A g^{-1} in the LIC pouch cells in the voltage range of 2.2–3.8 V.

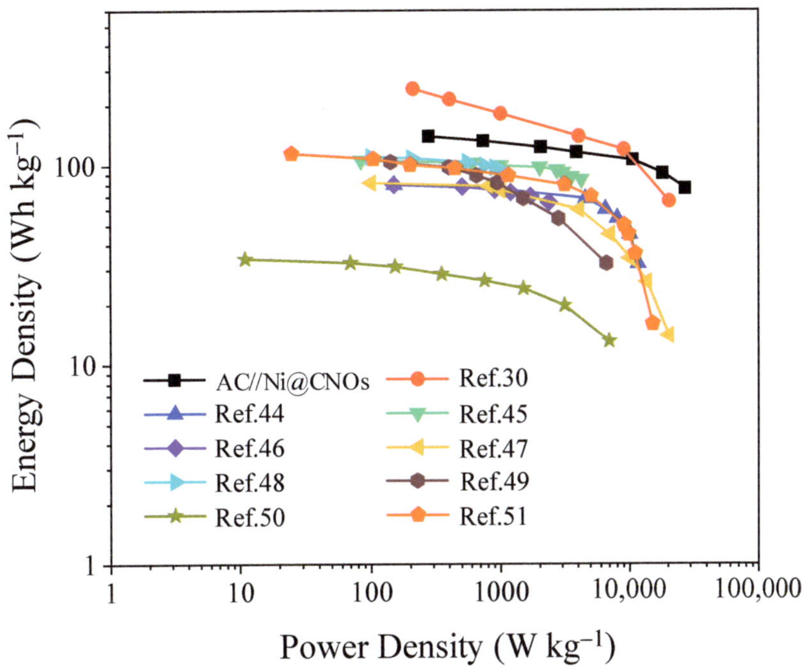

Figure 8. Ragone plots comparison of AC//Ni@CNOs with other recently reported LIC pouch cells systems (based on the mass of electrode materials).

Table 1. Literature comparison showing several reported LIC systems with carbonaceous anodes.

Ref	No. Anode//Cathode	Working Potential Range (vs. Li/Li$^+$)	Max. Energy Density Wh kg^{-1}/Power Density W kg^{-1}	Max. Power Density W kg^{-1}/Energy Density Wh kg^{-1}	Cyclability Cycles @ Current Density A g^{-1}
This work	Ni@CNOs//AC	2.0–4.2	140.1/275	27,000/76.6	40,000 cycles (94.09%) @ 2 A g^{-1} in the voltage range of 2.2–3.8 V
[30]	OLC-B//AC	2.0–4.0	243 @ 211	20,149 @ 66	10,000 cycles (78%)
[44]	Graphite//functionalized graphene	2.0–4.0	106/84	4200/85	100% over 1000 cycles
[45]	Hard carbon//activated carbon	1.4–4.3	80/150	2350/65	83% over 10,000 cycles
[46]	Hard carbon//activated carbon	2.0–4.0	82/100	20,000/14	97% over 600 cycles
[47]	Graphdiyne//AC	2.0–4.0	110.7/100.3	1000.4/95.1	1000 cycles @ 0.2 A g^{-1}
[48]	Sisal fiber-derived graphitic carbon//Sisal fiber AC	2.0–4.0	104/143	6628/32	3000 cycles @ 1 A g^{-1}
[49]	N-doped hard carbon//activated carbon	2.0–4.0	28.5/348	6940/13.1	97% over 5000 cycles
[50]	Soft carbon//activated carbon	0–4.4	115/25	15,000/16	63% over 15,000 cycles
[51]	Graphene//activated carbon	2.0–4.0	95/27	222.2/61.7	74% over 300 cycles

4. Conclusions

In this work, the effects of Ni@CNOs as pre-lithiated anode materials on the electrochemical performances of LICs were investigated. Due to their large specific surface area, high conductivity, and the presence of metal Ni increasing the electronic conductivity, Ni@CNOs exhibit excellent electron and ion transport for rapid electrochemical interactions with Li^+ and a significant improvement in the rate performance of LICs. For the Ni@CNOs as anodes vs. Li/Li^+, when the current density was increased to 20 A g^{-1}, the reversible specific capacity remained at 111 mAh g^{-1}. Furthermore, pouch-type AC//Ni@CNOs LICs with a high energy density and a high-power density were successfully fabricated. According to the experimental results, the AC//Ni@CNOs LIC achieves an energy density of up to 140.1 Wh Kg^{-1} at a power density of 275 W kg^{-1}. Even more, at an ultra-high power density of 27 kW kg^{-1}, the electrode can deliver an energy density of 76.6 Wh kg^{-1}, which is superior to the energy density of carbonaceous anodes and the power density of commercial supercapacitors. These capacitors also demonstrate excellent cycling stability, with a 94.09% capacitance retention over 40,000 cycles. Hence, the Ni@CNOs demonstrate potential for application in practical LICs. In future work, Ni will be removed from the CNOs, and then the pure CNOs' electrochemical performance and their application in the field of LICs will be tested.

Supplementary Materials: The following supporting information can be downloaded at: https://www.mdpi.com/article/10.3390/batteries9020102/s1, Table S1. The productivity of CNOs synthesized at different temperatures; Table S2. The productivity of CNOs synthesized with different flows; Figure S1. (a) The XPS survey spectrum and (b) high-resolution XPS spectrum of the C 1 s region in Ni@CNOs; Figure S2. TEM image of Ni@CNOs; Figure S3. The energy spectra of Ni@CNOs.

Author Contributions: Conceptualization, X.Z. (Xiaohu Zhang), X.Z. (Xiong Zhang) and Y.M.; methodology, X.Z. (Xiong Zhang), K.Z., W.Z., X.Z. (Xiaohu Zhang) and Y.M.; software, X.Z. (Xiaohu Zhang), K.Z. and X.Z. (Xiong Zhang); validation, X.Z. (Xiong Zhang), X.Z. (Xiaohu Zhang), K.W. and Y.M.; formal analysis, X.Z. (Xiaohu Zhang) and X.Z. (Xiong Zhang); investigation, X.Z. (Xiong Zhang), K.Z. and C.L.; resources, X.Z. (Xiong Zhang) and Y.M.; data curation, X.Z. (Xiong Zhang), K.Z., L.W., X.S., Y.A. and X.Z. (Xiaohu Zhang); writing—original draft preparation, X.Z. (Xiong Zhang), K.Z., W.Z. and X.Z. (Xiaohu Zhang); writing—review and editing, X.Z. (Xiong Zhang), K.Z., W.Z. and X.Z. (Xiaohu Zhang); visualization, X.Z. (Xiong Zhang), K.Z., W.Z. and X.Z. (Xiaohu Zhang); supervision, X.S., K.W. and Y.M.; project administration, X.Z. (Xiaohu Zhang), X.Z. (Xiong Zhang) and Y.M.; funding acquisition, X.Z. (Xiong Zhang) and Y.M. All authors have read and agreed to the published version of the manuscript.

Funding: This research was funded by the National Natural Science Foundation of China, grant numbers 52077207, 51822706, and 51777200, Beijing Natural Science Foundation, grant number JQ19012, and the Dalian National Laboratory for Clean Energy (DNL) Cooperation Fund, CAS grant numbers DNL201912 and DNL201915.

Institutional Review Board Statement: Not applicable.

Informed Consent Statement: Not applicable.

Data Availability Statement: Not applicable.

Conflicts of Interest: The authors declare no conflict of interest.

References

1. Dunn, B.; Kamath, H.; Tarascon, J.-M. Electrical energy storage for the grid: A battery of choices. *Science* **2011**, *334*, 928–935. [CrossRef]
2. Simon, P.; Gogotsi, Y. Perspectives for electrochemical capacitors and related devices. *Nat. Mater.* **2020**, *19*, 1151–1163. [CrossRef] [PubMed]
3. Zuo, W.; Li, R.; Zhou, C.; Li, Y.; Xia, J.; Liu, J. Battery-supercapacitor hybrid devices: Recent progress and future prospects. *Adv. Sci.* **2017**, *4*, 1600539. [CrossRef] [PubMed]
4. Sun, Y.; Liu, N.; Cui, Y. Promises and challenges of nanomaterials for lithium-based rechargeable batteries. *Nat. Energy* **2016**, *1*, 16071. [CrossRef]

5. Ock, I.W.; Lee, J.; Kang, J.K. Metal-organic framework-derived anode and polyaniline chain networked cathode with mesoporous and conductive pathways for high energy density, ultrafast rechargeable, and long-life hybrid capacitors. *Adv. Energy Mater.* **2020**, *10*, 2001851. [CrossRef]

6. Xu, J.; Gao, B.; Huo, K.F.; Chu, P.K. Recent progress in electrode materials for nonaqueous lithium-ion capacitors. *J. Nanosci. Nanotechnol.* **2020**, *20*, 2652–2667. [CrossRef]

7. Zheng, J.P. Theoretical energy density for electrochemical capacitors with intercalation electrodes. *J. Electrochem. Soc.* **2005**, *152*, A1864–A1869. [CrossRef]

8. Wang, H.W.; Zhu, C.R.; Chao, D.L.; Yan, Q.Y.; Fan, H.J. Nonaqueous hybrid lithium-ion and sodium-ion capacitors. *Adv. Mater.* **2017**, *29*, 1702093. [CrossRef]

9. Karimi, D.; Behi, H.; Van Mierlo, J.; Berecibar, M. A comprehensive review of lithium-ion capacitor technology: Theory, development, modeling, thermal management systems, and applications. *Molecules* **2022**, *27*, 3119. [CrossRef]

10. Ma, Y.F.; Chang, H.C.; Zhang, M.; Chen, Y.S. Graphene-based materials for lithium-ion hybrid supercapacitors. *Adv. Mater.* **2015**, *27*, 5296–5308.

11. Zhang, X.H.; Sun, X.Z.; An, Y.B.; Zhang, X.; Li, C.; Zhang, K.L.; Song, S.; Wang, K.; Ma, Y.W. Design of a fast-charge lithium-ion capacitor pack for automated guided vehicle. *J. Energy Storage* **2022**, *48*, 104045. [CrossRef]

12. Omonayo, B.; Annadanesh, S.; Zheng, J.P. Lithium-ion capacitor safety testing for commercial application. *Batteries* **2019**, *5*, 74.

13. Barcellona, S.; Ciccarelli, F.; Iannuzzi, D.; Piegari, L. Overview of lithium-ion capacitor applications based on experimental performances. *Electr. Power Compon. Syst.* **2016**, *44*, 1248–1260. [CrossRef]

14. Liu, W.J.; Zhang, X.; Xu, Y.N.; Li, C.; Wang, K.; Sun, X.Z.; Su, F.Y.; Chen, C.M.; Liu, F.Y.; Wu, Z.S.; et al. Recent advances of carbon-based materials for high performance lithium-ion capacitors. *Batter. Supercaps* **2021**, *4*, 407–428. [CrossRef]

15. Wang, Q.; Jiang, X.; Tong, Q.; Li, H.; Li, J.; Yang, W. Continuously interconnected N-doped porous carbon for high-performance lithium-ion capacitors. *Nanoenergy Adv.* **2022**, *2*, 303–315. [CrossRef]

16. Wang, Q.; Liu, F.; Jin, Z.; Qiao, X.; Huang, H.; Chu, X.; Xiong, D.; Zhang, H.; Liu, Y.; Yang, W. Hierarchically divacancy defect building dual-activated porous carbon fibers for high-performance energy-storage devices. *Adv. Funct. Mater.* **2020**, *30*, 2002580. [CrossRef]

17. Banerjee, A.; Upadhyay, K.K.; Puthusseri, D.; Aravindan, V.; Madhavi, S.; Ogale, S. MOF-derived crumpled-sheet-assembled perforated carbon cuboids as highly effective cathode active materials for ultra-high energy density Li-ion hybrid. *Nanoscale* **2014**, *6*, 4387–4394. [CrossRef]

18. Stoller, M.D.; Murali, S.; Quarles, N.; Zhu, Y.; Potts, J.R.; Zhu, X.; Ha, H.W.; Ruoff, R.S. Activated graphene as a cathode material for Li-ion hybrid supercapacitors. *Phys. Chem. Chem. Phys.* **2012**, *14*, 3388–3391. [CrossRef]

19. Lee, J.H.; Shin, W.H.; Lim, S.Y.; Kim, B.G.; Choi, J.W. Modified graphite and graphene electrodes for high-performance lithium ion hybrid capacitors. *Mater. Renew. Sustain. Energy* **2014**, *3*, 22. [CrossRef]

20. Yi, S.; Wang, L.; Zhang, X.; Li, C.; Liu, W.J.; Wang, K.; Sun, X.Z.; Xu, Y.N.; Yang, Z.X.; Cao, Y.S.; et al. Cationic intermediates assisted self-assembly two-dimensional Ti3C2Tx/rGO hybrid nanoflakes for advanced lithium-ion capacitors. *Sci. Bull.* **2021**, *66*, 914. [CrossRef]

21. Zeiger, M.; Jäckel, N.; Mochalin, V.N.; Presser, V. Review: Carbon onions for electrochemical energy storage. *J. Mater. Chem. A* **2016**, *4*, 3172–3196. [CrossRef]

22. Dhand, V.; Yadav, M.; Kim, S.H.; Rhee, K.Y. A comprehensive review on the prospects of multi-functional carbon nano onions as an effective, high- performance energy storage material. *Carbon* **2021**, *175*, 534–575. [CrossRef]

23. Tan, Z.-L.; Wei, J.-X.; Liu, Y.; Zaman, F.U.; Rehman, W.; Hou, L.-R.; Yuan, C.-Z. V2CTx MXene and its derivatives: Synthesis and recent progress in electrochemical energy storage applications. *Rare Metals* **2022**, *41*, 775. [CrossRef]

24. Sun, X.Z.; Geng, L.B.; Yi, S.; Li, C.; An, Y.B.; Zhang, X.H.; Zhang, X.; Wang, K.; Ma, Y.W. Effects of carbon black on the electrochemical performances of SiOx anode for lithium-ion capacitors. *J. Power Source* **2021**, *499*, 229936. [CrossRef]

25. Akintola, T.; Shellikeri, A.; Akintola, T.; Zheng, J.P. The influence of Li4Ti5O12 preparation method on lithium-ion capacitor performance. *Batteries* **2021**, *7*, 33. [CrossRef]

26. An, Y.B.; Liu, T.Y.; Li, C.; Zhang, X.; Hu, T.; Sun, X.Z.; Wang, K.; Wang, C.D.; Ma, Y.W. A general route for the mass production of graphene-enhanced carbon composites toward practical pouch lithium-ion capacitors. *J. Mater. Chem. A* **2021**, *9*, 15654. [CrossRef]

27. Portet, C.; Yushin, G.; Gogotsi, Y. Electrochemical performance of carbon onions, nanodiamonds, carbon black and multiwalled nanotube in electrical double layer capacitors. *Carbon* **2007**, *45*, 2511–2518. [CrossRef]

28. Chen, J.; Yang, B.; Li, H.; Ma, P.; Lang, J.; Yan, X. Candle soot: Onion-like carbon, an advanced anode material for a potassium-ion hybrid capacitor. *J. Mater. Chem. A* **2019**, *7*, 9247–9252. [CrossRef]

29. Jäckel, N.; Weingarth, D.; Zeiger, M.; Aslan, M.; Grobelsek, I.; Presser, V. Comparison of carbon onions and carbon blacks as conductive additives for carbon supercapacitors in organic electrolytes. *J. Power Source* **2014**, *272*, 1122–1133. [CrossRef]

30. Shi, K.Y.; Liu, J.W.; Chen, R. Nitrogen-Doped nano-carbon onion rings for energy storage in Lithium-ion capacitors. *J. Energy Storage* **2020**, *31*, 101609. [CrossRef]

31. Permana, A.D.C.; Ding, L.; Gonzalez-Martinez, I.G.; Hantusch, M.; Nielsch, K.; Mikhailova, D.; Omar, A. Comparative study of onion-like carbons prepared from different synthesis routes towards li-ion capacitor application. *Batteries* **2022**, *8*, 160. [CrossRef]

32. Permana, A.D.C.; Omar, A.; Gonzalez-Martinez, L.G.; Oswald, S.; Giebeler, L.; Nielsch, K.; Mikhailova, D. MOF-derived onion-like carbon with superior surface area and porosity for high performance lithium-ion capacitors. *Batter. Supercaps* **2022**, *5*, e202100353.

33. Aref, A.R.; Chen, S.W.; Rajagopalan, R.; Randall, C. Bimodal porous carbon cathode and prelithiated coalesced carbon onion anode for ultrahigh power energy efficient lithium ion capacitors. *Carbon* **2019**, *152*, 89–97. [CrossRef]
34. Zhang, L.; Zhou, Q.; Zhao, H.T.; Ruan, C.; Wang, Y.J.; Li, Z.J.; Lian, Y.F. The arc-discharged Ni-cored carbon onions with enhanced microwave absorption performances. *Mater. Lett.* **2020**, *265*, 127408. [CrossRef]
35. Hechmann, A.; Fromm, O.; Rodehorst, U.; Münster, P.; Winter, M.; Placke, T. New Insights into Electrochemical Anion Intercalation into Carbonaceous Materials for Dual-Ion Batteries: Impact of Graphitization Degree. *Carbon* **2018**, *131*, 201–212. [CrossRef]
36. Ma, Y.; Wang, K.; Xu, N.Y.; Zhang, D.X.; Peng, F.Q.; Li, N.S.; Zhang, X.; Sun, Z.X.; Ma, W.Y. Dehalogenation produces graphene wrapped carbon cages as fast-kinetics and large-capacity anode for lithium-ion capacitors. *Carbon* **2023**, *202*, 175–185. [CrossRef]
37. Li, C.; Zhang, X.; Wang, K.; Sun, Z.X.; Ma, W.Y. High-power and long-life lithium-ion capacitors constructed from N-doped hierarchical carbon nanolayer cathode and mesoporous graphene anode. *Carbon* **2018**, *140*, 237–248. [CrossRef]
38. Chen, C.J.; Wang, Z.G.; Miao, L.; Cai, J.; Peng, L.F.; Huang, Y.Y.; Zhang, L.N.; Xie, J. Nitrogen-rich hard carbon as a highly durable anode for high-power potassium-ion batteries. *Energy Storage Mater.* **2017**, *8*, 161–168. [CrossRef]
39. Yue, Y.; Juarez-Robles, D.; Mukherjee, P.; Liang, H. Superhierarchical Nickel–Vanadia Nanocomposites for Lithium Storage. *ACS Appl. Energy Mater.* **2018**, *1*, 2056–2066. [CrossRef]
40. Chen, Y.; Chen, H.Y.; Du, F.H.; Shen, X.P.; Ji, Z.Y.; Zhou, H.B.; Yuan, A.H. In-situ construction of nano-sized Ni-NiO-MoO2 heterostructures on holey reduced graphene oxide nanosheets as high-capacity lithium-ion battery anodes. *J. Alloys Compd.* **2022**, *926*, 166847. [CrossRef]
41. Jin, L.M.; Guo, X.; Shen, C.; Qin, N.; Zheng, J.S.; Wu, Q.; Zhang, C.M.; Zheng, J.P. A universal matching approach for high power-density and high cycling-stability lithium ion capacitor. *J. Power Source* **2019**, *441*, 227211. [CrossRef]
42. Naderi, R.; Shellikeri, A.; Hagen, M.; Cao, W.; Zheng, J.P. The influence of anode/cathode capacity ratio on cycle life and potential variations of lithium-ion capacitors. *J. Electrochem. Soc.* **2019**, *166*, A2610–A2617. [CrossRef]
43. Yu, X.L.; Zhan, C.Z.; Lv, R.T.; Bai, Y.; Lin, Y.X.; Huang, Z.H.; Shen, W.C.; Qiu, X.P.; Kang, F.Y. Ultrahigh-rate and high-density lithium-ion capacitors through hybriding nitrogen-enriched hierarchical porous carbon cathode with prelithiated microcrystalline graphite anode. *Nano Energy* **2015**, *5*, 43–53. [CrossRef]
44. Lee, J.H.; Shin, W.H.; Ryon, M.H.; Jin, J.K.; Kim, J.Y.; Choi, J.W. Functionalized graphene for high performance lithium ion capacitors. *ChemSusChem* **2012**, *5*, 2328–2333. [CrossRef] [PubMed]
45. Kim, J.H.; Kin, J.S.; Lim, Y.G.; Lee, J.G.; Kim, Y.J. Effect of carbon types on the electrochemical properties of negative electrodes for Li-ion capacitors. *J. Power Source* **2011**, *196*, 10490–10495. [CrossRef]
46. Cao, W.J.; Zheng, J.P. Li-ion capacitors with carbon cathode and hard carbon/stabilized lithium metal powder anode electrodes. *J. Power Source* **2012**, *213*, 180–185. [CrossRef]
47. Du, H.P.; Yang, H.; Huang, C.S.; He, J.J.; Liu, H.B.; Li, Y.L. Graphdiyne applied for lithium-ion capacitors displaying high power and energy densities. *Nano Energy* **2016**, *22*, 615–622. [CrossRef]
48. Yang, Z.W.; Guo, H.J.; Li, X.H.; Wang, Z.X.; Wang, J.X.; Wang, Y.S.; Yan, Z.L.; Zhang, D.C. Graphitic carbon balanced between high plateau capacity and high rate capability for lithium ion capacitors. *J. Mater. Chem. A* **2017**, *5*, 15302–15309. [CrossRef]
49. Han, X.Q.; Han, P.X.; Yao, J.H.; Zhang, S.; Cao, X.Y.; Xiong, J.W.; Zhang, J.N.; Cui, G.L. Nitrogen-doped carbonized polyimide microsphere as a novel anode material for high performance lithium-ion capacitors. *Electrochim. Acta* **2016**, *196*, 603–610. [CrossRef]
50. Schroeder, M.; Winter, M.; Passerini, S.; Balducci, A. On the cycling stability of lithium-ion capacitors containing soft carbon as anodic material. *J. Power Source* **2013**, *238*, 388–394. [CrossRef]
51. Ren, J.J.; Su, L.W.; Qin, X.; Yang, M.; Wei, J.P.; Zhou, Z.; Shen, P.W. Pre-lithiated graphene nanosheets as negative electrode materials for Li-ion capacitors with high power and energy density. *J. Power Source* **2014**, *264*, 108–113. [CrossRef]

Review

Strategies and Challenge of Thick Electrodes for Energy Storage: A Review

Junsheng Zheng [1,*], Guangguang Xing [1], Liming Jin [1,*], Yanyan Lu [1], Nan Qin [1], Shansong Gao [2] and Jim P. Zheng [3]

1 Clean Energy Automotive Engineering Center and School of Automotive Studies, Tongji University, Shanghai 201804, China
2 China Shenhua Coal to Liquid and Chemical Shanghai Research Institute, Shanghai 201108, China
3 Department of Electrical Engineering, University at Buffalo, The State University of New York, Buffalo, NY 14260, USA
* Correspondence: jszheng@tongji.edu.cn (J.Z.); limingjin@tongji.edu.cn (L.J.)

Abstract: In past years, lithium-ion batteries (LIBs) can be found in every aspect of life, and batteries, as energy storage systems (ESSs), need to offer electric vehicles (EVs) more competition to be accepted in markets for automobiles. Thick electrode design can reduce the use of non-active materials in batteries to improve the energy density of the batteries and reduce the cost of the batteries. However, thick electrodes are limited by their weak mechanical stability and poor electrochemical performance; these limitations could be classified as the critical cracking thickness (CCT) and the limited penetration depth (LPD). The understanding of the CCT and the LPD have been proposed and the recent works on breaking the CCT and improving the LPD are listed in this article. By comprising these attempts, some thick electrodes could not offer higher mass loading or higher accessible areal capacity that would defeat the purpose.

Keywords: thick electrodes; critical cracking thickness; limited penetration depth; mass loading; area capacity

Citation: Zheng, J.; Xing, G.; Jin, L.; Lu, Y.; Qin, N.; Gao, S.; Zheng, J.P. Strategies and Challenge of Thick Electrodes for Energy Storage: A Review. *Batteries* **2023**, *9*, 151. https://doi.org/10.3390/batteries9030151

Academic Editor: Palani Balaya

Received: 18 December 2022
Revised: 14 February 2023
Accepted: 24 February 2023
Published: 27 February 2023

1. Introduction

Over the past few decades, lithium-ion batteries (LIBs) have attracted more attention as energy storage systems (ESSs) due to the drive for a greener future. LIBs are electrochemical ESSs that supply energy by electrochemical reactions occurring in porous electrodes. The introduction of LIBs into vehicles has required more demands for advanced batteries. A typical areal capacity for hybrid electrical vehicles (HEVs) is about 2 mAh·cm^{-2}, while for typical electrical vehicles (EVs), it is 4 mAh·cm^{-2} for state-of-the-art LIBs [1]. Energy-to-weight ratio is a critical issue for ESSs, and a battery-level specific energy of ~225 Wh·kg^{-1} is widely accepted; it combines the weight and driving mileage for EVs [2]. To offer competitive advantages for EVs in market, the US Department of Energy (US DOE) and the Advanced Battery Consortium (USABC) held that the EVs should provide a range of at least 500 km, while batteries as ESSs need to possess high energy density of approximately 235 Wh·kg^{-1} and 500 Wh·L^{-1} at battery pack level [3,4].

For a greener future, the China Society of Automotive Engineers (CSAE) with the guidance from the Ministry of Industry and Information Technology of China (China IIT) publishes the Technology Roadmap of Clean and New Energy Vehicles 2.0. In addition, a higher demand of 500 Wh·kg^{-1} is proposed for advanced batteries for EVs in 2035 [5]. The US DOE organized a *Battery 500 Consortium* that aims to achieve battery energy above 500 Wh·kg^{-1}, and some feasible strategies (see Figure 1a) are given for lithium metal batteries. In general, advanced strategies proposed to obtain high energy storage systems include: (1) to study the new electrochemical energy storage mechanisms [6]; (2) to broaden the cell potential window [7]; (3) to develop electrode materials with high specific capacity [8]; and (4) to design electrodes with high mass loading [9]. There are lots of studies that focus on developing next-generation high-energy batteries, such as Li-oxygen and Li-sulfur batteries. A high energy density lithium-oxygen battery based on a reversible four-electron conversion to lithium oxide was reported [6]. However, most studies have focused on energy density on materials level while the specific energy density of the represented battery packs is nearly four times smaller than the energy density on the material level (Figure 1b) [10]. Much non-active but indispensable components are introduced into batteries due to the need for its operation and management. Therefore, the improvement on the material or electrode level could not completely transferred to the battery level. Narrowing the gap in specific energy density between the material layer and the battery layer brings a brighter prospect for improving the battery energy density. The fourth path of this paper is based on this idea. Furthermore, the first three ways make efforts on the material or electrode level and need to put much energy on new mechanisms while the fourth path does not. Additionally, the fourth path could further improve energy density on the battery pack level based on the first three ways.

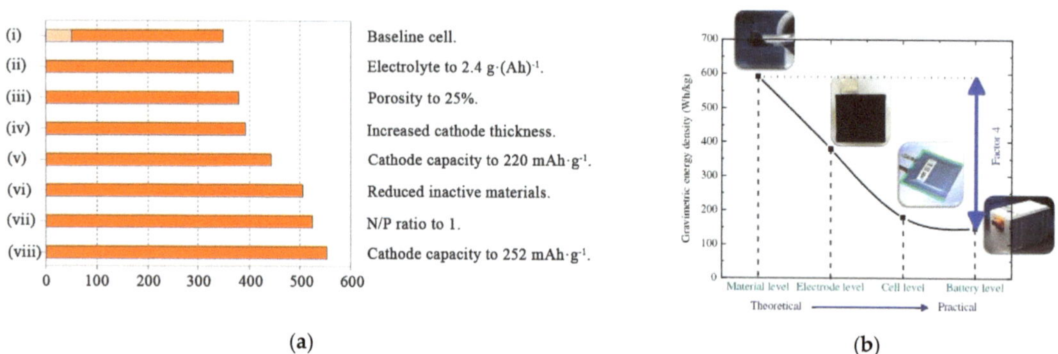

Figure 1. (**a**) Calculated cell−level specific energy based on different strategies [11]. (**b**) The specific energy density based on a standard 52Ah pouch cell and state−of−the−art packs [10].

The fourth path to design electrodes with high mass loading is thick electrodes. The principle of thick electrodes design is that higher mass loading could decrease the share of non-active materials like current collectors and separators. However, it does not mean that the thicker the electrodes' thickness the higher battery energy density. Figure 2a gives the advantage of thick electrodes, and the trend of curves supports that point. To further vividly explain this principle, taking the commercial laminated batteries for example (see Figure 2b), a repeating unit that contains electrode materials and current collectors and separators could be extracted. When using thick electrodes to replace the conventional electrodes in the repeating unit, the ratio of non-active materials in batteries is significantly decreased. The strategy of thick electrodes is to minimize the use of non-active materials to improve the battery energy density. Additionally, from Figure 2b, the use of non-active materials in batteries constructed by thick electrodes is already too low, which means that there is not more space for improving battery energy density from increasing electrode thickness. It is agreed with the second half curves in Figure 2a. Therefore, there is an optimal interval for the thickness or mass loading of the electrodes, as shown in Figure 2c. However, the concrete values are related various factors and not given.

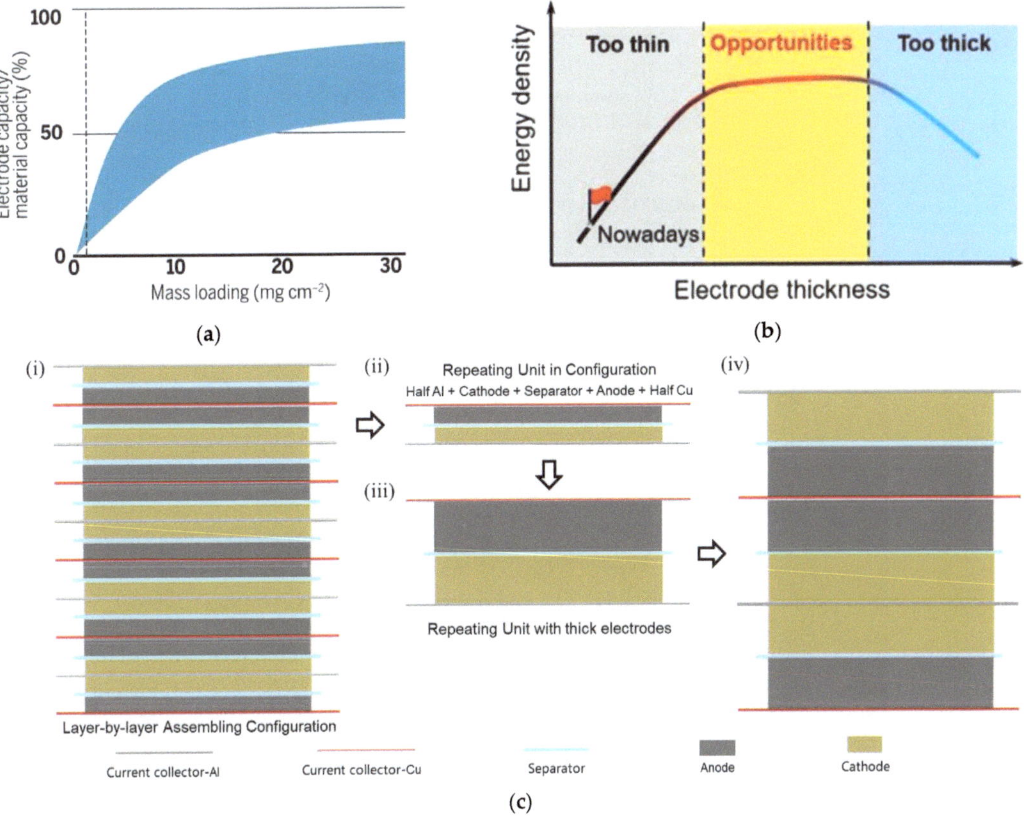

Figure 2. (**a**) The capacity of an electrode is positively proportional to the mass loading of active materials on the electrode [12]. (**b**) A detailed explanation about thick electrode design. (**c**) The opportunities for thick electrode design [13].

In presently commercial LIBs, the thickness of electroactive components including the cathode and the anode are both limited between 50 and 100 μm [14,15]. The design of thick electrodes is not a novel strategy, and its application is restricted by two serious obstacles: weak mechanical stability in production and poor electrochemical performance in working. It has been acknowledged in academia that there are two critical thickness for battery electrodes with high mass loading. One is the critical cracking thickness (CCT) about mechanical stability [16–19]; the other is the limited penetration depth (LPD) for electrolyte transport in the electrode [2,20–22].

In past years, much of the studies were devoted to new electrode design with high mass loading to boost the development of LIBs. To overcome the limitations of CCT, the plentiful works draw the support of three-dimensional frameworks to offer mechanical stability [9,23–27]. Electrodes with a thickness up to 850 μm and an aerial mass of 55 mg·cm⁻² have been constructed with the aid of wood templates [28]. For improving the limitation of LPD, to construct the ordered pores to decrease tortuosity is the most popular choice [29–31]. Thickness-independent electrodes constructed by vertical alignment of two-dimensional flakes could enable directional ions transport [32].

It is noteworthy that the limitations of ion diffusion in the liquid electrolyte deteriorate the rate capabilities of thick electrode design [2,33]. Therefore, introducing thick electrode design into all-solid-state lithium batteries (ASSLBs) could be helpful due to the uniform transference number of inorganic solid electrolytes [34]. Moreover, the current ASSLBs are produced by expensive and multi-step processes based on thin-layer-deposition techniques, and much efforts have been invested in increasing electrode thickness [35]. The combination of the two concepts could offer higher energy density and power density for lithium batteries. Hong et al. developed a new solvent-free dry technique to produce thick electrodes for ASSLBs which uses a Li^+-conducting ionomer as a binder to form fibrous linear binding [36].

This paper will describe and discuss the latest advances in thick, porous electrodes from various perspectives, including the understanding of the two critical thicknesses, break-throughs in the two critical thicknesses, and the relevant comparison between these studies.

2. The Challenge of Thick Electrodes

To obtain high energy density of 500 Wh·kg^{-1} for advanced batteries is the shared goal for China and US governments where are the largest automotive markets in the world. The *Battery 500 Consortium* proposed pathways to 500 Wh·kg^{-1} practical cells and an essential requirement is increasing the cathode thickness [11]. LIBs constructed by thick electrodes with high mass loading can benefit both vehicular range and unit cost in the application for EVs [37,38]. The superiority of thick electrodes design has been discovered formerly and the electrode thickness has been increased to over 100 μm in commercial batteries. Therefore, there must be some problems hindering the realization of thicker electrodes.

2.1. The Critical Cracking Thickness (CCT)

During the drying of wet films, cracks were observed in diverse systems such as desiccated soil, concrete casting, ceramics slips and model colloidal dispersions [39,40]. Figure 3a give the diagram about the drying process [41]. It has been widely accepted that the capillary stresses during drying process are the cause of cracking formation [16,42–44]. When the slurry containing suspended particles is dried, capillary stresses are generated between particles in the air–solvent interface. If the particles are soft, they can remove stresses. However, in reverse case, the stresses are released by cracks formation when the particles are hard [16].

It has been observed that the CCT would increase with the increase of particle size and be not affected by the drying speed [16,19,41]. It is noteworthy that lower drying speed has positive effects on the fracture toughness but not the film thickness [43]. It has also proven that the drying rate actually has no impact on the CCT but crack size [45]. Additionally, decreasing film thickness and increasing particle shear modulus would increase the critical capillary pressure for cracking formation. Singh et al. analyzed the influencing factors of the capillary stresses and established a formula about the CCT:

$$h_{max} = 0.41 \left(\frac{GM\varnothing_{rcp}R^3}{2\gamma} \right)^{1/2} \tag{1}$$

where h_{max} is the CCT, G is the shear modulus of the particles, M is the coordination number, $\varnothing_{rcp}$ is the particle volume fraction at random close packing, R is the particle radius, and γ is the air–solvent interfacial tension [16].

The traditional technology in electrode manufacture is to mix active materials with conductive additives and binders in organic solvents, and then to coat this slurry on current collectors like Al or Cu film. Additionally, a requisite step in this manufacture is to evaporate the solvents. The cracks also have been observed on battery electrodes, cracks were generated in NMC electrodes (NMC811:PVDF:CB = 90:5:5, wt.%) at a thickness above 175 μm and any crack-free μ-Si electrodes (μ-Si:PAA:CB = 80:10:10, wt. %) could not be fabricated at a thickness above 100 μm, as depicted in Figure 3b [46].

2.2. The Limited Penetration Depth (LPD)

High mass loading electrodes would increase the thickness of electrode films, which increases the diffusion distance of charges in electrodes. In the electrochemical process, the longer the charge diffusion distance, the lower the mass transfer efficiency [21]. Additionally, due to the sluggish ions transport kinetics, not all active materials in thick electrodes could be used in high C-rates. It has been proven that the limited diffusion of Li$^+$ inside the porous electrode leads to the under-utilization of the active material [14,47–49].

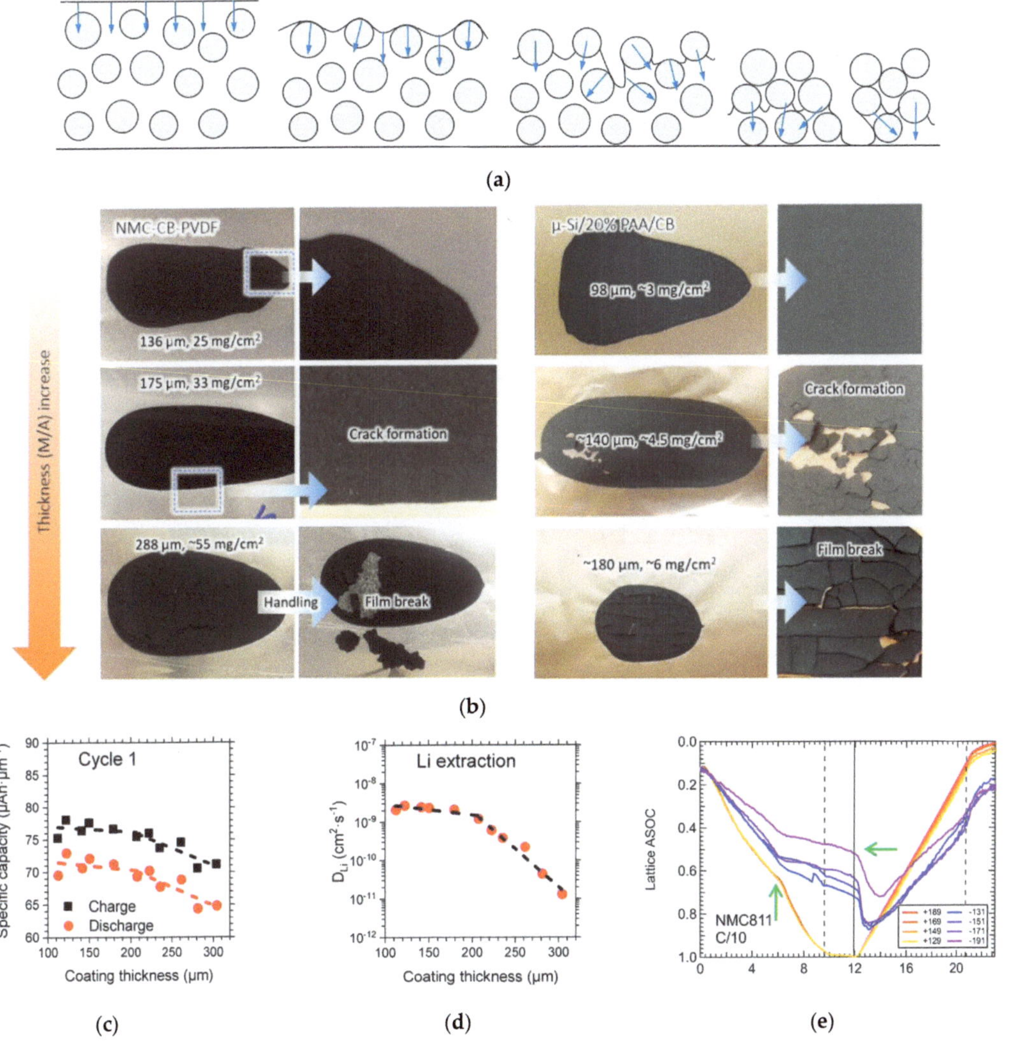

Figure 3. (**a**) Illustration of the drying process [41]. (**b**) Cracks in NMC electrodes (NMC811:PVDF:CB = 90:5:5, wt.%) and µ−Si electrodes (µ-Si:PAA:CB = 80:10:10, wt. %) [46]. (**c**) First cycle thickness−normalized specific capacity as a function of electrode thickness [47]. (**d**) D_{Li}^+ as a function of electrode thickness [47]. (**e**) An inhomogeneity in different thicknesses across an NMC811 electrode during the charging and discharging process [50].

When the LPD is greater than the designed electrodes thickness, mass transport in the electrolyte would not be the limited factors of the full utilization of active materials in electrodes [2]. To balance the migration of anions that do not participate in the electrochemical process, a lithium salt concentration gradient would form: that is what the LPD means [20]. The LPD is in inverse proportion to C-rates and tortuosity [22]. Additionally, a simple analytical equation for the LPD for electrolyte transport in the electrode is given:

$$L_d = \frac{\varepsilon}{T} \frac{D_0 c_0 F}{(1 - t_+)I} \tag{2}$$

where L_d is the LPD, ε is the porosity of the electrode, T is the tortuosity factor of the pore matrix, D_0 is the diffusion coefficient of the lithium salt species in the electrolyte, c_0 is the initial concentration of electrolyte, t_+ is the transference number of Li$^+$, I is the applied current density and F is the Faraday constant [2].

From experiments on NMC622, electrodes with different thicknesses, their specific capacities at C/50 rate were obtained and these values were normalized by their coating thickness [47]. An obvious turn at ~200 μm was shown in the plot of the normalized capacity as the function of thickness, as shown in Figure 3c. Moreover, the lithium diffusion coefficient ($D_{Li}{}^+$) that calculated from GITT is a related parameter of Li$^+$ diffusion behavior. The same turn at ~200 μm were observed in the plot of $D_{Li}{}^+$ as the function of thickness (see Figure 3d). Additionally, the operando studies about the depth-dependent inhomogeneity are supported in this suspect, as shown in Figure 3e [50]. The same result has been proposed that the maximum film thickness limited by ions diffusion is approximately 200 μm [32].

The ESSs for EVs are required in excellent performance in terms of energy and power. To fabricate battery electrodes with high mass loading needs to break the limit of the CCT, and the LPD is the obstacle that stands in the way of battery electrodes with high accessible areal capacity. According to the understanding of the CCT and the LPD, various ways have been developed to solve the problems.

3. Strategies for Increasing Electrode Thickness

Making thick electrodes a reality needs to break the limitations of CCT and LPD. Additionally, how to fabricate crack-free electrodes is the first concern, which means the solution could be two-step or one-step. The two-step solution is to solve those two problems separately, for example, a thick and free-standing electrode was constructed with two-dimensional nanomaterials that breaks the CCT and then a laser drilling technique was adopted to fabricate a micro-hole array in this electrode to increase the LPD [51]. The one-step solution is to solve those two problems simultaneously. For example, the ultra-thick electrodes made of wood formwork improve the mechanical stability of the frame. Additionally, the gap between the active materials and the carbon frames improve the diffusion kinetics during the drying process [52].

3.1. Increasing the CCT

From the formation mechanism, the cause of cracking is generated stresses during the drying process [16,19,41], and even is the traditional technology (the wet-slurry casting technology) from a higher vision. Therefore, there are three strategies to construct thicker electrodes. One is making efforts to decrease the generated stresses during the drying process, and another is using three-dimensional (3D) frameworks to offer mechanical stability while the other is taking new manufacture technologies beyond the traditional technology.

3.1.1. Decreasing Generated Stresses

According to $h_{max} \sim (1/\gamma)^{1/2}$ from Equation (1), the CCT could be improved by decreasing the surface tension. Thus, making efforts to decrease the surface tension is a straightforward method to alleviate or eliminate cracks. According to this point, Du et al. have successfully constructed a NMC532 electrode with areal loading above 25 mg·cm^{-2} (~4 mAh·cm^{-2}) by introducing isopropyl alcohol (IPA) into aqueous solvent systems to decrease the surface tension, as shown in Figure 4a [41].

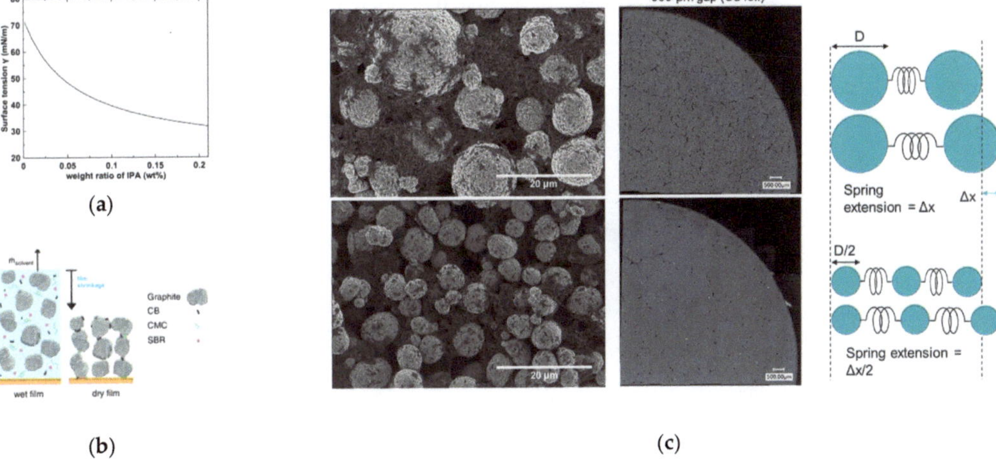

Figure 4. (**a**) Calculated surface tension of IPA–water mixture versus composition [41]. (**b**) Schematic of the wet film and dry film [45]. (**c**) SEM and optical microscope images of aqueous-processed cathode coatings (500 μm coating wet gap) on copper foil. Additionally, a simple ball and spring model is given [53].

As shown in Figure 4b, the remaining particles are forced to fill a certain ratio of the formed void when the solvent is evaporating during the drying process. Additionally, cracking would occur when above a certain value. Therefore, the other way to reduce mechanical stresses is to reduce the amount of solvent which needs to be inevitably evaporated during drying process [44]. By increasing the solid content of the electrode slurry with using a new binder system to 65 wt.%, crack-free NMC111 electrodes have been constructed with an active mass loading of up to 60 mg·cm^{-2} and an average dry film thickness of (322 ± 9) μm [44].

Moreover, the observed results of NMC electrodes with different particle sizes are contrary to the estimate of $h_{max} \sim R^{3/2}$ [53]. There is no contradiction, since the stresses are removed by the particles themselves when particles are soft, or by cracks when particles are hard. Additionally, there are more than one kind of particle in the electrode slurry. The NMC particles are hard, so the stresses are released by softer binders and conductive additives and even cracks. Additionally, a simple ball and spring model is given in Figure 4c, to reduce the size of active particles, which would decrease stresses in binders and conductive additives. Therefore, the formation of cracks was reformed when NMC811 particles of smaller size were used.

3.1.2. Utilizing 3D Frameworks

The mechanical stability of thick electrodes could draw support from 3D frameworks. Several thick and ultra-thick electrodes have been successfully fabricated with the aid of carbon frameworks [9,23–25,27], metal foams [15,54–56], and 3D conductive textile [57]. Besides, some 3D frameworks could offer convenient ion and electron channels to improve the limited diffusion kinetics due to the increased thickness.

Carbon nanotubes (CNT) and carbon nanofibers (CNF) are easy to form crosslinked networks and provide features as both binders and conductive additives. Park et al. constructed a high-performance electrode with the thickness of up to 800 μm through segregated CNT networks, as shown in Figure 5a [46]. Due to the improved mechanical robustness through these networks, the extremely thick electrodes could be constructed. With the help of graphite fibers (GF) bonded with pyrolytic carbon (PC) and graphite nanoplatelets (GNP), an ultra-thick electrode with the thickness of 17 mm performed the reversible capacity of 11.63 mAh·cm^{-2} (as shown in Figure 5b) [24]. Additionally, some carbon networks could undertake the functions of binders and conductive additives. It may be profitable for decreasing the ratio of non-active materials in battery electrodes.

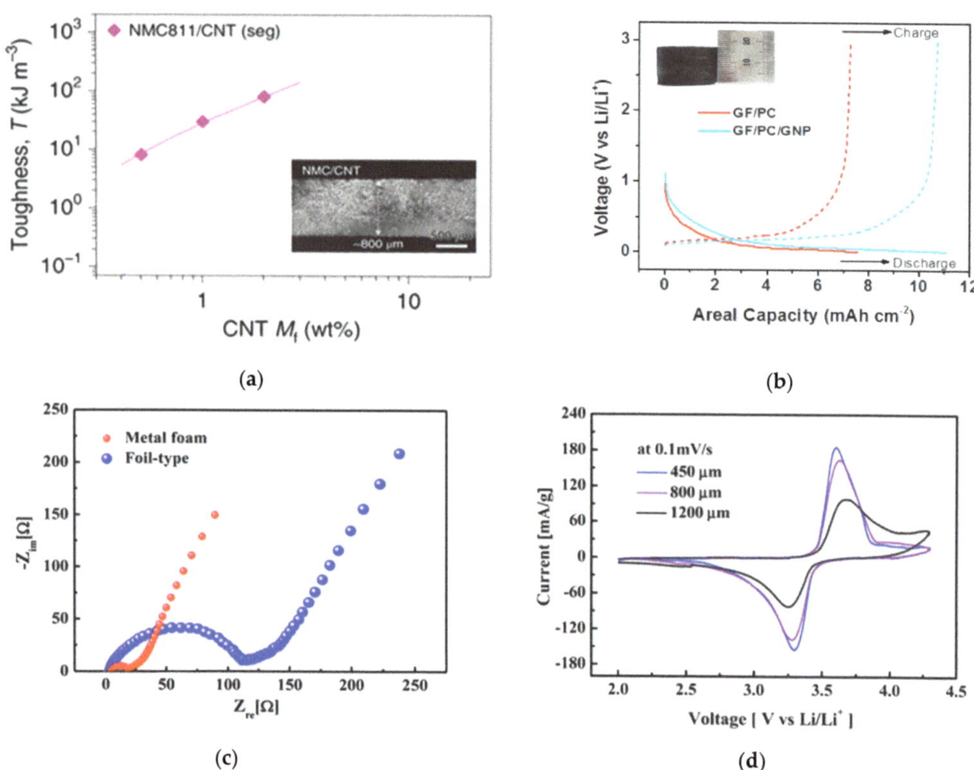

Figure 5. (**a**) Tensile toughness of NMC811/CNT electrodes plotted versus CNT content [46]. (**b**) The reversible capacity of the 3D thick all−carbon frameworks [24]. (**c**) Comparison of the impedance curves for electrodes constructed by the metal foam and the foil [54]. (**d**) Comparison of the cyclic voltametric curves for the electrodes using different cell size of metal foams [56].

Additionally, the adhesion properties between the active material and traditional current collectors are constant when increase the mass loading of electrodes, so the possibility of delamination at the current-collector/active-material interface became larger. 3D current collectors could provide greater adhesion than the foil-type current collectors [54]. Thicker electrodes could be constructed when all particles are not more than 50 μm from the nearest current collector. Metal foams have attracted more attention since they are the most suitable to introduce into the slurry casting process to fabricate electrodes. A graphite electrode was constructed with the thicknesses of 0.6 and 1.2 mm by using Cu foams as the current collector [15]. Additionally, Yang et al. fabricated a 450 μm thick graphene electrode with mass loading of 10~15 mg·cm^{-2} by using Al foams [55].

The 3D current collector also could improve kinetics of electrodes. From the comparison of foam-collector-type and foil-collector-type electrodes with similar mass loading of active materials, the charge transfer resistance was seven times less for the foam than for the foil, at values of 15 and 110 Ω (as shown in Figure 5c) [54], respectively. Furthermore, by comparing the cyclic voltametric curves for the cells using different sizes of metal foams, as shown in Figure 5d, a remarkable result was that the peak became lower and wider for the cell using 1200 μm size of metal foams [56]. It indicated that using a 3D current collector to fabricate electrodes could settle mechanical, but not completely electrochemical, problems in thicker electrodes.

Other 3D frameworks are also used to fabricated thick electrodes. The 3D conductive textiles have been introduced batteries electrodes due to its stable potential range in organic electrolyte and electrodes have been fabricated with a thickness of ~600 μm through 3D conductive textiles, which are 8–12 times higher of those on metal collector [57].

3.1.3. Taking New Technology

Since the cracks occur during drying process, taking some manufacture technologies without the evaporation process are possible ways to construct crack-free thick electrodes. Some technologies have been successfully introduced into battery electrodes like spraying [58–60], sintering [61], 3D printing [62–65], powder extrusion moulding (PEM) [66–68] and dry powder coating [69,70]. The solvent-free process is an ideal alternative solution to replace the wet slurry casting process. The related stresses generated during the drying process would no longer be considerations. Moreover, manufacture costs would be cut as the elimination of using and removing solvents.

A solvent-free technology is spray deposition that has been used in coating industries for over 30 years to create functional paints. Recently, this technology has been introduced into producing battery electrodes. The main processes of this technology are mixing, dry-spraying and hot-pressing. Figure 6a gives the optimized technology for using in battery electrodes. The fabricated LTO electrode has interconnected particles that facilitate electron and ion transport via shortened pathways within the electrodes [59]. Additionally, the film thickness could be controlled just by adjusting the spraying time [58].

Another solvent-free technology is dry-coating technology that does not introduce any solvents into the craft, as shown in Figure 6b. This technology maintains a liquid-free state in the full process from the raw materials to finished products. Free-standing electrodes with thicknesses between 50 microns to about 1 mm could be easily fabricated by this craft. It has been proven that the bonding strength between dry-deposited particles and current collectors could be greater than slurry-cast electrodes [70].

PEM is a cost-effective manufacturing method to produce battery electrode. The main steps of PEM are mixing, extrusion, de-binding and sintering, as shown in Figure 6c. Thick electrodes constructed by PEM technology shows better mechanical properties than alternative technologies [66]. The $Li_4Ti_5O_{12}$ (LTO) anodes and the $LiFePO_4$ (LFP) cathodes have been produced with the thickness as high as 500 μm through this method [66,67]. Additionally, a LIB was assembled with this LTO anode and this LFP cathode and possessed mass loading of ~100 mg·cm^{-2} that is better than the current one [68]. Just as important, this technology is also an environmentally friendly technology for electrode manufacturing.

Due to the low-cost and simple manufacturability, extrusion-based 3D printing is a potential craft for electrode fabrication. The thickness of thick electrodes could be well controlled thanks to the layer-by-layer additive technique [71]. Sun et al. constructed an ultra-thick LTO electrode by using 3D printing with the thickness of 1500 μm [62]. Additionally, 3D-printed electrodes with highly interconnected networks could offer ion- and electron-transport paths, which indicated a better electrochemical performance [63].

Additionally, a new design called fiber-aligned thick electrodes has been used to construct thick electrodes. The composite membrane contained aligned carbon fibers that could provide low tortuosity, high conductivity, and mechanically strong features for high mass loading electrodes, and this novel design is shown in Figure 6d [9,72].

There are many attempts to improve mechanical stability for producing thick crack-free electrodes. In this part, we mainly focus on the growth of thickness. Some technologies also bring other gains, 3D current collectors could serve as both the support and the collector that improve construction stability and electron conductivity. The thick electrodes fabricated by 3D printing technology have advanced ion transport pathways due to their highly ordered structure. Additionally, solvent-free technologies no longer need the using and removing of solvents, as well the corresponding costs.

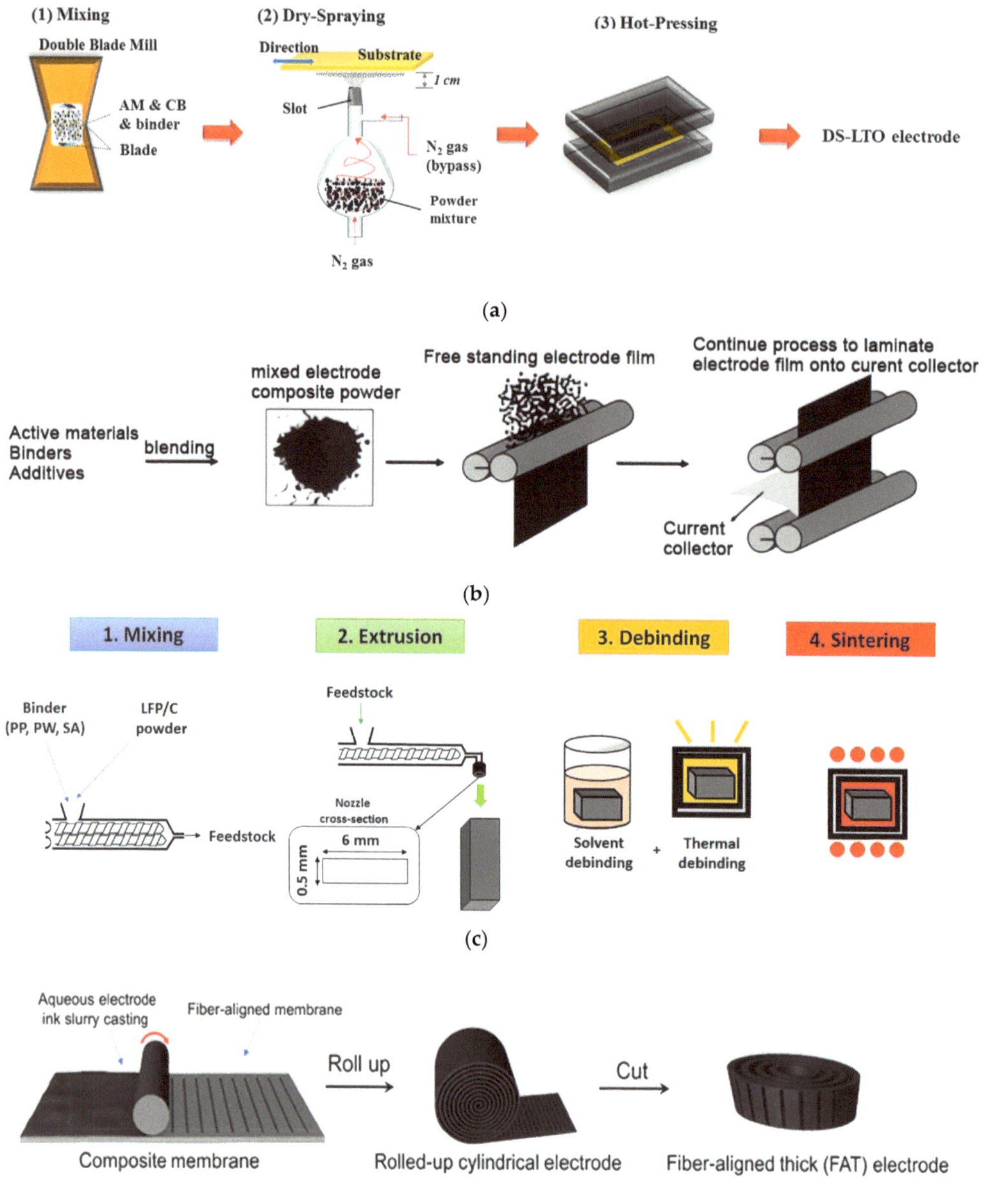

Figure 6. (**a**) A schematic of the dry spraying process [59]. (**b**) A schematic of the dry coating process. (**c**) Schematic illustrations of powder extrusion moulding [67]. (**d**) Schematic fabrication of fiber-aligned thick electrode [9].

3.2. Increasing the LPD

Improving mechanical stability to fabricate crack-free electrodes is just the first step to get the target on energy density of 500 Wh·kg^{-1}. When the porosity of thick electrodes is below 30%, it is found that ionic conduction within such a thick and dense electrode becomes a main reason that causes poor rate performances [2,50,73]. For thick electrodes, the ionic resistance in pores (R_{ion}) is higher than the charge-transfer resistance for Li intercalation (R_{ct}), as shown in Figure 7a, so there are limited ion diffusion behaviors across thick electrodes [74].

According to Equation (2) of the LPD, the LPD is proportional to the porosity of the electrode and inversely proportional to the tortuosity of the electrode. The diffusion coefficient of Li$^+$ is directly related to the diffusion behaviors of Li$^+$ in porous electrodes. Additionally, a simplified expression could be inferred from Equation (2):

$$D_{eff} = \frac{\varepsilon}{\tau} D_{Li} \tag{3}$$

where D_{eff} is the effective ionic conductivity, ε is the porosity of the electrode, τ is the tortuosity of the electrode and D_{Li} is the intrinsic ions conductivity [75–77]. Additionally, the tortuosity reflects the ions diffusion length in the electrodes.

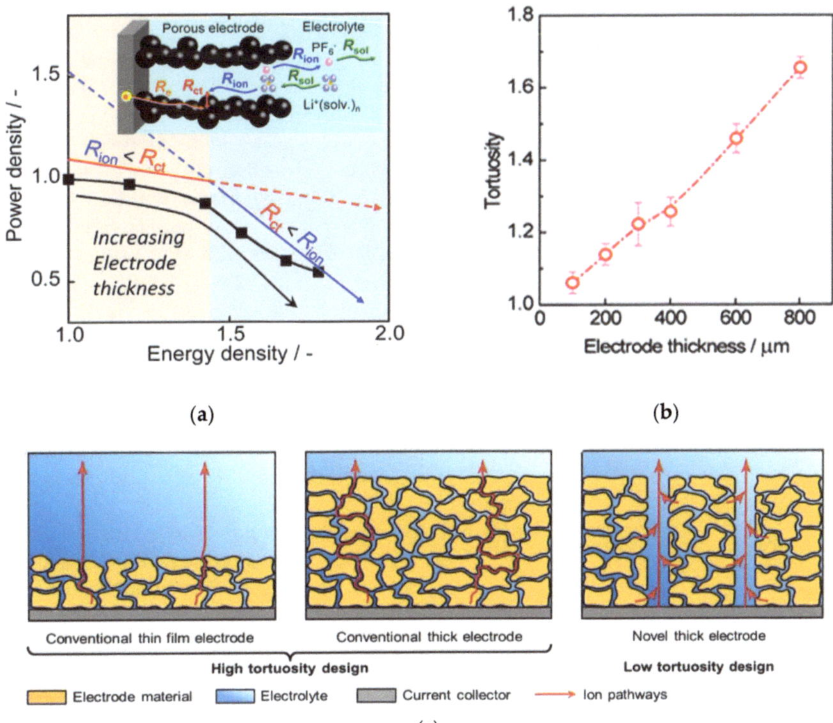

Figure 7. (**a**) The changes of R_{ion} and R_{ct} and their magnitude show opposite trends with respect to electrode thickness [74]. (**b**) Higher tortuosity is present in a thicker electrode [78]. (**c**) Illustration about the value of the low−tortuosity design for thick electrodes [13].

Therefore, there are two strategies to improve ion diffusion behaviors. One is to optimize porosity of the electrode to get better rate performance, and the other is to construct vertical pathways to current collectors for fabricating low-tortuosity electrodes. However, a high porosity would reduce the ratio of active materials in electrodes, and it goes against the design of thick electrodes. Additionally, the tortuosity reflects the ions diffusion length in the electrodes. Additionally, Figure 7b clearly shows that the tortuosity of the electrode increases with the growth of electrode thickness [78]. Therefore, the low-tortuosity design with building ions transport pathways paralleled to the ion transport direction has become a key principle for thick electrodes [29–31]. Additionally, the superiority of low-tortuosity design in thick electrodes could be obtained from Figure 7c [13].

It is noteworthy that the low-tortuosity design would increase the porosity of the electrode, but this increase would bring more effective promotion on ions diffusion behavior. Moreover, there is an optimal design of the oriented porosity ε_0 and matrix porosity ε_m, which could balance the ion transport kinetics along the channels and in the matrix. Additionally, it has been given an optimal value ($\varepsilon_0 \approx 0.11$ at $\varepsilon_0 + \varepsilon_m = 0.42$) at which the oriented-pore achieved the best rate capability without a sacrifice of the energy density [30].

3.2.1. Optimizing Electrode Porosity

When thick electrodes are discussed, not much attention has been paid to how dense the electrode is, since most of the testing end at coin cell level. However, the practical use of thick electrodes not only needs to consider their mass loading: porosity is equally important [79]. Figure 8a revealed the role of the porosity in the evolution of the rate-limiting step in thick electrodes [48]. The porosity is critical for performing electrochemical properties of electrodes. Additionally, some studies have proven that the electrochemical performance of thick electrodes could be improved through the porosity gradient [80,81].

To put kinetics into perspective, the porosity across the electrode showing a gradient increase is beneficial from the region near to the current collector to the region close to the separator. The reason is that electrodes with large porosity near separators could facilitate fast ion transport while electrodes with small porosity close current collectors could ensure optimum electronic contact [80,82–84]. Based on the electrochemical porous-electrode model, two teams obtained the same results via distinct simulation methods. The result is a higher porosity in the electrode near the separator, which can hold more electrolytes and minimize resistance at the electrode [80,82]. Another report also supports this suspect, but it puts forward that it is not much use to construct graded electrode beyond two layers for reducing the resistance across the electrodes. Beyond simulations, the electrodes with staged porosity in order to improve ions migration has been constructed by applying the capillary suspension concept [84]. This strategy is under patent-pending protection application [85].

However, an opposite viewpoint has been given. The highest stresses are located in the electrodes near the current collectors and by increasing the porosity of this region the maximum stress could be reduced. Furthermore, the lower porosity of electrodes near separators would contribute to maintain a higher potential during discharge [81]. Moreover, the comparison of electrochemical performance between directional ice templating with low porosity nearest the separator (DIT LP-S) and directional ice templating with low porosity nearest the current collector (DIT LP-CC) shows that the former performed better rate capability (Figure 8b,c) [86]. Similarly, this strategy also has patent protection [87].

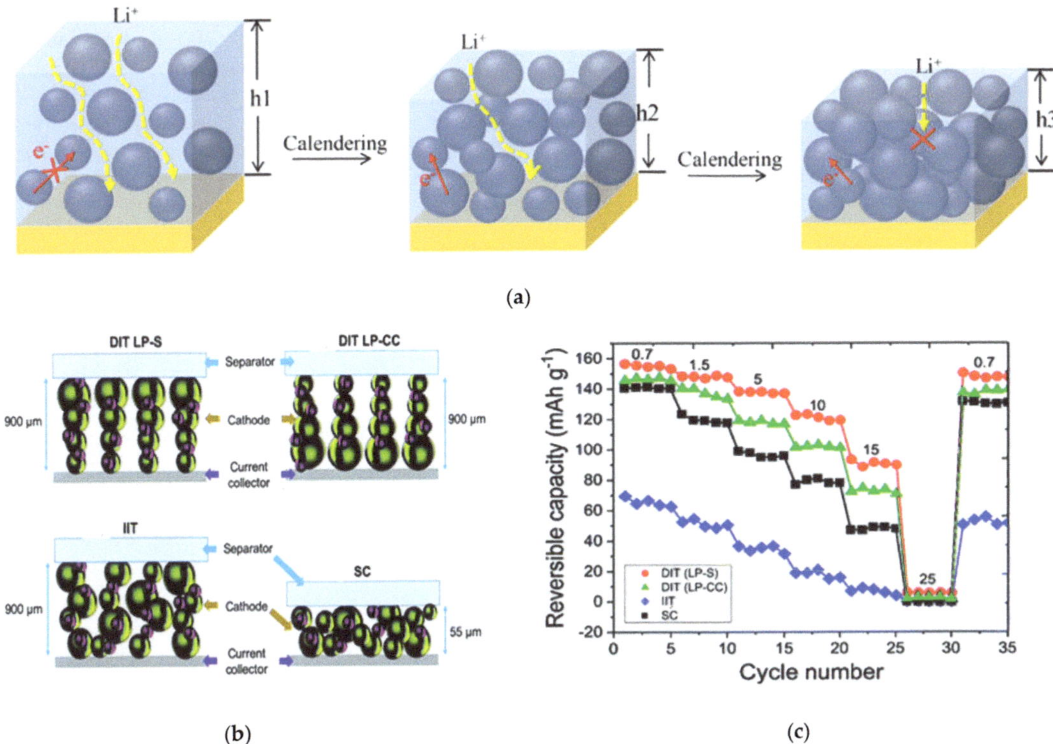

Figure 8. (**a**) The charge transport in electrodes with different porosity [48]. (**b**) Schematics of the four electrode types fabricated for performance comparison: (i) directional ice templating with low porosity nearest the separator (DIT LP−S); (ii) directional ice templating with low porosity nearest the current collector (DIT LP−CC); (iii) isotropic ice templating (IIT); and (iv) slurry casting (SC) [86]. (**c**) Reversible capacities of the four types of these electrodes at different current densities [86].

3.2.2. Decreasing Electrode Tortuosity

Figure 8c also gives other sights that both DIT LP-S and DIT LP-CC electrodes performed better rate capability than isotropic ice templating (IIT) electrodes and the IIT electrodes also had poorer rate performance than the traditional slurry casting (SC) electrode. Combining with Figure 8b, these sights indicate that the increased thickness of electrodes would deteriorate the rate performance and decreasing electrode tortuosity could improve electrochemical performance in thick electrodes. Furthermore, more general and sustainable approaches are highly desired to fabricate uniformly aligned microchannels in the electrode.

Nature provides many templates and inspirations for improving electrode performance and building hierarchical microstructures [88]. The wood has open channels to transport water, ions and other components along the growth direction. Additionally, the structure provides guidance for making ordered electrodes. Thick electrodes constructed via wood template inherit directional porous structure where it provides pathways for easy ion and electron transport across the entire electrode [25,28,89–91]. As shown in Figure 9a, the interconnected carbon framework could provide a conductive network for electron transport while the gap between active materials and frameworks could offer low-tortuosity pathways for ion transport [52]. To further illustrate the value of wood-template, Lu and coworkers constructed the traditional LiCoO2 cathode (control LCO) and the wood-template LiCoO2 cathodes (LCO-1 and LCO-2, the LCO-2 for higher mass loading underwent the double processes of LCO-1) and measured their related properties [92]. The results are given in Figure 9b,c, and the tortuosity of wood-template cathodes is approximately close to 1 due to ordered microchannels and smaller than the control LCO. The ion conductivity and electron conductivity of wood-template cathodes are larger than the control LCO. Through comparing the porosity and ion conductivity between LCO-1 and LCO-2, a larger porosity is not always positive to improve ions diffusion or the LPD.

The ice-template technique is also used to build orientational pores for ion migration [64,86,93]. The ordered pores could be well controlled by adjusting the freezing and sintering parameters. This technique creates electrodes that combine the energy and the power [93]. Moreover, Miller et al. successfully constructed thick electrodes with the thickness-independent electrochemical performance by aligning vertical two-dimensional flakes [72]. Additionally, Figure 9d gives a scheme of thick electrodes constructed by this guidance, and approximately 200 µm is the maximum thickness of each film due to the ion transport limitations [32]. It is another proof of the LPD.

Commercial availability is also a critical factor for building directional aligned pores in electrodes. Additionally, a simple, up-scalable and inexpensive technique to construct pathways parallel to the direction of the Li+ migration has been proposed by applying a magnetic field during the electrode fabrication (Figure 9e) [94–99]. A magnetic control method based on sacrificial features was reported, magnetized nylon rods or magnetic emulsions were used as a template to fabricate directionally aligned pore arrays in the thick electrode [95]. Additionally, it is noteworthy that the aligned pore structure performs higher areal capacity under the same conditions (at constant thickness and total porosity) and smaller pore spacings are superior, as shown in Figure 9f,g.

Moreover, Figure 9f,g give a critical viewpoint that the pore parameters (like diameter and spacing) play a pivotal key on rate performance of electrodes. Therefore, a more precise technique for constructing vertically aligned pores is requested and essential. The laser-based manufacturing process attracts more attention in the low-tortuosity design for thick electrodes. This concept involves high-precision ablation of a small fraction of the active material from the initial coating and generating additional diffusion pathways (Figure 9h) [10,100]. The laser processing on electrodes is inevitably accompanied by slight capacity loss but significant improvements in rate performance [101–103]. By using 200 ns-laser radiation and a pitch distance of 200 µm, the loss in active material can reach values of about 30 wt.%, while a pitch distance of 600 µm would reduce the material loss below 10 wt.% [104]. Additionally, the rate performance in different pitch distances were given in Figure 9i [105].

Furthermore, the ultrashort pulse duration in the femtosecond (fs) or picosecond (ps) range is less than the heat diffusion time. Additionally, the ablation volume is emitted before any heat diffusion or thermal damage occurs and side effects are significantly reduced [106]. Additionally, due to the cold nature of fs-laser ablation, a high-aspect ratio of approximately 15 could be reached, which related to the capacity loss. Therefore, it can construct fine and accurate alignment hole array. Additionally, the ideal pore parameters for maximizing the rate capability could be identified by evaluating the ions distribution in electrodes [107]. Furthermore, the surface on laser-generated structures leads to an accelerated and homogenous wetting of the electrodes with liquid electrolyte [103,104].

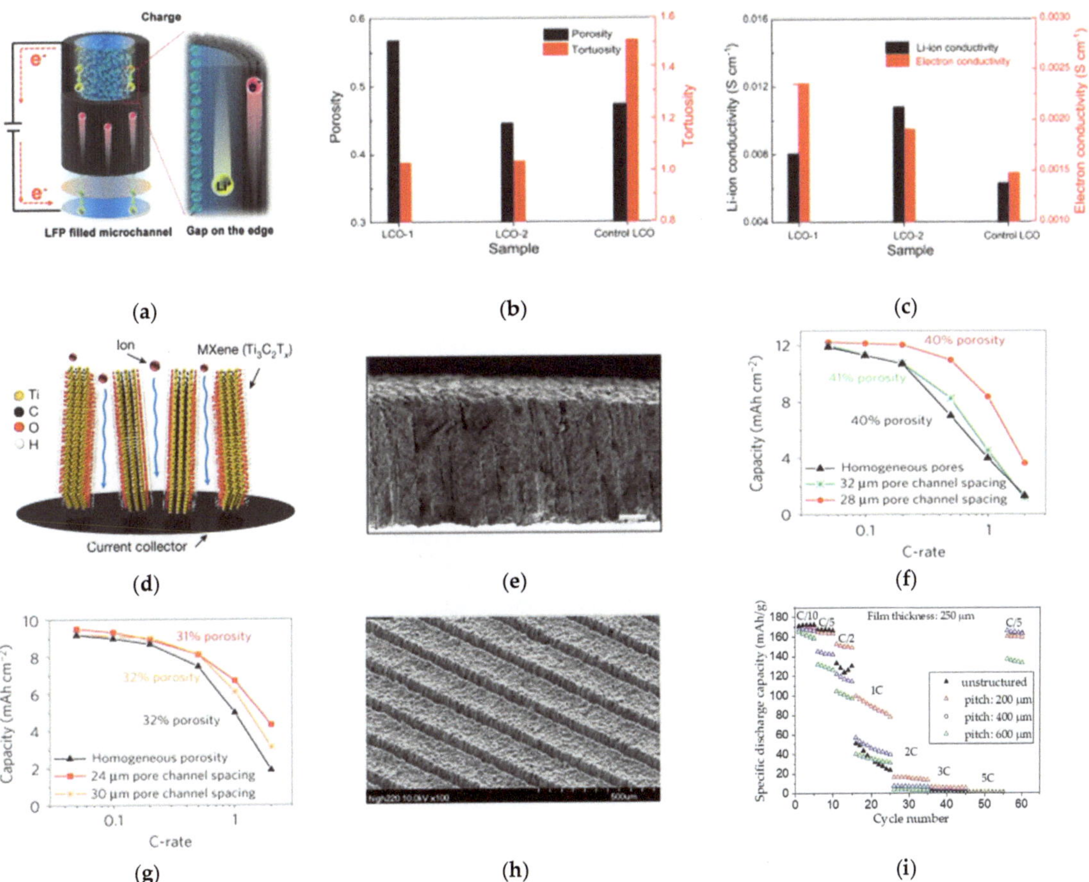

Figure 9. (**a**) Graphical illustrations of the ion and electron transportation behavior occurring in the wood-templated electrodes [52]. The comparison of LCO-1, LCO-2, and control LCO in terms of (**b**) tortuosity and porosity, (**c**) Li-ion and electron conductivities [92]. (**d**) Schematic illustration of ion transport in vertically aligned $Ti_3C_2T_x$ MXene films [32]. (**e**) Cross-sectional SEM image of a low tortuosity LCO electrode constructed by magnetic method [94]. Areal capacity versus C-rate for LCO electrodes with pore channels: (**f**) 310-μm-thick electrodes with 39–42% porosity and (**g**) 220-μm-thick electrodes with 30–32% porosity [95]. (**h**) SEM images for laser-structured electrodes (thickness ≈ 210 μm) [101]. (**i**) Specific discharge capacity of cells containing 250 μm thick electrodes with different pitch distances [105].

3.3. Summary

This chapter gives the recent efforts on improving the CCT and the LPD to construct thicker electrodes. It must be emphasized again that the first law of thick electrodes is to obtain high performance electrodes with high mass loading and high accessible areal capacity, not thicker electrodes. Some thick electrodes constructed by the above methods show lower volumetric specific capacity that have no competition with the traditional electrodes. Additionally, some design of thick electrodes cannot be transferred directly to state-of-the-art large-scale cell manufacturing processes in industry. Table 1 gives the summary of the thick electrodes. Additionally, the volumetric specific capacity of thick electrodes is not below 400 mAh·cm^{-3} which is the current level of traditional electrodes.

Table 1. Summary of recently reported thick electrodes.

Active Materials	Thickness/μm	Mass Loading/mg·cm⁻²	Areal Capacity/ mAh·cm⁻²@mA·cm⁻²	Volumetric Capacity/mAh·cm⁻³	Reference
NMC622	154	37.6	6.58@0.38	427.3	[2]
Graphite	182	23.4	7.84@0.82	430.8	
LFP	1000	128	19.6@1	196	[9]
Graphite	1200	50	17.25@0.93	143.8	[15]
NMC532	240	30	5.84@1.15	228.3	[30]
NMC111	320	72	9.86@1.12	308.3	[33]
Graphite	320	43	11.23@1.62	352.1	
LCO	600	115.4	15.7@0.5	261.7	[34]
NMC111	322	60	5.1@1.8	158.4	[44]
NMC811	740	155	29@1.47	391.9	[46]
2 μm Si	210	15	45@1.79	2142	
LFP	800	60	5.7@1	71.3	[52]
LFP/C	240	12	1.86@1.02	77.5	[54]
LFP	430	46.5	7.2@1	167.4	[56]
LTO	600	168	26.5@1.68	441.7	[57]
LTO	~1500	30	4.74@1.06	31.6	[62]
LTO	475	138	15.2@1.02	319	[66]
LTO	550	110	11.11@1.6	202	[68]
LFP	500	90	11.07@1.23	221.4	
Graphite	240	16.5	3.79@1.23	158.1	[84]
LCO	1500	206	24.5@1.44	163.8	[92]
NCA	600	73.8	13@1.48	216.7	[93]
LCO	440	100.5	13.6@1.41	309.1	[94]
S	300	6	6.9@1	230	[97]
LCO	700	172	20.1@1.21	287.1	[102]
NMC622	250	51.7	8.79@0.93	351.6	[105]

4. Conclusions and Outlook

Recently moving towards carbon neutrality has become a global consensus and green transportation attracts more attention. The EVs with the driving range of above 500 km could become more competitive to survive in the market for automobiles. Additionally, a practicable way to reach this value is to adopt thick electrodes with high mass loading and high area capacity. The obstacles that stand in the way of using thick electrodes are weak mechanical stability and poor electrochemical performance, or are limited by the CCT and the LPD. Here, the understanding of these mechanisms and the recent efforts on breaking the limitations are given.

The design of thick electrodes is aimed at obtaining higher energy density of LIBs at battery-pack level, but not larger thickness for electrodes. Some thick electrodes with large porosity show lower volumetric specific capacity that go against the original intention of this design. Therefore, two corresponding parameters to evaluate the thick electrode is needed, such as thickness and porosity, areal capacity and volumetric capacity.

The design of thick electrodes is aimed at applications for practical use like EVs, and the preparation process must be suitable for a large-scale use. Some methods for fabricating thick electrodes are only for a laboratory scale. Some technologies with drying process may cause the microstructural heterogeneity in the electrodes due to the drying-induced migration of the binder to the electrode surface. Additionally, combining with the requirement of low-cost solution, some solvent-free manufacturing technologies are appropriate, like the dry-coated technique, which is suitable for mass manufacture and cuts the cost of using and removing the toxic solvents. Additionally, an environmentally friendly industry is more easily accepted by local government and residents.

The electrochemical performance is the most basic requirements for thick electrodes. Decreasing the tortuosity of thick electrodes by fabricating ordered microchannels paralleling the diffusion of Li⁺ has been considered as the most effective ways to help thick electrodes in performing better properties. Because the low-tortuosity design resulted in active mass loss, it is essential to precisely construct vertical channels. From the studies on constructing aligned pathways, the laser-ablation process could be better controlled to precisely adjust parameters of the vertical channels.

Author Contributions: Conceptualization, J.Z., L.J. and J.P.Z.; validation, Y.L.; investigation, G.X.; resources, L.J.; data curation, G.X.; writing—original draft preparation, G.X.; writing—review and editing, Y.L. and N.Q.; visualization, N.Q.; supervision, J.Z. and J.P.Z.; project administration, L.J. and S.G.; funding acquisition, J.Z. and S.G. All authors have read and agreed to the published version of the manuscript.

Funding: The work is supported by China Shenhua Coal to Liquid and Chemical Shanghai Research Institute, Grant No. SAC-B-CT-TECH-2022-04-25, and National Science Foundation of China, Grant No. 51777140, and the Fundamental Research Funds for the Central Universities at Tongji University, Grant No. 17002150019.

Institutional Review Board Statement: Not applicable.

Informed Consent Statement: Not applicable.

Data Availability Statement: No new data were created in this paper.

Conflicts of Interest: The funders had no role in the design of the study; in the collection, analyses, or interpretation of data; in the writing of the manuscript; or in the decision to publish the results.

References

1. Rempel, J. Vehicle Technologies Office Merit Review 2015: High Energy High Power Battery Exceeding PHEV-40 Requirements. Available online: https://www.energy.gov/sites/prod/files/2015/06/f23/es209_rempel_2015_p.pdf (accessed on 30 January 2023).
2. Gallagher, K.G.; Trask, S.E.; Bauer, C.; Woehrle, T.; Lux, S.F.; Tschech, M.; Lamp, P.; Polzin, B.J.; Ha, S.; Long, B.; et al. Optimizing Areal Capacities through Understanding the Limitations of Lithium-Ion Electrodes. *J. Electrochem. Soc.* **2015**, *163*, A138–A149. [CrossRef]
3. Gröger, O.; Gasteiger, H.A.; Suchsland, J.-P. Review—Electromobility: Batteries or Fuel Cells? *J. Electrochem. Soc.* **2015**, *162*, A2605–A2622. [CrossRef]
4. Andre, D.; Kim, S.-J.; Lamp, P.; Lux, S.F.; Maglia, F.; Paschos, O.; Stiaszny, B. Future generations of cathode materials: An automotive industry perspective. *J. Mater. Chem. A* **2015**, *3*, 6709–6732. [CrossRef]
5. China Society of Automotive Engineers. Technology Roadmap of Clean and New Energy Vehicles 2.0. Available online: http://zhishi.sae-china.org/ppt.html?id=2100 (accessed on 30 January 2023).
6. Xia, C.; Kwok, C.; Nazar, L. A high-energy-density lithium-oxygen battery based on a reversible four-electron conversion to lithium oxid. *Science* **2018**, *361*, 777–781. [CrossRef]
7. Zhang, J.; Wang, P.F.; Bai, P.; Wan, H.; Liu, S.; Hou, S.; Pu, X.; Xia, J.; Zhang, W.; Wang, Z.; et al. Interfacial Design for a 4.6 V High-Voltage Single-Crystalline LiCoO2 Cathode. *Adv. Mater.* **2022**, *34*, e2108353. [CrossRef]
8. Du, K.; Tao, R.; Guo, C.; Li, H.; Liu, X.; Guo, P.; Wang, D.; Liang, J.; Li, J.; Dai, S.; et al. In-situ synthesis of porous metal fluoride@carbon composite via simultaneous etching/fluorination enabled superior Li storage performance. *Nano Energy* **2022**, *103*, 107862. [CrossRef]
9. Shi, B.; Shang, Y.; Pei, Y.; Pei, S.; Wang, L.; Heider, D.; Zhao, Y.Y.; Zheng, C.; Yang, B.; Yarlagadda, S.; et al. Low Tortuous, Highly Conductive, and High-Areal-Capacity Battery Electrodes Enabled by Through-thickness Aligned Carbon Fiber Framework. *Nano Lett.* **2020**, *20*, 5504–5512. [CrossRef] [PubMed]
10. Pfleging, W. A review of laser electrode processing for development and manufacturing of lithium-ion batteries. *Nanophotonics* **2018**, *7*, 549–573. [CrossRef]
11. Liu, J.; Bao, Z.; Cui, Y.; Dufek, E.J.; Goodenough, J.B.; Khalifah, P.; Li, Q.; Liaw, B.Y.; Liu, P.; Manthiram, A.; et al. Pathways for practical high-energy long-cycling lithium metal batteries. *Nat. Energy* **2019**, *4*, 180–186. [CrossRef]
12. Cheng, H.; Li, K. Charge delivery goes the distance. *Science* **2017**, *356*, 582–583. [CrossRef] [PubMed]
13. Kuang, Y.; Chen, C.; Kirsch, D.; Hu, L. Thick Electrode Batteries: Principles, Opportunities, and Challenges. *Adv. Energy Mater.* **2019**, *9*, 1901457. [CrossRef]
14. Zheng, H.; Li, J.; Song, X.; Liu, G.; Battaglia, V. A comprehensive understanding of electrode thickness effects on the electrochemical performances of Li-ion battery cathodes. *Electrochim. Acta* **2012**, *71*, 258–265. [CrossRef]
15. Wang, J.S.; Liu, P.; Sherman, E.; Verbrugge, M.; Tataria, H. Formulation and characterization of ultra-thick electrodes for high energy lithium-ion batteries employing tailored metal foams. *J. Power Sources* **2011**, *196*, 8714–8718. [CrossRef]
16. Singh, K.; Tirumkudulu, M. Cracking in drying colloidal films. *Phys. Rev. Lett.* **2007**, *98*, 218302. [CrossRef] [PubMed]
17. Chiu, R.; Garino, T.; Cima, M. Drying of Granular Ceramic Films: I, Effect of Processing Variables on Cracking Behavior. *J. Am. Ceram. Soc.* **1993**, *76*, 2257–2264. [CrossRef]

18. Slowik, V.; Ju, J. Discrete modeling of plastic cement paste subjected to drying. *Cem. Concr. Compos.* **2011**, *33*, 925–935. [CrossRef]
19. Tirumkudulu, M.; Russel, W. Cracking in drying latex films. *Langmuir* **2005**, *21*, 4938–4948. [CrossRef] [PubMed]
20. Tambio, S.; Cadiou, F.; Maire, E.; Besnard, N.; Deschamps, M.; Lestriez, B. The Concept of Effective Porosity in the Discharge Rate Performance of High-Density Positive Electrodes for Automotive Application. *J. Electrochem. Soc.* **2020**, *167*, 160509. [CrossRef]
21. Lu, C.; Huang, Q.; Chen, X. High-performance silicon nanocomposite based ionic actuators. *J. Mater. Chem. A* **2020**, *8*, 9228–9238. [CrossRef]
22. Yan, L.; Yudong, L.; Ting'an, Z.; Naixiang, F. Research on the Penetration Depth in Aluminum Reduction Cell with New Type of Anode and Cathode Structures. *JOM* **2014**, *66*, 1202–1209. [CrossRef]
23. Fang, R.; Zhao, S.; Hou, P.; Cheng, M.; Wang, S.; Cheng, H.; Liu, C.; Li, F. 3D Interconnected Electrode Materials with Ultrahigh Areal Sulfur Loading for Li-S Batteries. *Adv. Mater.* **2016**, *28*, 3374–3382. [CrossRef] [PubMed]
24. Li, G.; Ouyang, T.; Xiong, T.; Jiang, Z.; Adekoya, D.; Wu, Y.; Huang, Y.; Balogun, M. All-carbon-frameworks enabled thick electrode with exceptional high-areal-capacity for Li-Ion storage. *Carbon* **2021**, *174*, 1–9. [CrossRef]
25. Woodward, R.; Markoulidis, F.; Luca, F.D.; Anthony, D.; Malko, D.; McDonald, T.; Shaffer, M.; Bismarck, A. Carbon foams from emulsion-templated reduced graphene oxide polymer composites: Electrodes for supercapacitor devices. *J. Mater. Chem. A* **2018**, *6*, 1840–1849. [CrossRef]
26. Sun, H.; Mei, L.; Liang, J.; Zhao, Z.; Lee, C.; Fei, H.; Ding, M.; Lau, J.; Li, M.; Wang, C.; et al. Three-dimensional holey-graphene/niobia composite architectures for ultrahigh-rate energy storage. *Science* **2017**, *356*, 599–604. [CrossRef]
27. Kang, J.; Pham, H.Q.; Kang, D.-H.; Park, H.-Y.; Song, S.-W. Improved rate capability of highly loaded carbon fiber-interwoven LiNi 0.6 Co 0.2 Mn 0.2 O 2 cathode material for high-power Li-ion batteries. *J. Alloys Compd.* **2016**, *657*, 464–471. [CrossRef]
28. Shen, F.; Luo, W.; Dai, J.; Yao, Y.; Zhu, M.; Hitz, E.; Tang, Y.; Chen, Y.; Sprenkle, V.L.; Li, X.; et al. Ultra-Thick, Low-Tortuosity, and Mesoporous Wood Carbon Anode for High-Performance Sodium-Ion Batteries. *Adv. Energy Mater.* **2016**, *6*, 1600377. [CrossRef]
29. Ebner, M.; Chung, D.-W.; García, R.E.; Wood, V. Tortuosity Anisotropy in Lithium-Ion Battery Electrodes. *Adv. Energy Mater.* **2014**, *4*, 1301278. [CrossRef]
30. Xiong, R.; Zhang, Y.; Wang, Y.; Song, L.; Li, M.; Yang, H.; Huang, Z.; Li, D.; Zhou, H. Scalable Manufacture of High-Performance Battery Electrodes Enabled by a Template-Free Method. *Small Methods* **2021**, *5*, 2100280. [CrossRef]
31. Zhang, L.; Pan, Y.; Chen, Y.; Li, M.; Liu, P.; Wang, C.; Wang, P.; Lu, H. Designing vertical channels with expanded interlayers for Li-ion batteries. *Chem. Commun.* **2019**, *55*, 4258–4261. [CrossRef] [PubMed]
32. Xia, Y.; Mathis, T.S.; Zhao, M.Q.; Anasori, B.; Dang, A.; Zhou, Z.; Cho, H.; Gogotsi, Y.; Yang, S. Thickness-independent capacitance of vertically aligned liquid-crystalline MXenes. *Nature* **2018**, *557*, 409–412. [CrossRef]
33. Singh, M.; Kaiser, J.; Hahn, H. Thick Electrodes for High Energy Lithium Ion Batteries. *J. Electrochem. Soc.* **2015**, *162*, A1196–A1201. [CrossRef]
34. Kato, Y.; Shiotani, S.; Morita, K.; Suzuki, K.; Hirayama, M.; Kanno, R. All-Solid-State Batteries with Thick Electrode Configurations. *J. PhysIcal Chem. Lett.* **2018**, *9*, 607–613. [CrossRef] [PubMed]
35. Kubanska, A.; Castro, L.; Tortet, L.; Dollé, M.; Bouchet, R. Effect of composite electrode thickness on the electrochemical performances of all-solid-state li-ion batteries. *J. Electroceram.* **2017**, *38*, 189–196. [CrossRef]
36. Hong, S.-B.; Lee, Y.-J.; Kim, U.-H.; Bak, C.; Lee, Y.M.; Cho, W.; Hah, H.J.; Sun, Y.-K.; Kim, D.-W. All-Solid-State Lithium Batteries: Li+-Conducting Ionomer Binder for Dry-Processed Composite Cathodes. *ACS Energy Lett.* **2022**, *7*, 1092–1100. [CrossRef]
37. Kukay, A.; Sahore, R.; Parejiya, A.; Hawley, W.; Li, J.; Wood, D. Aqueous Ni-rich-cathode dispersions processed with phosphoric acid for lithium-ion batteries with ultra-thick electrodes. *J. Colloid Interface Sci.* **2021**, *581*, 635–643. [CrossRef]
38. Li, H. Practical Evaluation of Li-Ion Batteries. *Joule* **2019**, *3*, 911–914. [CrossRef]
39. Shin, H.; Santamarina, J. Desiccation cracks in saturated fine-grained soils: Particle-level phenomena and effective-stress analysis. *Géotechnique* **2011**, *61*, 961–972. [CrossRef]
40. Lura, P.; Pease, B.; Mazzotta, G.B.; Rajabipour, F.; Weiss, J. Influence of Shrinkage-Reducing Admixtures on Development of Plastic Shrinkage Cracks. *ACI Mater. J.* **2007**, *104*, 187–194. [CrossRef]
41. Du, Z.; Rollag, K.M.; Li, J.; An, S.J.; Wood, M.; Sheng, Y.; Mukherjee, P.P.; Daniel, C.; Wood, D.L. Enabling aqueous processing for crack-free thick electrodes. *J. Power Sources* **2017**, *354*, 200–206. [CrossRef]
42. Dufresne, E.R.; Corwin, E.I.; Greenblatt, N.A.; Ashmore, J.; Wang, D.Y.; Dinsmore, A.D.; Cheng, J.X.; Xie, X.S.; Hutchinson, J.W.; Weitz, D.A. Flow and fracture in drying nanoparticle suspensions. *Phys. Rev. Lett.* **2003**, *91*, 224501. [CrossRef]
43. Birk-Braun, N.; Yunus, K.; Rees, E.J.; Schabel, W.; Routh, A.F. Generation of strength in a drying film: How fracture toughness depends on dispersion properties. *Phys. Rev. Lett.* **2017**, *95*, 022610. [CrossRef] [PubMed]
44. Ibing, L.; Gallasch, T.; Schneider, P.; Niehoff, P.; Hintennach, A.; Winter, M.; Schappacher, F.M. Towards water based ultra-thick Li ion battery electrodes—A binder approach. *J. Power Sources* **2019**, *423*, 183–191. [CrossRef]
45. Kumberg, J.; Müller, M.; Diehm, R.; Spiegel, S.; Wachsmann, C.; Bauer, W.; Scharfer, P.; Schabel, W. Drying of Lithium-Ion Battery Anodes for Use in High-Energy Cells: Influence of Electrode Thickness on Drying Time, Adhesion, and Crack Formation. *Energy Technol.* **2019**, *7*, 1900722. [CrossRef]
46. Park, S.-H.; King, P.J.; Tian, R.; Boland, C.S.; Coelho, J.; Zhang, C.; McBean, P.; McEvoy, N.; Kremer, M.P.; Daly, D.; et al. High areal capacity battery electrodes enabled by segregated nanotube networks. *Nat. Energy* **2019**, *4*, 560–567. [CrossRef]
47. Gao, H.; Wu, Q.; Hu, Y.; Zheng, J.P.; Amine, K.; Chen, Z. Revealing the Rate-Limiting Li-Ion Diffusion Pathway in Ultrathick Electrodes for Li-Ion Batteries. *J. Phys. Chem. Lett.* **2018**, *9*, 5100–5104. [CrossRef] [PubMed]

48. Hu, J.; Wu, B.; Cao, X.; Bi, Y.; Chae, S.; Niu, C.; Xiao, B.; Tao, J.; Zhang, J.; Xiao, J. Evolution of the rate-limiting step: From thin film to thick Ni-rich cathodes. *J. Power Sources* **2020**, *454*, 227966. [CrossRef]

49. Appiah, W.; Park, J.; Song, S.; Byun, S.; Ryou, M.; Lee, Y. Design optimization of LiNi$_{0.6}$Co$_{0.2}$Mn$_{0.2}$O$_2$/graphite lithium-ion cells based on simulation and experimental data. *J. Power Sources* **2016**, *319*, 147–158. [CrossRef]

50. Li, Z.; Yin, L.; Mattei, G.S.; Cosby, M.R.; Lee, B.-S.; Wu, Z.; Bak, S.-M.; Chapman, K.W.; Yang, X.-Q.; Liu, P.; et al. Synchrotron Operando Depth Profiling Studies of State-of-Charge Gradients in Thick Li(Ni0.8Mn0.1Co0.1)O2 Cathode Films. *Chem. Mater.* **2020**, *32*, 6358–6364. [CrossRef]

51. Xu, C.; Li, Q.; Wang, Q.; Kou, X.; Fang, H.-T.; Yang, L. Femtosecond laser drilled micro-hole arrays in thick and dense 2D nanomaterial electrodes toward high volumetric capacity and rate performance. *J. Power Sources* **2021**, *492*, 229638. [CrossRef]

52. Chen, C.; Zhang, Y.; Li, Y.; Kuang, Y.; Song, J.; Luo, W.; Wang, Y.; Yao, Y.; Pastel, G.; Xie, J.; et al. Highly Conductive, Lightweight, Low-Tortuosity Carbon Frameworks as Ultrathick 3D Current Collectors. *Adv. Energy Mater.* **2017**, *7*, 1700595. [CrossRef]

53. Sahore, R.; Wood, D.L.; Kukay, A.; Grady, K.M.; Li, J.; Belharouak, I. Towards Understanding of Cracking during Drying of Thick Aqueous-Processed LiNi0.8Mn0.1Co0.1O2 Cathodes. *ACS Sustain. Chem. Eng.* **2020**, *8*, 3162–3169. [CrossRef]

54. Yang, G.F.; Song, K.Y.; Joo, S.K. A metal foam as a current collector for high power and high capacity lithium iron phosphate batteries. *J. Mater. Chem. A* **2014**, *2*, 19648–19652. [CrossRef]

55. Yang, Z.; Tian, J.; Ye, Z.; Jin, Y.; Cui, C.; Xie, Q.; Wang, J.; Zhang, G.; Dong, Z.; Miao, Y.; et al. High energy and high power density supercapacitor with 3D Al foam-based thick graphene electrode: Fabrication and simulation. *Energy Storage Mater.* **2020**, *33*, 18–25. [CrossRef]

56. Yang, G.-F.; Song, K.-Y.; Joo, S.-K. Ultra-thick Li-ion battery electrodes using different cell size of metal foam current collectors. *RSC Adv.* **2015**, *5*, 16702–16706. [CrossRef]

57. Hu, L.; La Mantia, F.; Wu, H.; Xie, X.; McDonough, J.; Pasta, M.; Cui, Y. Lithium-Ion Textile Batteries with Large Areal Mass Loading. *Adv. Energy Mater.* **2011**, *1*, 1012–1017. [CrossRef]

58. Al-Shroofy, M.; Zhang, Q.; Xu, J.; Chen, T.; Kaur, A.P.; Cheng, Y.-T. Solvent-free dry powder coating process for low-cost manufacturing of LiNi1/3Mn1/3Co1/3O2 cathodes in lithium-ion batteries. *J. Power Sources* **2017**, *352*, 187–193. [CrossRef]

59. Park, D.-W.; Cañas, N.A.; Wagner, N.; Friedrich, K.A. Novel solvent-free direct coating process for battery electrodes and their electrochemical performance. *J. Power Sources* **2016**, *306*, 758–763. [CrossRef]

60. Qin, X.; Wang, X.; Xie, J.; Wen, L. Hierarchically porous and conductive LiFePO4 bulk electrode: Binder-free and ultrahigh volumetric capacity Li-ion cathode. *J. Mater. Chem.* **2011**, *21*, 12444–12448. [CrossRef]

61. Elango, R.; Demortière, A.; Andrade, V.; Morcrette, M.; Seznec, V. Thick Binder-Free Electrodes for Li–Ion Battery Fabricated Using Templating Approach and Spark Plasma Sintering Reveals High Areal Capacity. *Adv. Energy Mater.* **2018**, *8*, 1703031. [CrossRef]

62. Sun, C.; Liu, S.; Shi, X.; Lai, C.; Liang, J.; Chen, Y. 3D printing nanocomposite gel-based thick electrode enabling both high areal capacity and rate performance for lithium-ion battery. *Chem. Eng. J.* **2020**, *381*, 122641. [CrossRef]

63. Tang, X.; Zhou, H.; Cai, Z.; Cheng, D.; He, P.; Xie, P.; Zhang, D.; Fan, T. Generalized 3D Printing of Graphene-Based Mixed-Dimensional Hybrid Aerogels. *ACS Nano* **2018**, *12*, 3502–3511. [CrossRef]

64. Gao, X.; Yang, X.; Sun, Q.; Luo, J.; Liang, J.; Li, W.; Wang, J.; Wang, S.; Li, M.; Li, R.; et al. Converting a thick electrode into vertically aligned "Thin electrodes" by 3D-Printing for designing thickness independent Li-S cathode. *Energy Storage Mater.* **2020**, *24*, 682–688. [CrossRef]

65. Sajadi, S.M.; Enayat, S.; Vásárhelyi, L.; Alabastri, A.; Lou, M.; Sassi, L.M.; Kutana, A.; Bhowmick, S.; Durante, C.; Kukovecz, Á.; et al. Three-dimensional printing of complex graphite structures. *Carbon* **2021**, *181*, 260–269. [CrossRef]

66. Sotomayor, M.; Torre-Gamarra, C.; Bucheli, W.; Amarilla, J.; Varez, A.; LevenfelD, B.; Sanchez, J. Additive-free Li$_4$Ti$_5$O$_{12}$ thick electrodes for Li-ion batteries with high electrochemical performance. *J. Mater. Chem. A* **2018**, *6*, 5952–5961. [CrossRef]

67. de la Torre-Gamarra, C.; Sotomayor, M.E.; Sanchez, J.-Y.; Levenfeld, B.; Várez, A.; Laïk, B.; Pereira-Ramos, J.-P. High mass loading additive-free LiFePO$_4$ cathodes with 500 μm thickness for high areal capacity Li-ion batteries. *J. Power Sources* **2020**, *458*, 228033. [CrossRef]

68. Sotomayor, M.E.; Torre-Gamarra, C.d.l.; Levenfeld, B.; Sanchez, J.-Y.; Varez, A.; Kim, G.-T.; Varzi, A.; Passerini, S. Ultra-thick battery electrodes for high gravimetric and volumetric energy density Li-ion batteries. *J. Power Sources* **2019**, *437*, 226923. [CrossRef]

69. Duong, H.; Shin, J.; Yudi, Y. *Dry Electrode Coating Technology*; Maxwell Technologies, Inc.: San Diego, CA, USA, 2018.

70. Ludwig, B.; Zheng, Z.; Shou, W.; Wang, Y.; Pan, H. Solvent-Free Manufacturing of Electrodes for Lithium-ion Batteries. *Sci. Rep.* **2016**, *6*, 23150. [CrossRef]

71. Fu, K.; Yao, Y.; Dai, J.; Hu, L. Progress in 3D Printing of Carbon Materials for Energy-Related Applications. *Adv. Mater.* **2017**, *29*, 1603486. [CrossRef]

72. Yoon, Y.; Lee, K.; Kwon, S.; Seo, S.; Yoo, H.; Kim, S.; Shin, Y.; Park, Y.; Kim, D.; Choi, J.Y.; et al. Vertical alignments of graphene sheets spatially and densely piled for fast ion diffusion in compact supercapacitors. *ACS Nano* **2014**, *8*, 4580–4590. [CrossRef]

73. Johns, P.A.; Roberts, M.R.; Wakizaka, Y.; Sanders, J.H.; Owen, J.R. How the electrolyte limits fast discharge in nanostructured batteries and supercapacitors. *Electrochem. Commun.* **2009**, *11*, 2089–2092. [CrossRef]

74. Ogihara, N.; Itou, Y.; Sasaki, T.; Takeuchi, Y. Impedance Spectroscopy Characterization of Porous Electrodes under Different Electrode Thickness Using a Symmetric Cell for High-Performance Lithium-Ion Batteries. *J. Phys. Chem. C* **2015**, *119*, 4612–4619. [CrossRef]

75. Vijayaraghavan, B.; Ely, D.R.; Chiang, Y.-M.; García-García, R.; Garcia, R.E. An Analytical Method to Determine Tortuosity in Rechargeable Battery Electrodes. *J. Electrochem. Soc.* **2012**, *159*, A548–A552. [CrossRef]

76. Harris, S.; Lu, P. Effects of Inhomogeneities—Nanoscale to Mesoscale—On the Durability of Li-Ion Batteries. *J. Phys. Chem. C* **2013**, *117*, 6481–6492. [CrossRef]

77. Ebner, M.; Wood, V. Tool for Tortuosity Estimation in Lithium Ion Battery Porous Electrodes. *J. Electrochem. Soc.* **2015**, *162*, A3064–A3070. [CrossRef]

78. Li, H.; Tao, Y.; Zheng, X.; Luo, J.; Kang, F.; Cheng, H.-M.; Yang, Q.-H. Ultra-thick graphene bulk supercapacitor electrodes for compact energy storage. *Energy Environ. Sci.* **2016**, *9*, 3135–3142. [CrossRef]

79. Xiao, Q.; Gu, M.; Yang, H.; Li, B.; Zhang, C.; Liu, Y.; Liu, F.; Dai, F.; Yang, L.; Liu, Z.; et al. Inward lithium-ion breathing of hierarchically porous silicon anodes. *Nat. Commun.* **2015**, *6*, 8844. [CrossRef]

80. Ramadesigan, V.; Methekar, R.N.; Latinwo, F.; Braatz, R.D.; Subramanian, V.R. Optimal Porosity Distribution for Minimized Ohmic Drop across a Porous Electrode. *J. Electrochem. Soc.* **2010**, *157*, A1328–A1334. [CrossRef]

81. Golmon, S.; Maute, K.; Dunn, M.L. Multiscale design optimization of lithium ion batteries using adjoint sensitivity analysis. *Int. J. Numer. Methods Eng.* **2012**, *92*, 475–494. [CrossRef]

82. Golmon, S.; Maute, K.; Dunn, M. A design optimization methodology for Li$^+$ batteries. *J. Power Sources* **2014**, *253*, 239–250. [CrossRef]

83. Qi, Y.; Jang, T.; Ramadesigan, V.; Schwartz, D.T.; Subramanian, V.R. Is There a Benefit in Employing Graded Electrodes for Lithium-Ion Batteries? *J. Electrochem. Soc.* **2017**, *164*, A3196–A3207. [CrossRef]

84. Bitsch, B.; Gallasch, T.; Schroeder, M.; Börner, M.; Winter, M.; Willenbacher, N. Capillary suspensions as beneficial formulation concept for high energy density Li-ion battery electrodes. *J. Power Sources* **2016**, *328*, 114–123. [CrossRef]

85. Wang, C.P.; Lopatin, S.D.; Bachrach, R.Z.; Sikha, G. Graded Electrode Technologies for High Energy Lithium-Ion Batteries. U.S. Patent 20110168550, 21 July 2011.

86. Huang, C.; Dontigny, M.; Zaghib, K.; Grant, P.S. Low-tortuosity and graded lithium ion battery cathodes by ice templating. *J. Mater. Chem. A* **2019**, *7*, 21421–21431. [CrossRef]

87. Kolosnitsyn, V.; Karaseva, E. Improvements relating to electrode structures in batteries. European Patent 2006010894, 02 February 2006.

88. Huebsch, N.; Mooney, D.J. Inspiration and application in the evolution of biomaterials. *Nature* **2009**, *462*, 426–432. [CrossRef]

89. Ma, Y.; Yao, D.; Liang, H.; Yin, J.; Xia, Y.; Zuo, K.; Zeng, Y.-P. Ultra-thick wood biochar monoliths with hierarchically porous structure from cotton rose for electrochemical capacitor electrodes. *Electrochim. Acta* **2020**, *352*, 136452. [CrossRef]

90. Liu, K.; Mo, R.; Dong, W.; Zhao, W.; Huang, F. Nature-derived, structure and function integrated ultra-thick carbon electrode for high-performance supercapacitors. *J. Mater. Chem. A* **2020**, *8*, 20072–20081. [CrossRef]

91. Lv, Z.; Yue, M.; Ling, M.; Zhang, H.; Yan, J.; Zheng, Q.; Li, X. Controllable Design Coupled with Finite Element Analysis of Low-Tortuosity Electrode Architecture for Advanced Sodium-Ion Batteries with Ultra-High Mass Loading. *Adv. Energy Mater.* **2021**, *11*, 2003725. [CrossRef]

92. Lu, L.L.; Lu, Y.Y.; Xiao, Z.J.; Zhang, T.W.; Zhou, F.; Ma, T.; Ni, Y.; Yao, H.B.; Yu, S.H.; Cui, Y. Wood-Inspired High-Performance Ultrathick Bulk Battery Electrodes. *Adv. Mater.* **2018**, *30*, e1706745. [CrossRef]

93. Behr, S.; Amin, R.; Chiang, Y.M.; Tomsia, A.P. Highly structured, additive free lithium-ion cathodes by freeze-casting technology. *Ceram. Forum Int.* **2015**, *92*, E39–E43.

94. Li, L.; Erb, R.M.; Wang, J.; Wang, J.; Chiang, Y.M. Fabrication of Low-Tortuosity Ultrahigh-Area-Capacity Battery Electrodes through Magnetic Alignment of Emulsion-Based Slurries. *Adv. Energy Mater.* **2018**, *9*, 1802472. [CrossRef]

95. Sander, J.S.; Erb, R.M.; Li, L.; Gurijala, A.; Chiang, Y.M. High-performance battery electrodes via magnetic templating. *Nat. Energy* **2016**, *1*, 16099. [CrossRef]

96. Billaud, J.; Bouville, F.; Magrini, T.; Villevieille, C.; Studart, A.R. Magnetically aligned graphite electrodes for high-rate performance Li-ion batteries. *Nat. Energy* **2016**, *1*, 16097. [CrossRef]

97. Ma, J.; Qiao, Y.; Huang, M.; Shang, H.; Zhou, H.; Li, T.; Liu, W.; Qu, M.; Zhang, H.; Peng, G. Low tortuosity thick cathode design in high loading lithium sulfur batteries enabled by magnetic hollow carbon fibers. *Appl. Surf. Sci.* **2021**, *542*, 148664. [CrossRef]

98. Pan, G.; Hu, L.; Zhang, F.; Chen, Q. Out-of-Plane Alignment of Conjugated Semiconducting Polymers by Horizontal Rotation in a High Magnetic Field. *J. Phys. Chem. Lett.* **2021**, *12*, 3476–3484. [CrossRef] [PubMed]

99. Li, X.; Zhao, H.; Liu, C.; Cai, J.; Zhang, Y.; Jiang, Y.; Zhang, D. High-Efficiency Alignment of 3D Biotemplated Helices via Rotating Magnetic Field for Terahertz Chiral Metamaterials. *Adv. Opt. Mater.* **2019**, *7*, 1900247. [CrossRef]

100. Chen, K.-H.; Namkoong, M.J.; Goel, V.; Yang, C.; Kazemiabnavi, S.; Mortuza, S.M.; Kazyak, E.; Mazumder, J.; Thornton, K.; Sakamoto, J.; et al. Efficient fast-charging of lithium-ion batteries enabled by laser-patterned three-dimensional graphite anode architectures. *J. Power Sources* **2020**, *471*, 228475. [CrossRef]

101. Park, J.; Hyeon, S.; Jeong, S.; Kim, H.-J. Performance enhancement of Li-ion battery by laser structuring of thick electrode with low porosity. *J. Ind. Eng. Chem.* **2019**, *70*, 178–185. [CrossRef]

102. Park, J.; Jeon, C.; Kim, W.; Bong, S.; Jeong, S.; Kim, H. Challenges, laser processing and electrochemical characteristics on application of ultra-thick electrode for high-energy lithium-ion battery. *J. Power Sources* **2021**, *482*, 228948. [CrossRef]

103. Mangang, M.; Seifert, H.J.; Pfleging, W. Influence of laser pulse duration on the electrochemical performance of laser structured LiFePO4 composite electrodes. *J. Power Sources* **2016**, *304*, 24–32. [CrossRef]
104. Pfleging, W.; Pröll, J. A new approach for rapid electrolyte wetting in tape cast electrodes for lithium-ion batteries. *J. Mater. Chem. A* **2014**, *2*, 14918–14926. [CrossRef]
105. Zhu, P.; Seifert, H.J.; Pfleging, W. The Ultrafast Laser Ablation of Li(Ni0.6Mn0.2Co0.2)O2 Electrodes with High Mass Loading. *Appl. Sci.* **2019**, *9*, 4067. [CrossRef]
106. Mottay, E.; Liu, X.; Zhang, H.; Mazur, E.; Sanatinia, R.; Pfleging, W. Industrial applications of ultrafast laser processing. *MRS Bull.* **2016**, *41*, 984–992. [CrossRef]
107. Habedank, J.B.; Kraft, L.; Rheinfeld, A.; Krezdorn, C.; Jossen, A.; Zaeh, M.F. Increasing the Discharge Rate Capability of Lithium-Ion Cells with Laser-Structured Graphite Anodes: Modeling and Simulation. *J. Electrochem. Soc.* **2018**, *165*, A1563–A1573. [CrossRef]

batteries

MDPI

Article

SOC Estimation Based on Combination of Electrochemical and External Characteristics for Hybrid Lithium-Ion Capacitors

Xiaofan Huang [1], Renjie Gao [2], Luyao Zhang [3,4], Xinrong Lv [3,4], Shaolong Shu [2], Xiaoping Tang [1,*], Ziyao Wang [1] and Junsheng Zheng [3,4,*]

[1]   Huadian Electric Power Research Institute Co., Ltd., Hangzhou 310030, China
[2]   School of Electronics and Information Engineering, Tongji University, Shanghai 201804, China
[3]   Clean Energy Automotive Engineering Centre, Tongji University, Shanghai 201804, China
[4]   School of Automotive Studies, Tongji University, Shanghai 201804, China
*   Correspondence: xiaoping-tang@chder.com (X.T.); jszheng@tongji.edu.cn (J.Z.)

Abstract: Hybrid lithium-ion capacitors (HyLICs) have received considerable attention because of their ability to combine the advantages of high-energy lithium-ion batteries and high-power supercapacitors. State of charge (SOC) is the main factor affecting the practical application of HyLICs; therefore, it is essential to estimate the SOC accurately. In this paper, a partition SOC-estimation method that combines electrochemical and external characteristics is proposed. The discharge process of the HyLICs was divided into three phases based on test results of electrochemical characteristics. To improve the estimation accuracy and reduce the amount of calculation, the Extended Kalman Filter (EKF) method was applied for SOC estimation at the interval where the capacitor energy storage characteristics dominated, and the Ampere-hour (Ah) method was used to estimate the SOC at the interval where battery energy storage characteristics dominated. The proposed method is verified under different operating conditions. The experimental results show good agreement with the estimation results, which indicates that the proposed method can estimate the SOC of the HyLICs accurately.

Keywords: SOC estimation; electrochemical characteristic; hybrid lithium-ion capacitor

Citation: Huang, X.; Gao, R.; Zhang, L.; Lv, X.; Shu, S.; Tang, X.; Wang, Z.; Zheng, J. SOC Estimation Based on Combination of Electrochemical and External Characteristics for Hybrid Lithium-Ion Capacitors. *Batteries* **2023**, *9*, 163. https://doi.org/10.3390/batteries9030163

Academic Editor: Carlos Ziebert

Received: 17 December 2022
Revised: 18 February 2023
Accepted: 4 March 2023
Published: 9 March 2023

1. Introduction

Hybrid Lithium ion Capacitors (HyLICs) are a new type of devices combined with the battery-type negative electrodes and capacitor-type positive electrodes [1–3]. HyLICs have the characteristics of traditional lithium-ion batteries and supercapacitors [4]. State of charge (SOC) is a significant parameter that indicates the level of charge in the HyLICs [5,6]. However, the SOC is an inner state of the HyLICs that depends on temperature, material degradation, electrochemical reactions and aging cycles [7,8]. In addition, HyLICs have typically nonlinear and time-varying characteristics [9]. Consequently, based on the traditional SOC estimation method, it is difficult to estimate SOC for HyLICs [10].

The methods for the estimation of the SOC can be mainly classified into three categories: direct-measurement methods, data-driven methods and model-based methods [11,12]. Direct measurement methods include the ampere-hour counting method and the open-circuit voltage (OCV) method. The ampere-hour counting method is easy to implement, but the initial value is difficult to find and the error may accumulate [13,14]. Considering the relationship between OCV and SOC is simple and straight; the OCV method can meet the requirements [15]. However, this highly depends on the temperature, chemistry, state of health (SOH) and other factors that can easily affect how their relationship works, and it also varies when batteries age [16].

The data-driven method, such as neural networks [17], extreme learning machine [18] and support vector machine [19], takes advantage of advanced machine learning algorithms to achieve SOC estimation with available historic data, and it is not dependent on a model featuring degradation-dependent parameters [10,20]. Hussein et al. [21] presented two artificial neural networks (ANNs) for SOC estimation of an Li ion battery. However, the implementation cost rises due to the strictly necessary advanced hardware. Malkhandi et al. [22] estimated the SOC of an Li ion battery using the fuzzy logic without thoroughly eliminating the impact of incomplete charging. Yang et al. [23] proposed a stacked Long Short-Term Memory (LSTM) network for SOC estimation that can capture the nonlinear relationship among measured current, voltage, temperature and SOC. Xia et al. [24] proposed a hybrid intelligent method based on a Wavelet Neural Network (WNN) to estimate the SOC of lithium-ion batteries. The discrete wavelet transform and Levenberg–Marquardt are used in the data-training operations.

Nowadays, more work has focused on developing the SOC estimation methods to improve their accuracy based on different models. The model-based method usually starts with the construction of the battery degradation models. Input parameters such as load current, terminal voltage and temperature were taken into calculation for the equivalent model to estimate the SOC of a lead battery [25]. The electrochemical and equivalent circuit model are common equivalent models for batteries [26,27]. They are used to simulate the dynamic characteristics of batteries. Based on various types of battery models, some filtering algorithms derived from control theory are used to estimate the SOC [28]. Pan et al. [11] used the grey prediction model (GM) combined with an OCV model based on the piecewise cubic-Hermite interpolation to build a grey extended Kalman filter (GMEKF) for the SOC estimation of an Li ion battery. However, the estimation accuracy and robustness of these methods still mainly depend on the type of battery model [14].

The above methods are based on the external characteristics of the battery [29], and rarely take the internal electrochemical reaction of the battery into account. Therefore, the accuracy of the estimation results depends heavily on the model or algorithm selection. As a newly developed energy equipment, the property investigation of HyLICs is still at the very beginning compared to the broad study on the Li-ion battery or lead battery. When it comes to the SOC estimation performed in HyLICs, it is certain that our work is not the first time to complete it; previous work was not discussed but spent quite a quantity of space to analyze the characteristics of HyLICs. Based on their nonlinear and time-varying characteristics, existing SOC estimation methods that were mainly performed on the external characteristics of the equipment, as well as the previous little research on HyLICs, are not suitable for our work.

In this paper, we conduct our research based on the internal characteristics of the research object and take the electrochemical impedance of HyLICs, which can be obtained by an electrochemical characteristics test, as the key parameter of the study. The voltage of HyLICs was divided into three intervals based on the linearization result of the HyLICs' electrochemical impedance. The extended Kalman filter method and the ampere-hour integral method are used in the different intervals to estimate the SOC of HyLICs. The method is applied to different operating conditions, and the experimental results show that the method has higher estimation accuracy and stronger reliability.

2. Experiment and Electrochemical Characteristic Analysis

2.1. Hybrid Pulse Power Characterization (HPPC) Test

The hybrid pulse power characterization was recorded on battery tester BTS-60V20A produced by Neware Ins., Shenzhen. The specific procedure is as follows:

Step 1: The HyLICs were fully charged with constant voltage and standard current.

Step 2: The fully charged HyLICs were left to rest for 5 h until they reached the equilibrium state.

Step 3: The HPPC sequence was loaded before the constant current discharge was conducted. The HPPC discharge cycle was repeated at every 10% drop in SOC until the cycle ran at 10% SOC. The time interval of resting was 1 h.

Step 4: The test procedure stopped when the voltage of HyLICs reached its discharge cut-off voltage.

HPPC test data will be detailed in the following analysis of Section 4.1.

2.2. Electrochemical Characteristic Test

The electrochemical characteristic test mainly consists of two parts: cyclic voltammetry (CV) test and electrochemical impedance spectroscopy (EIS). The cyclic voltammetry (CV) profiles at different voltage ranges and scanning rates were recorded on the electrochemical workstation CHI660 produced by CH Instruments Ins., Shanghai. The EIS was tested at the open circuit voltage (OCV) within the frequency of 10^{-1}–10^{6} Hz on CHI660. The galvanostatic charge and discharge test was conducted on Neware battery tester. The specific procedure is as follows:

Step 1: The HyLICs were fully charged and discharged for 8 times with a constant current of 1.6 C.

Step 2: A CV test was performed on the HyLICs. A triangular wave voltage was then applied to the HyLICs. After that, the HyLICs were scanned forward with a scanning rate of 10^{-4} V/s from 2.2 V to 4.1 V at both charge and discharge periods.

Step 3: The HyLICs were tested with the interval of 0.1 V during the cycle from 2.2 V to 4.1 V and back to 2.2 V, to obtain the EIS of charge and discharge for each voltage state.

2.3. Electrochemical Characteristic Analysis

In order to obtain more detailed electrochemical information inside HyLICs, the Zview software was used to identify the parameters at the equivalent circuit diagram as shown in Figure 1, where R_1 represents the ohmic impedance generated by the contact of electrolyte diaphragm; R_2 represents the transfer resistance; C_1 represents the electric double layer generated at the boundary between the electrolyte and the electrode; R_3 and C_2 connected in parallel represent the impedance of the solid electrolyte interface membrane on the negative electrode of the battery; W_1 represents the ion diffusion resistance of the battery. The kinetic parameters of the reaction process were calculated by an EIS test, and the EIS test results are shown in Figure 2. Based on the results, the electronic impedance, ion impedance and total impedance can be further fitted into curves.

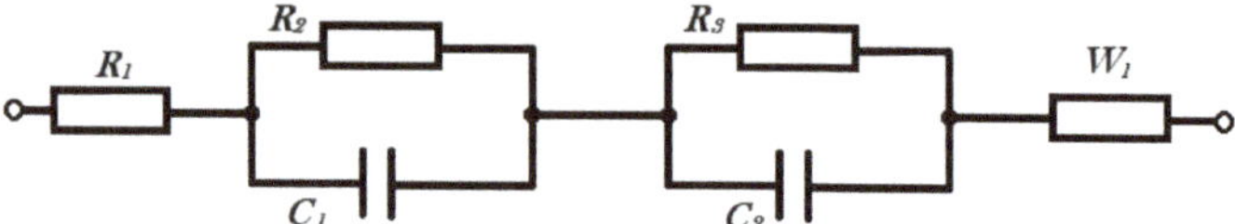

Figure 1. Electrochemical impedance equivalent circuit model.

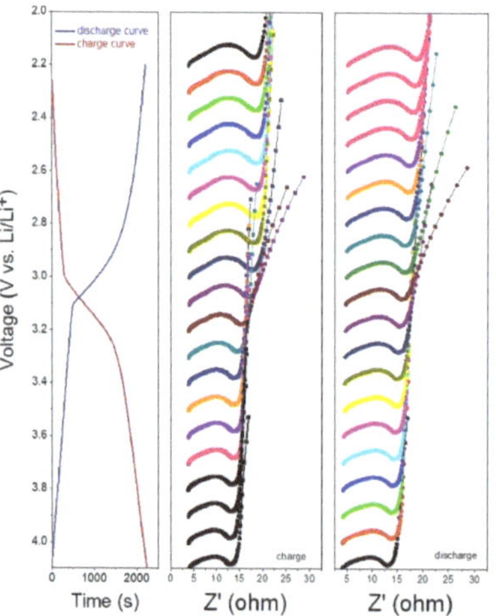

Figure 2. Electrochemical impedance spectroscopy of HyLICs.

The electronic impedance, ion impedance and total impedance of the HyLICs changed with terminal voltage are shown in Figure 3. It can be seen that for the total impedance of the HyLIC, its nonlinear and time-varying characteristics are quite obvious and it is inappropriate to adopt one method to estimate its SOC during different energy storage conditions. The standard Levenberg–Marquardt method [30] and general global optimization algorithm are used to perform piecewise linear fitting to the total impedance of the HyLICs, and the mathematical relationship is shown in Equation (1).

$$y = \begin{cases} 0.0357x + 34.125, & x < 2.8875 \\ -16.9x + 83.23, & 2.8875 \leq x < 3.3355 \\ -1.2976x + 30.7036, & x \geq 3.3355 \end{cases} \tag{1}$$

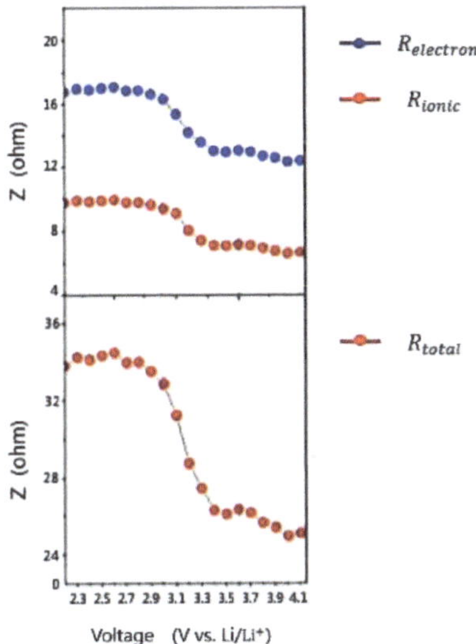

Figure 3. Capacitor impedance changes with terminal voltage of HyLICs.

After fitting, the mean square error MSE = 0.3025, the residual sum of squares RSS = 1.8308 and the correlation coefficient R = 0.9969 indicate that the model fits the data well. Intervals are separated by blue and green dotted lines in Figure 4. When the HyLICs voltage is between 2.2 V and 2.90 V, the corresponding SOC interval is 3–20%, the slope of the discharge OCV–SOC curve is large, and the OCV of the HyLICs changes drastically with the SOC value. The total impedance curve remains basically stable. In this interval, the capacitive energy storage characteristics of HyLICs occupy a dominant position. When the terminal voltage rises to 2.90–3.35 V, the corresponding SOC range is 20–80%, with an obviously smaller slope of the discharge OCV–SOC curve. As the SOC value increases greatly, the OCV value changes less and the total impedance curve drops sharply in this interval. They all show that the energy storage characteristics of lithium batteries dominate in this interval of HyLICs. When the terminal voltage reaches 3.35–3.8 V, the corresponding SOC interval increases to 80–100%, the slope of the discharge OCV–SOC curve increases again, and the total impedance of the capacitor is reduced and becomes stable. The above results prove that the capacitive energy-storage characteristic of HyLICs takes control again.

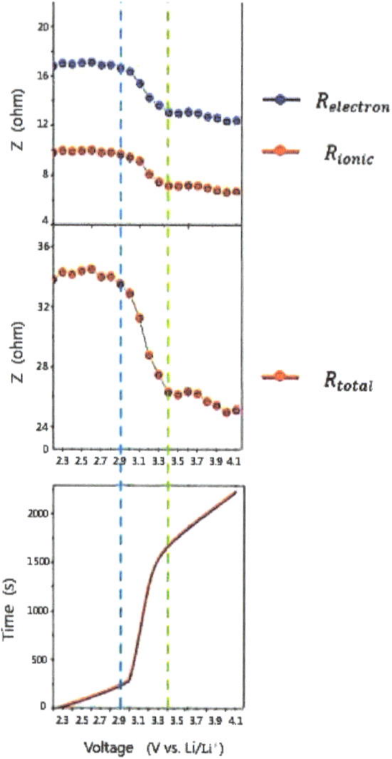

Figure 4. Classification results based on electrochemical characteristics of HyLICs.

3. SOC Partition Estimation Method Based on Electrochemical Characteristics

3.1. Battery Model and Parameter Identification

3.1.1. Equivalent Circuit Model

The equivalent circuit model [31] is a circuit network that can quantitatively describe the working characteristics of power batteries. Compared to other equivalent models [32], the equivalent circuit model has the advantages of strong applicability, high precision and easy quantitative analysis. Considering the real-time requirements of the Battery Management System for state information estimation and the computing ability of microprocessors, it is necessary to establish a lithium-ion capacitor model with as low complexity as possible on the premise of meeting the accuracy requirements. Comparing its model accuracy, complexity, computational complexity and other factors to the Rint model and Partnership for a New Generation Vehicles (PNGV) model, the first-order RC equivalent circuit model (Thevenin model) was selected for the establishment of an equivalent model of HyLICs [33]. The circuit diagram is shown in Figure 5, and the circuit equation is:

$$\begin{cases} \dot{U}_d = \dfrac{i_L}{C_d} - \dfrac{U_d}{R_d R_d} \\ U_t = U_{oc} - U_d - R_i \cdot i_L \end{cases} \tag{2}$$

where U_{oc} is the OCV, i_L is the load current, U_t is the terminal voltage, R_i is the ohmic internal resistance, and R_d and C_d are the polarization resistance and polarization capacitance of the RC network, respectively.

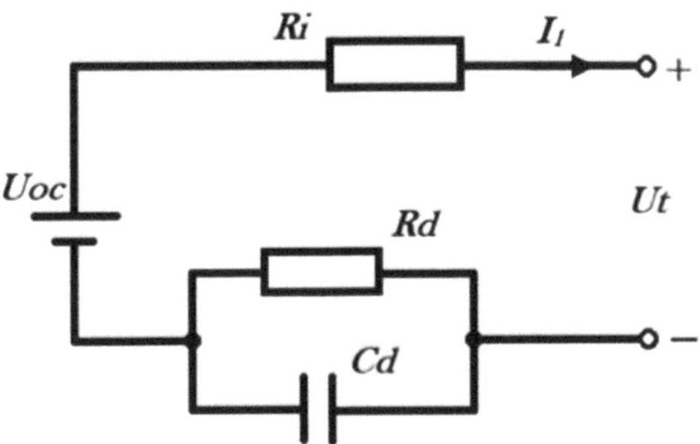

Figure 5. The first-order RC equivalent circuit model.

3.1.2. Parameter Identification

The parameters of the HyLICs change with the influence of factors, such as operating temperature and working conditions. The accuracy of the fixed parameter model will inevitably decrease with the change of battery life-cycle parameters, so that the battery state cannot be accurately estimated. Therefore, it is necessary to introduce an online update mechanism for battery-model parameters. According to the characteristics of rapid changes in battery-system status and slow changes in parameters, it is necessary to reduce the influence weight of old data on parameter estimation, while increasing the influence of new information on system-parameter identification, and the newer the data, the greater the weight [34]. For this reason, the Recursive Least Squares method with Forgetting Factor (FFRLS) [11] is used to identify the parameters of the first-order RC model of lithium-ion batteries R_d, R_i, C_d.

3.2. The EKF Method for SOC Estimation

The EKF [35,36] method is applied to estimate the SOC of the HyLICs. The terminal voltage sequence $U_{T,0}$, $U_{T,1}$, $U_{T,2}$, $\dots$, $U_{T,k}$ is the system input and the terminal current sequence $I_{L,0}$, $I_{L,1}$, $\dots$, $I_{L,k}$ is the system output. The model of the system is as follows:

$$\begin{cases} \begin{bmatrix} \dot{SOC} \\ \dot{U}_d \end{bmatrix} = \begin{bmatrix} 1 & 0 \\ 0 & -\frac{1}{R_d C_d} \end{bmatrix} \begin{bmatrix} SOC \\ U_d \end{bmatrix} + \begin{bmatrix} -\frac{1}{q_0} \\ \frac{1}{C_d} \end{bmatrix} I_L + W \\ U_t = OCV(SOC) - U_d - R_i I_L + V \end{cases} \tag{3}$$

where $U_{OC} = OCV(SOC)$ represents the nonlinear function between open-circuit voltage U_{OC} and the SOC of the HyLICs; W and V are white noise that obeys Gaussian distribution.

The linear variable system is as follows:

$$\begin{cases} \begin{bmatrix} \dot{SOC} \\ \dot{U}_d \end{bmatrix} = \begin{bmatrix} 1 & 0 \\ 0 & -\frac{1}{R_d C_d} \end{bmatrix} \begin{bmatrix} SOC \\ U_d \end{bmatrix} + \begin{bmatrix} -\frac{1}{q_0} \\ \frac{1}{C_d} \end{bmatrix} I_L + W \\ U_t = \begin{bmatrix} \frac{dOCV}{dSOC} & -1 \end{bmatrix} \begin{bmatrix} SOC \\ U_d \end{bmatrix} + OCV\left(\widehat{SOC}\right) - \widehat{U}_d - R_i I_L - \begin{bmatrix} \frac{dOCV}{dSOC} & -1 \end{bmatrix} \begin{bmatrix} \widehat{SOC} \\ \widehat{U}_d \end{bmatrix} + V \end{cases} \tag{4}$$

The discrete time model of the system is as follows:

$$\begin{cases} \begin{bmatrix} \dot{SOC} \\ \dot{U}_d \end{bmatrix} = \begin{bmatrix} 1 & 0 \\ 0 & e^{-\frac{1}{R_d C_d}} \end{bmatrix} \begin{bmatrix} SOC \\ U_d \end{bmatrix} + \begin{bmatrix} -\frac{1}{q_0} \\ R_d\left(1 - e^{-\frac{1}{R_d C_d}}\right) \end{bmatrix} I_{L,k} + W_{k-1} \\ U_{t,k} = \begin{bmatrix} \frac{dOCV}{dSOC} \\ -1 \end{bmatrix}^T \begin{bmatrix} SOC \\ U_d \end{bmatrix}_k + OCV_k\left(\widehat{SOC}_k\right) - \widehat{U}_{d,k} - R_i I_{L,k} - \begin{bmatrix} \frac{dOCV}{dSOC} \\ -1 \end{bmatrix}_k^T \begin{bmatrix} \widehat{SOC} \\ \widehat{U}_d \end{bmatrix}_k + V_k \end{cases} \tag{5}$$

where $x_k = \begin{bmatrix} SOC & U_d \end{bmatrix}_k^T$ is the state vector, $U_{t,k}$ is the observation vector, $I_{L,k}$ is the control vector, W_{k-1} and V_k are uncorrelated zero-mean Gaussian white noise. In addition, the SOC estimation can be achieved by applying the linearized system discrete-time model to the EKF algorithm.

3.3. SOC Partition Estimation Method Based on Electrochemical Characteristics

From the analysis in Section 2.2, it can be seen that when the terminal voltage is 2.2–2.9 V and 3.35–3.8 V, the capacitive energy-storage characteristics of HyLICs dominate. Capacitor energy storage has strong nonlinear characteristics and the EKF method is suitable for nonlinear systems. The EKF method is not sensitive to initial parameters and can reduce the accumulation of experimental errors. Therefore, the SOC value of these battery intervals will be estimated using the EKF method considering the complexity of the algorithm and the real-time of the system [37]. When the terminal voltage is between 2.90 V and 3.35 V, the battery-energy-storage characteristics are more apparent, and the ampere-hour method should be used to estimate the SOC value more accurately [38].

4. Results and Discussion

4.1. Analysis of the Model for Battery SOC Estimation

A comparison of the U_{OC} value obtained by parameter identification with the discharge OCV–SOC curve is shown in Figure 6. The average error is 0.02 V and the maximum error is 0.033 V. It can be seen that the parameter-identification result is true, the difference between the values is small and the identification accuracy is high.

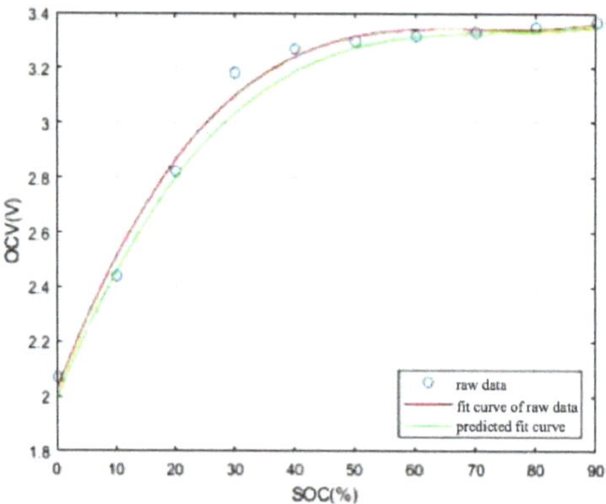

Figure 6. The OCV–SOC curve after the parameter identification.

Combining the common SOC estimation methods described in this chapter, the analysis of the electrochemical characteristics of lithium-ion capacitors, and the methods selected for different intervals, the flow chart of the SOC partition-estimation method based on the electrochemical characteristics of lithium-ion capacitors can be obtained as following and as shown in Figure 7:

1. The historical information stored in the system should be queried, and the SOC value obtained at the last time of the last operation of the HyLIC should be used as the initial value of the SOC estimation algorithm;
2. The real-time operating condition data of the HyLIC would be acquired, including terminal voltage, working current, etc., the terminal voltage as the characteristic value of the SOC electrochemical characteristic zone would be used, and the chemical characteristic interval of the lithium ion capacitor would be judged according to the terminal voltage value;

3. The current SOC of the HyLIC would be estimated according to the current interval of the lithium-ion capacitor and the corresponding method previously determined, and the result would be saved;
4. Determine whether the work is over. If it is, jump out of the loop; otherwise, return to 2.

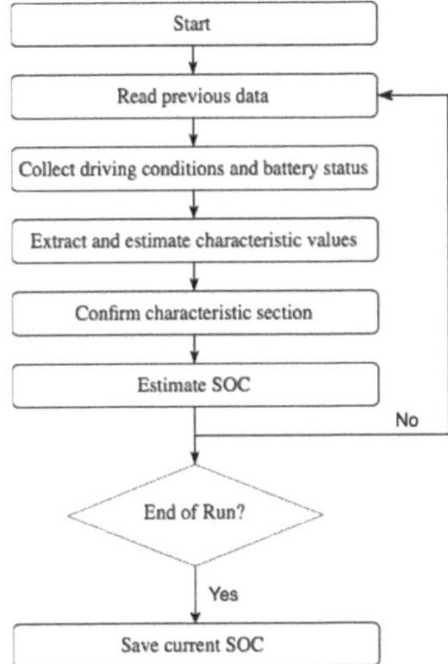

Figure 7. SOC partition-estimation method based on electrochemical characteristics.

According to the parameter-identification results, the Thevenin model of the HyLICs shown in Figure 8 is established in Simulink. Load the HPPC test current at the input terminal of the model, and compare the output terminal voltage value HPPC test results. As shown in Figure 9, the average error is 0.176 V, and the maximum error is 0.431 V. In the middle- and high-voltage range in Figure 8, the estimated value of the Thevenin model is in good agreement with the measured value, while the error becomes larger in the low-voltage range. The reason may be that the charging and discharging principle of HyLICs is more complicated, but the accuracy of the model is still acceptable.

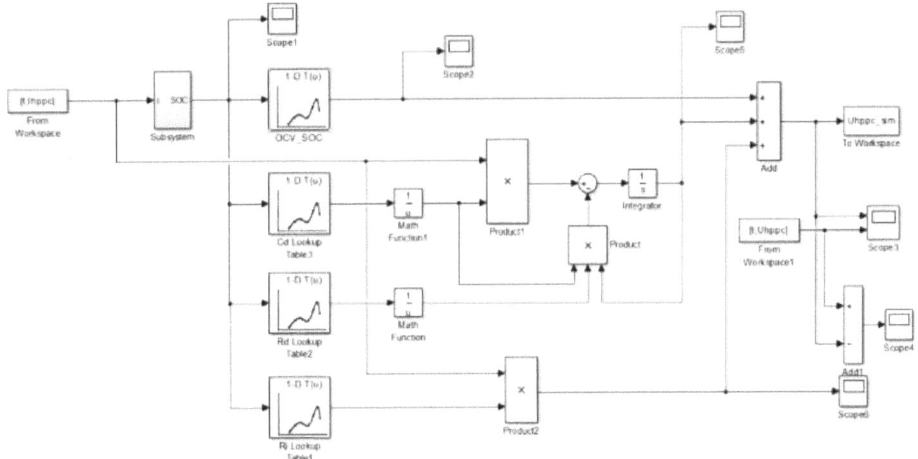

Figure 8. HyLICs Simulink model.

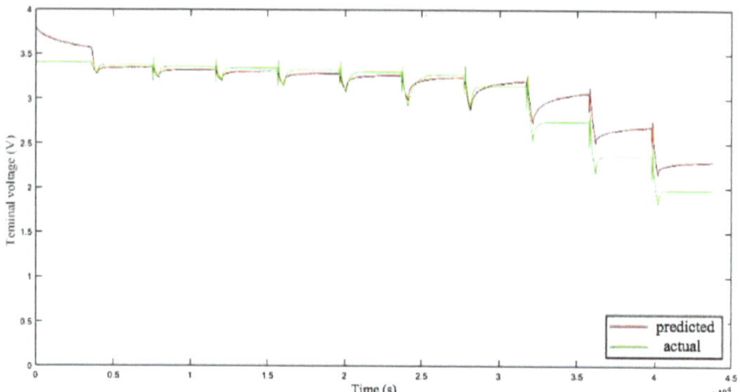

Figure 9. Comparison chart of predicted terminal voltage and actual value.

4.2. Algorithm Verification

4.2.1. Discharge OCV Test and Verification

The relevant data is recorded during the test of 1C rate in the discharge open-circuit voltage at room temperature of the HyLICs. The voltage and current data in the discharge open-circuit voltage test are used as the input of the algorithm and implemented in MATLAB. The estimated SOC value obtained by the partition-estimation algorithm and the SOC value measured in the laboratory are shown in Figure 10a. The difference between them is shown in Figure 10b.

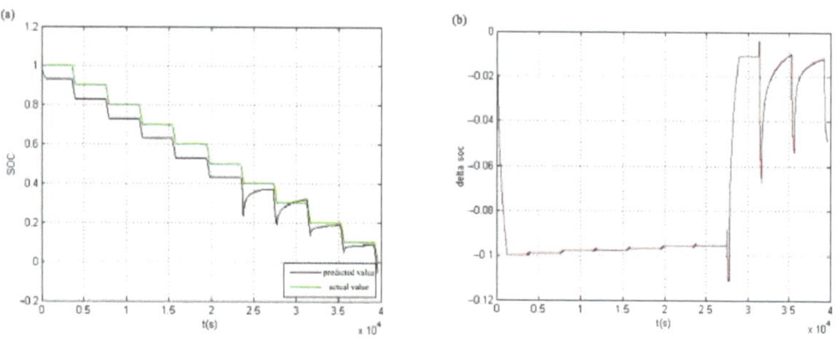

Figure 10. Comparison chart of (**a**) predicted SOC and actual value; (**b**) the difference through discharge OCV test.

It can be seen from Figure 10 that this method is close to the data measured in the laboratory in the low SOC interval, and the accuracy is a little lower in the medium and high SOC intervals. The reason is that the SOC value of the HyLICs is greater than 80% when the system is started and the EKF method is used in the interval. Before the estimation result of the EKF method converges to a stable value, the algorithm switches to the stable region. During that period, the ampere-hour method is used for SOC estimation, and the initial value of this method has a great effect on it. The initial value estimation error will accumulate as the test proceeds. It can be seen from Figure 10b that in the discharge open-circuit-voltage test, the error between the predicted and actual value is less than 10%, which is still within an acceptable range. Regardless of whether the error is feasible for other circumstances, there is no denying that the error-correcting process still requires deeper and more qualified work to optimize the current estimation method and to reduce error.

4.2.2. HPPC Test and Verification

In order to verify the estimation accuracy of the model under different voltage and current, the model was placed under HPPC experimental conditions and an HPPC test of HyLICs was conducted at room temperature [39]. The voltage and current data in the HPPC test are used as the algorithm input. The test was implemented in MATLAB. Figure 11a,b show the SOC estimation value obtained by the partition-estimation algorithm, the SOC value measured by the laboratory and the error of them, respectively.

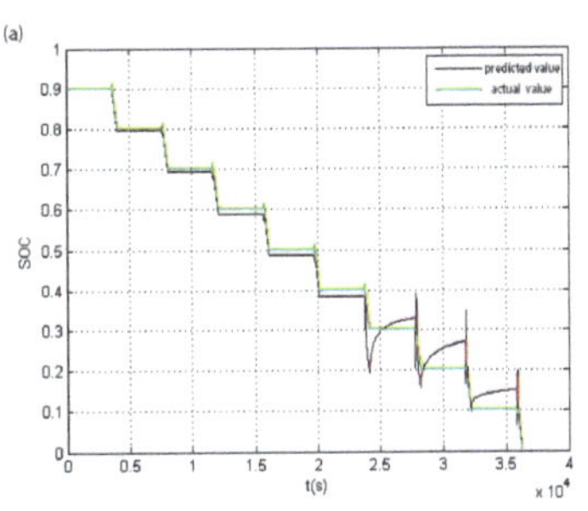
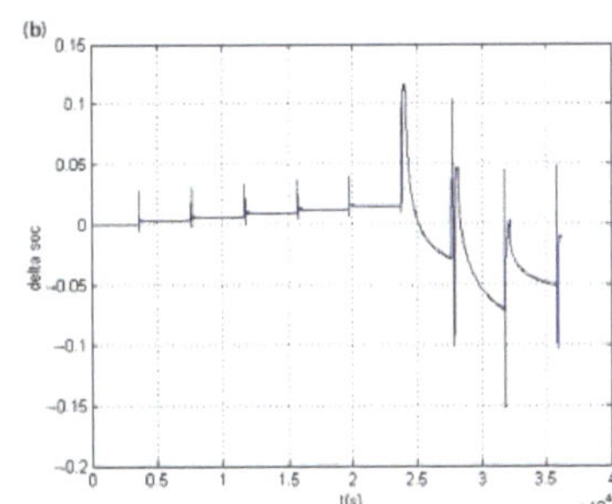

Figure 11. Comparison chart of (**a**) predicted SOC and actual value; (**b**) the difference through HPPC test.

It can be seen from Figure 11 that during the entire HPPC test process, the error between the predicted value of the partition-estimation algorithm and the actual measurement value is very small. It is only when the current pulses are too large, short in duration and recover quickly that large differences occur. Fast convergence proves that the algorithm has a good suppression effect on this disturbance. In addition, the average error between the predicted value and the actual value is 0.034 V, showing a high estimation and a strong algorithm reliability.

4.2.3. NEDC Cycle Verification

Tests in Sections 4.2.1 and 4.2.2 were based on working conditions in a laboratory environment. However, in real vehicle-mounted working conditions, the operation of HyLICs is more complicated. To further verify the accuracy and reliability of the partition estimation method, applying the method to actual standard operating conditions, comparing the estimated results with the standard values of the conditions and analyzing the pros and cons of the algorithm seem quite necessary.

In order to exert the advantages of high power density and long cycle life of a lithium-ion capacitor, the energy power device of the 48 V vehicle start–stop system was adopted. The parallel mild hybrid system was formed with the internal combustion engine to achieve high energy-saving efficiency at relatively low cost, provide sufficient power when the vehicle starts and maximize the energy recovered during braking. With its typical characteristics of a nonlinear and time-varying energy-storage pattern, HyLICs used in this paper as start–stop power supply devices are mainly for the needs of urban operating conditions in China. Therefore, NEDC operating conditions are selected as the standard comprehensive operating conditions [40].

Through a NEDC cycle test, the NEDC operating-condition simulation in the 48 V hybrid-electric vehicle model was established, and the I-t, U-t, P-t and other relational curves of the battery working process under this operating condition were derived in the AVL-cruise software. The I-t curve is shown in Figure 12. According to the ampere-hour method, the relationship between the SOC of the power battery and the time under NEDC operating conditions can be calculated. The voltage and current values under NEDC operating conditions are used as the input of the partition-estimation algorithm in MATLAB. The estimated SOC value, the measured value and the difference between them are shown in Figure 13a,b. The average error between the estimated value and the measured value is 0.031 V, and the maximum error is 0.057 V. The error of this algorithm is within the accuracy range required by practical application, so it has high reliability.

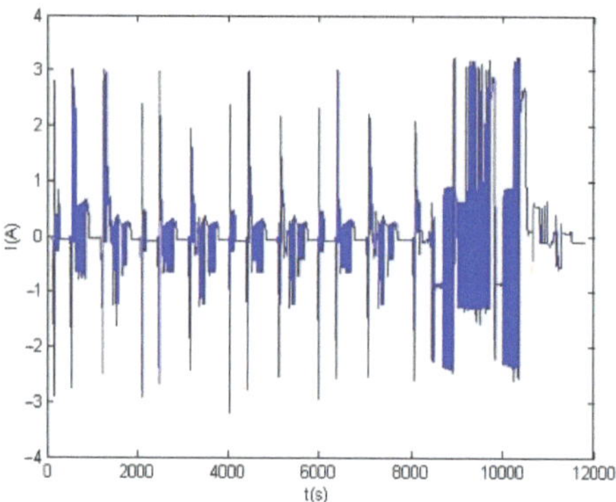

Figure 12. The I-t curve at the loading profiles of NEDC.

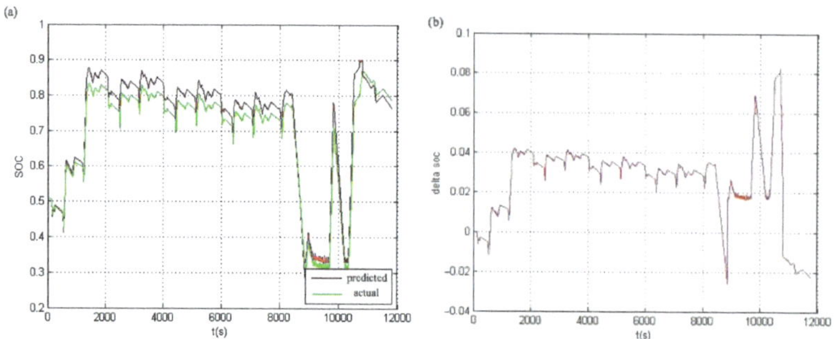

Figure 13. Comparison chart of (**a**) predicted terminal voltage and actual value; (**b**) the difference through NEDC cycle test.

5. Conclusions

In this article, from the perspective of the electrochemical-reaction principle of HyLICs, an SOC partition-estimation method based on its electrochemical characteristics was proposed. The accuracy was increased and the amount of calculation was deduced by combining the Extended Kalman filter (EKF) method and the ampere-hour (Ah) method at different phases. The proposed method was verified with the data under different working conditions by a valid battery model. The results show that the method can accurately estimate the SOC of the HyLICs. Based on this method's accurate estimation, HyLICs-loaded vehicles are more competitive in the automotive industry's development.

When selecting an equivalent-circuit model for lithium-ion capacitors, the Thevenin equivalent circuit model was chosen in consideration of factors such as model accuracy, real-time calculation and complexity of algorithm. However, during model validation, it was found that the Thevenin model was not sufficient to fully describe the external characteristics of lithium-ion capacitors in the low SOC range, which might affect the accuracy of the SOC estimation algorithm based on the model. Therefore, it was necessary to study and establish the HyLICs equivalent model superior to the Thevenin model while considering the model accuracy, computational real-time and algorithm complexity. The discharge open-circuit voltage test data and HPPC test data used in the verification of the SOC estimation algorithm in the paper were measured under laboratory conditions. The NEDC cycle working-condition data used were obtained by establishing a 48 V hybrid-electric vehicle model in AVL-cruise software and running the NEDC working condition simulation. Compared with the external characteristic data of the capacitor collected in real time under the actual operating condition, these data have less interference from pollution or noise. In later research, the real data collected to verify the feasibility of the SOC estimation algorithm will make the verification results more convincing.

Author Contributions: Conceptualization, X.H.; methodology, R.G.; software, X.H. and Z.W.; validation, R.G.; formal analysis, R.G.; investigation, L.Z. and X.T.; resources, Z.W.; data curation, S.S.; writing—original draft, L.Z.; writing—review and editing, L.Z. and X.L.; visualization, X.T.; supervision, S.S. and J.Z.; project administration, J.Z.; funding acquisition, X.T. and J.Z. All authors have read and agreed to the published version of the manuscript.

Funding: The work is supported by Key scientific and technological projects of China Huadian Corporation Ltd. (Grant No. CHDKJ20-01-111 and CHDER/MYZX-CG-2022-0022).

Institutional Review Board Statement: Not applicable.

Informed Consent Statement: Not applicable.

Data Availability Statement: Data is unavailable due to privacy.

Conflicts of Interest: The authors declare no conflict of interest.

References

1. Liang, T.; Wang, H.W.; Fei, R.X.; Wang, R.; He, B.B.; Gong, Y.S.; Yan, C.J. A high-power lithium-ion hybrid capacitor based on a hollow N-doped carbon nanobox anode and its porous analogue cathode. *Nanoscale* **2019**, *11*, 20715–20724. [CrossRef]
2. Yuan, J.; Qin, N.; Lu, Y.; Jin, L.; Zheng, J.; Zheng, J.P. The effect of electrolyte additives on the rate performance of hard carbon anode at low temperature for lithium-ion capacitor. *Chin. Chem. Lett.* **2022**, *33*, 3889–3893. [CrossRef]
3. Zhang, J.; Wang, J.; Shi, Z.; Xu, Z. Mesoporous carbon material as cathode for high performance lithium-ion capacitor. *Chin. Chem. Lett.* **2018**, *29*, 620–623. [CrossRef]
4. An, Z.X.; Fang, W.Y.; Xu, J.Q.; Zhang, J.J. Comparative analysis of electrochemical performances and capacity degrading behaviors in lithium-ion capacitors based on different anodic materials. *Ionics* **2019**, *25*, 3277–3285. [CrossRef]
5. Navid, Q.; Hassan, A. An Accurate and Precise Grey Box Model of a Low-Power Lithium-Ion Battery and Capacitor/Supercapacitor for Accurate Estimation of State-of-Charge. *Batteries* **2019**, *5*, 50. [CrossRef]
6. Kurzweil, P.; Shamonin, M. State-of-Charge Monitoring by Impedance Spectroscopy during Long-Term Self-Discharge of Supercapacitors and Lithium-Ion Batteries. *Batteries* **2018**, *4*, 35. [CrossRef]
7. Liu, K.L.; Shang, Y.L.; Ouyang, Q.; Widanage, W.D. A Data-Driven Approach With Uncertainty Quantification for Predicting Future Capacities and Remaining Useful Life of Lithium-ion Battery. *IEEE Trans. Ind. Electron.* **2021**, *68*, 3170–3180. [CrossRef]
8. El Ghossein, N.; Sari, A.; Venet, P. Lifetime Prediction of Lithium-Ion Capacitors Based on Accelerated Aging Tests. *Batteries* **2019**, *5*, 28. [CrossRef]
9. Lin, C.; Mu, H.; Xiong, R.; Shen, W.X. A novel multi-model probability battery state of charge estimation approach for electric vehicles using H-infinity algorithm. *Appl. Energy* **2016**, *166*, 76–83. [CrossRef]
10. Lipu, M.S.H.; Hannan, M.A.; Hussain, A.; Ayob, A.; Saad, M.H.M.; Karim, T.F.; How, D.N.T. Data-driven state of charge estimation of lithium-ion batteries: Algorithms, implementation factors, limitations and future trends. *J. Clean. Prod.* **2020**, *277*, 124110. [CrossRef]
11. Pan, H.H.; Lu, Z.Q.; Lin, W.L.; Li, J.Z.; Chen, L. State of charge estimation of lithium-ion batteries using a grey extended Kalman filter and a novel open-circuit voltage model. *Energy* **2017**, *138*, 764–775. [CrossRef]
12. El Ghossein, N.; Sari, A.; Venet, P. Development of a Capacitance versus Voltage Model for Lithium-Ion Capacitors. *Batteries* **2020**, *6*, 54. [CrossRef]
13. Zheng, Y.J.; Ouyang, M.G.; Han, X.B.; Lu, L.G.; Li, J.Q. Investigating the error sources of the online state of charge estimation methods for lithium-ion batteries in electric vehicles. *J. Power Sources* **2018**, *377*, 161–188. [CrossRef]
14. Liu, S.Q.; Wang, J.H.; Liu, Q.S.; Tang, J.; Liu, H.L.; Zhou, Y.; Pan, X.Y. A Novel Discharge Mode Identification Method for Series-Connected Battery Pack Online State-of-Charge Estimation Over A Wide Life Scale. *IEEE Trans. Power Electron.* **2021**, *36*, 326–341. [CrossRef]
15. He, H.W.; Zhang, X.W.; Xiong, R.; Xu, Y.L.; Guo, H.Q. Online model-based estimation of state-of-charge and open-circuit voltage of lithium-ion batteries in electric vehicles. *Energy* **2012**, *39*, 310–318. [CrossRef]
16. Li, Y.; Guo, H.; Qi, F.; Guo, Z.P.; Li, M.Y. Comparative Study of the Influence of Open Circuit Voltage Tests on State of Charge Online Estimation for Lithium-Ion Batteries. *IEEE Access* **2020**, *8*, 17535–17547. [CrossRef]
17. Wei, M.; Ye, M.; Li, J.B.; Wang, Q.; Xu, X.X. State of Charge Estimation of Lithium-Ion Batteries Using LSTM and NARX Neural Networks. *IEEE Access* **2020**, *8*, 189236–189245. [CrossRef]
18. Ma, Y.Y.; Wu, L.F.; Guan, Y.; Peng, Z. The capacity estimation and cycle life prediction of lithium-ion batteries using a new broad extreme learning machine approach. *J. Power Sources* **2020**, *476*, 228581. [CrossRef]
19. Li, J.B.; Ye, M.; Meng, W.; Xu, X.X.; Jiao, S.J. A Novel State of Charge Approach of Lithium Ion Battery Using Least Squares Support Vector Machine. *IEEE Access* **2020**, *8*, 195398–195410. [CrossRef]

20. Chen, L.; Chen, J.; Wang, H.M.; Wang, Y.J.; An, J.J.; Yang, R.; Pan, H.H. Remaining Useful Life Prediction of Battery Using a Novel Indicator and Framework With Fractional Grey Model and Unscented Particle Filter. *IEEE Trans. Power Electron.* **2020**, *35*, 5850–5859. [CrossRef]

21. Hussein, A.A. Derivation and Comparison of Open-Loop and Closed-Loop Neural Network Battery State-of-Charge Estimators. In Proceedings of the 7th International Conference on Applied Energy (ICAE), Abu Dhabi, United Arab Emirates, 28–31 March 2015; pp. 1856–1861.

22. Malkhandi, S. Fuzzy logic-based learning system and estimation of state-of-charge of lead-acid battery. *Eng. Appl. Artif. Intell.* **2006**, *19*, 479–485. [CrossRef]

23. Yang, F.F.; Song, X.B.; Xu, F.; Tsui, K.L. State-of-Charge Estimation of Lithium-Ion Batteries via Long Short-Term Memory Network. *IEEE Access* **2019**, *7*, 53792–53799. [CrossRef]

24. Xia, B.Z.; Cui, D.Y.; Sun, Z.; Lao, Z.Z.; Zhang, R.F.; Wang, W.; Sun, W.; Lai, Y.Z.; Wang, M.W. State of charge estimation of lithium-ion batteries using optimized Levenberg-Marquardt wavelet neural network. *Energy* **2018**, *153*, 694–705. [CrossRef]

25. Waag, W.; Fleischer, C.; Sauer, D.U. Critical review of the methods for monitoring of lithium-ion batteries in electric and hybrid vehicles. *J. Power Sources* **2014**, *258*, 321–339. [CrossRef]

26. Mousavi, G.S.M.; Nikdel, M. Various battery models for various simulation studies and applications. *Renew. Sustain. Energy Rev.* **2014**, *32*, 477–485. [CrossRef]

27. Zou, Y.; Hu, X.S.; Ma, H.M.; Li, S.E. Combined State of Charge and State of Health estimation over lithium-ion battery cell cycle lifespan for electric vehicles. *J. Power Sources* **2015**, *273*, 793–803. [CrossRef]

28. Seaman, A.; Dao, T.S.; McPhee, J. A survey of mathematics-based equivalent-circuit and electrochemical battery models for hybrid and electric vehicle simulation. *J. Power Sources* **2014**, *256*, 410–423. [CrossRef]

29. Yu, Y.X.; Mao, J.S.; Chen, X.X. Comparative analysis of internal and external characteristics of lead-acid battery and lithium-ion battery systems based on composite flow analysis. *Sci. Total Environ.* **2020**, *746*, 140763. [CrossRef]

30. Lourakis, M.I.A.; Argyros, A.A. Is Levenberg-Marquardt the Most Efficient Optimization Algorithm for Implementing Bundle Adjustment? In Proceedings of the 10th IEEE International Conference on Computer Vision (ICCV 2005), Beijing, China, 17–20 October 2005; pp. 1526–1531.

31. Hu, X.S.; Li, S.B.; Peng, H. A comparative study of equivalent circuit models for Li-ion batteries. *J. Power Sources* **2012**, *198*, 359–367. [CrossRef]

32. He, H.W.; Xiong, R.; Peng, J.K. Real-time estimation of battery state-of-charge with unscented Kalman filter and RTOS mu COS-II platform. *Appl. Energy* **2016**, *162*, 1410–1418. [CrossRef]

33. Chen, B.; Ma, H.D.; Fang, H.Z.; Fan, H.Z.; Luo, K.; Fan, B. An Approach for State of Charge Estimation of Li-ion Battery Based on Thevenin Equivalent Circuit model. In Proceedings of the Prognostics and System Health Management Conference (PHM-Hunan), Lab Sci & Technol Integrated Logist Support, Zhangjiajie, China, 24–27 August 2014; pp. 647–652.

34. Huo, Y.T.; Hu, W.; Li, Z.; Rao, Z.H. Research on parameter identification and state of charge estimation of improved equivalent circuit model of Li-ion battery based on temperature effects for battery thermal management. *Int. J. Energy Res.* **2020**, *44*, 11583–11596. [CrossRef]

35. He, H.W.; Qin, H.Z.; Sun, X.K.; Shui, Y.P. Comparison Study on the Battery SoC Estimation with EKF and UKF Algorithms. *Energies* **2013**, *6*, 5088–5100. [CrossRef]

36. Yang, Y.H.; Nagayama, T.; Xue, K. Structure system estimation under seismic excitation with an adaptive extended kalman filter. *J. Sound Vib.* **2020**, *489*, 115690. [CrossRef]

37. Xu, J.; Gao, M.; He, Z.; Han, Q.; Wang, X. State of Charge Estimation Online Based on EKF-Ah Method for Lithium-Ion Power Battery. In Proceedings of the 2nd International Congress on Image and Signal Processing, Tianjin, China, 17–19 October 2009; pp. 4120–4124.

38. Liu, Z.X.; Li, Z.; Zhang, J.B.; Su, L.S.; Ge, H. Accurate and Efficient Estimation of Lithium-Ion Battery State of Charge with Alternate Adaptive Extended Kalman Filter and Ampere-Hour Counting Methods. *Energies* **2019**, *12*, 757. [CrossRef]

39. He, L.; Hu, M.K.; Wei, Y.J.; Liu, B.J.; Shi, Q. State of charge estimation by finite difference extended Kalman filter with HPPC parameters identification. *Sci. China Technol. Sci.* **2020**, *63*, 410–421. [CrossRef]

40. Shim, B.J.; Park, K.S.; Koo, J.M.; Jin, S.H. Work and speed based engine operation condition analysis for new European driving cycle (NEDC). *J. Mech. Sci. Technol.* **2014**, *28*, 755–761. [CrossRef]

Review

Review of Energy Storage Capacitor Technology

Wenting Liu [1,2], Xianzhong Sun [1,2,3,4,*], Xinyu Yan [5], Yinghui Gao [1,2,3,4], Xiong Zhang [1,2,3,4], Kai Wang [1,2,3,4] and Yanwei Ma [1,2,3,4,*]

1 Institute of Electrical Engineering, Chinese Academy of Sciences, Beijing 100190, China; liuwenting23@mails.ucas.ac.cn (W.L.); gyh@mail.iee.ac.cn (Y.G.); zhangxiong@mail.iee.ac.cn (X.Z.); wangkai@mail.iee.ac.cn (K.W.)
2 University of Chinese Academy of Sciences, Beijing 100049, China
3 Key Laboratory of High Density Electromagnetic Power and Systems (Chinese Academy of Sciences), Haidian District, Beijing 100190, China
4 Shandong Key Laboratory of Advanced Electromagnetic Conversion Technology, Institute of Electrical Engineering and Advanced Electromagnetic Drive Technology, Qilu Zhongke, Jinan 250013, China
5 TBEA Sunoasis Co., Ltd., Xi'an 830011, China; 13649270712@163.com
* Correspondence: xzsun@mail.iee.ac.cn (X.S.); ywma@mail.iee.ac.cn (Y.M.)

Abstract: Capacitors exhibit exceptional power density, a vast operational temperature range, remarkable reliability, lightweight construction, and high efficiency, making them extensively utilized in the realm of energy storage. There exist two primary categories of energy storage capacitors: dielectric capacitors and supercapacitors. Dielectric capacitors encompass film capacitors, ceramic dielectric capacitors, and electrolytic capacitors, whereas supercapacitors can be further categorized into double-layer capacitors, pseudocapacitors, and hybrid capacitors. These capacitors exhibit diverse operational principles and performance characteristics, subsequently dictating their specific application scenarios. To make informed decisions in selecting capacitors for practical applications, a comprehensive knowledge of their structure and operational principles is imperative. Consequently, this review delved into the structure, working principles, and unique characteristics of the aforementioned capacitors, aiming to clarify the distinctions between dielectric capacitors, supercapacitors, and lithium-ion capacitors.

Keywords: film capacitors; ceramic dielectric capacitors; electrolytic capacitors; double-layer capacitors; pseudocapacitors; lithium-ion capacitors; structure; operational principles

Citation: Liu, W.; Sun, X.; Yan, X.; Gao, Y.; Zhang, X.; Wang, K.; Ma, Y. Review of Energy Storage Capacitor Technology. *Batteries* **2024**, *10*, 271. https://doi.org/10.3390/batteries10080271

Academic Editor: Johan E. ten Elshof

Received: 23 June 2024
Revised: 21 July 2024
Accepted: 25 July 2024
Published: 29 July 2024

1. Introduction

Due to global economic growth and expanding population, there has been a consistent and unwavering increase in the demand for energy [1]. The extensive exploitation of fossil fuels, however, has resulted in numerous challenges, including global warming and environmental pollution [2]. These issues not only pose a direct threat to human health but also inflict significant damage on our ecosystems. Furthermore, given the limited reserves of fossil fuels, the issue of energy scarcity is rapidly escalating [3]. Consequently, there is an urgent need to move away from traditional fossil fuels and explore renewable energy sources. However, renewable energy sources such as solar energy, wind energy, tidal energy, and geothermal energy are inherently intermittent and unstable, posing challenges to their utilization [4,5]. To enhance the utilization of renewable energy, it is imperative to transform it into other forms, primarily electricity, for storage. Consequently, the advancement of energy storage technology holds immense significance in optimizing energy structures, enhancing energy efficiency, safeguarding energy security, and fostering sustainable energy development.

For over two centuries, batteries have been extensively utilized for energy storage purposes and continue to be so today. In recent years, lithium-ion batteries with polymer solid-state electrolytes have received increasing attention due to their inherent safety and excellent thermal stability, promising large-scale applications [6]. Furthermore, various other technologies such as solid oxide fuel cells (SOFCs), electrochemical capacitors (ECs), superconducting magnetic energy storage (SMES) systems, flywheel energy storage systems, and dielectric capacitors are also commonly employed for storing energy [7]. Capacitors possess higher charging/discharging rates and faster response times compared with other energy storage technologies, effectively addressing issues related to discontinuous and uncontrollable renewable energy sources like wind and solar [3]. Furthermore, they can tackle challenges such as peak shaving, frequency regulation, and intelligent power supply within the power grid, thereby enhancing the efficiency of multi-energy coupling and promoting energy conservation and emission reduction [8]. With the rapid development of the electronics industry, capacitors have undergone an evolution from relatively primitive forms such as air-dielectric capacitors, mica-dielectric capacitors, and paper-dielectric capacitors to ceramic-dielectric capacitors and electrolytic capacitors [9]. The advent of diverse dielectric materials, especially organic media, combined with sophisticated manufacturing techniques, has led to a significant reduction in capacitors' overall size and a remarkable boost in performance. In recent years, the emergence of the double-layer theory has fueled the rise of supercapacitors, with double-layer supercapacitors and hybrid supercapacitors experiencing rapid growth and exhibiting promising applications [10]. The performance of different capacitors is shown in Table 1, and the comparison chart of energy density and power density for different capacitors is shown in Figure 1.

Table 1. Performance comparison of different capacitors.

Type	High Capacitance	High Voltage	Rate Performance	Cycle Stability	Cost	Polarity	Life Time	Main Purpose	Electrode Material
Film capacitor	✖	□	□	○	△	non-existent	□	improve frequency and absorb power supply noise	metal foil or metalized film
Aluminum electrolytic capacitor	○	○	△	△	□	exist	△	smoothing and decoupling	aluminum foil (cathode/anode)
tantalum electrolytic capacitor	○	△	○	△	△	exist	○	coupling and decoupling	Ta (anode) MnO_2 (cathode)
Ceramic capacitor	△	○	□	□	○	non-existent	□	coupling and decoupling	metals such as silver and copper
EDLC	□	✖	✖	△	✖	exist	△	backup power	carbon-based materials (anode/cathode)
Lithium-ion Capacitor	□	✖	✖	△	✖	exist	✖	backup power and energy storage applications	carbon-based materials (positive/negative electrode)

□: excellent; ○: good; △: not good; ✖: bad.

As new energy technology and capacitor energy storage continue to evolve, users may encounter numerous questions related to capacitors. To make informed decisions about their selection and usage, it is imperative to gain a comprehensive understanding of capacitors' structure and operating principles. Furthermore, there are some new researchers in the realm of capacitor energy storage who lack a thorough comprehension of capacitors' classification, structure, and operational mechanisms, which can readily lead to confusion regarding the various types of capacitors. Hence, this review endeavors to offer a comprehensive overview of the structures and operational principles of various capacitor types. Its objective is to illuminate the distinctions between dielectric capacitors, supercapacitors, and lithium-ion capacitors, thereby facilitating a thorough understanding of these capacitor categories among the readership.

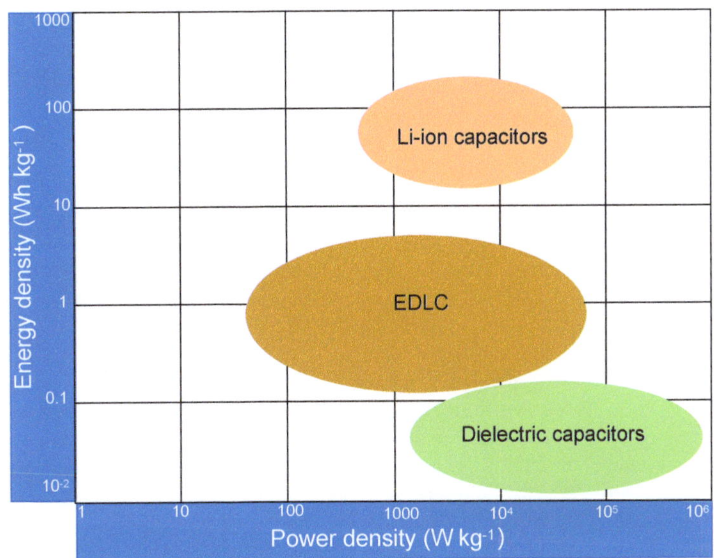

Figure 1. The comparison chart of energy density and power density for different capacitors. Reproduced from Ref. [11] with permission.

2. Dielectric Capacitor

The development of commercial dielectric capacitors can be traced back to 1876 when Fitzgerald invented the wax-impregnated paper dielectric capacitor equipped with foil electrodes [12]. This innovation was swiftly adopted in early radio-receiving equipment, significantly advancing radio communication technology. Subsequently, in 1909, William Dubilier introduced the mica dielectric capacitor, injecting new vitality into the field of radio transmission. Around 1926, capacitors based on titanium dioxide hit the market, further diversifying the types and enhancing the performance of capacitors. Then, in 1936, Cornell-Dubilier successfully commercialized aluminum electrolytic capacitors, launching a series of commercial products that marked a new era in the development of aluminum electrolytic capacitor technology. With the discovery of barium titanate in 1941, researchers embarked on the study of barium titanate-based dielectric capacitors, further advancing capacitor technology. During World War II, significant breakthroughs were made in capacitor manufacturing technology. Bosch leveraged lacquered paper and vacuum metallization techniques to mature the metalized paper capacitor, significantly enhancing its performance [13]. In 1954, Bell Labs successfully developed the first metalized polymer film capacitor, rejuvenating the capacitor industry [14]. From the 1970s to the 1980s, significant progress was made in the manufacture of multilayer ceramic capacitors (MLCCs) through the adoption of tape-casting technology and co-firing of ceramic electrodes. This resulted in a substantial increase in capacitance values and voltage withstand capabilities, vastly expanding the application areas of ceramic capacitors and making them indispensable components in numerous electronic devices. Currently, the primary types of dielectric capacitors include film capacitors, electrolytic capacitors, and ceramic capacitors [15].

The capacitance value of a capacitor is typically determined using Equation (1) [16]:

$$C = \frac{\varepsilon_0 \varepsilon_r A}{D} \tag{1}$$

where ε_r is the dielectric constant with respect to the electrolyte utilized, ε_0 is the permittivity of the vacuum, A is the surface area of the electrode material accessible to the electrolyte ions, and D is the effective thickness (charge separation distance) between the electrodes.

Therefore, the capacitance value can be increased by increasing the surface area of the electrodes and decreasing the distance between them.

In comparison to various electrical storage devices like batteries, dielectric capacitors possess the capability to discharge stored energy in an extremely brief timeframe (microseconds), resulting in the generation of substantial power pulses [17]. Their rapid charging and discharging rates render them ideally suited for high-power/pulse power systems, including medical defibrillators, pulsed lasers, power conditioning systems, and advanced electromagnetic emission systems [18–21]. Additionally, they are effective in harnessing energy from intermittent renewable sources [22]. However, the relatively low energy density of dielectric capacitors poses significant constraints on the miniaturization and lightweighting of equipment [23]. If the energy density of dielectric capacitors could be enhanced, it would lead to a substantial broadening of their application scope in the realm of energy storage.

The energy storage mechanism of a dielectric relies on its polarization process triggered by an electric field [24]. When an electric field is applied, the dielectric becomes polarized, leading to the accumulation of equal amounts of positive and negative charges on its surface. Consequently, energy is stored within the dielectric in the form of an electric field, as shown in Figure 2. The mechanism behind energy storage and release in dielectrics is elucidated through the electric displacement (D)-electric field (E) loop. As an electric field is applied, dielectrics become polarized due to the relative displacement of oppositely charged particles within their dipoles. Conversely, upon the removal of the electric field, depolarization occurs, causing the oppositely charged centers to tend toward overlap [25].

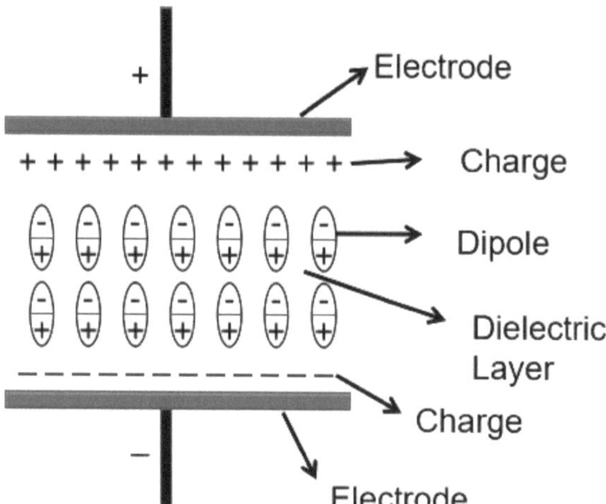

Figure 2. A schematic depiction of an electrically charged dielectric capacitor reveals how the charges of oriented electric dipoles, under the influence of a bias, contribute to binding the opposite charges at their respective electrode interfaces.

Dielectric capacitors can be categorized into several types, including film capacitors without electrolytes, electrolytic capacitors that utilize electrolytes, and ceramic capacitors. Film capacitors are made by depositing metal layers with different patterns on both sides of a thin film. While they can achieve voltages of tens of thousands of volts per unit, their capacitance is relatively small, necessitating parallel connection for high-power applications. Electrolytic capacitors can be classified into aluminum electrolytic capacitors (with an anode made of aluminum oxide, a separator of fibrous paper for insulation and electrolyte absorption, and an electrolyte typically composed of boric acid, ammonia water, and ethylene glycol) and tantalum electrolytic capacitors (with an anode made of tantalum pentoxide and a separator of fibrous paper). Typically, the voltage rating of a single unit is ≤100 V (low-voltage electrolytic capacitor) or ≥100 V (high-voltage electrolytic capacitor). Under high voltage conditions, they need to be used in series. Ceramic capacitors can be categorized into ceramic disc capacitors and multilayer ceramic capacitors. These capacitors are compact and cost-effective and possess excellent electrical properties, leading to their widespread application. In particular, high-capacity multilayer ceramic capacitors have been used to replace more expensive tantalum capacitors.

2.1. Film Capacitor

Film capacitors, alternatively known as plastic film capacitors, frequently employ metal foil as electrodes and plastic film as the dielectric. These capacitors often incorporate artificially synthesized polymer materials as dielectrics, enabling the selection of appropriate dielectric materials based on specific requirements, thereby exhibiting immense development potential. Recently, research in the field of film capacitors has focused on optimizing their dielectric properties. Meng et al. [26] proposed a hierarchical structural design approach to enhance the dielectric properties of metalized polypropylene (PP) film capacitors by utilizing terephthalaldehyde. This method offers insights into improving the dielectric properties of PP films by regulating charge transport behavior. Hu et al. [27] developed an ultra-high-temperature capacitor with relaxor ferroelectric (RFE) films by controlling polarization behavior. They further suggested that adjusting the intrinsic/extrinsic polarization ratio can enhance energy storage performance, providing a feasible approach to improving the high-temperature performance and dielectric strength of capacitors.

Film capacitors can be classified based on their structure, type of dielectric, and electrode formation method. To begin with, film capacitors are produced either in the form of winding utilizing a capacitor winding machine or as stacks of dielectric films [28]. These two distinct manufacturing methods are commonly referred to as coil technology and the stacking technique, respectively [29,30]. Therefore, based on their structure, film capacitors can be broadly classified into two types: "wound type" and "stacked type". The digital images are shown in Figure 3. The wound type involves winding and stamping polymer films and then inserting them into a shell. This type can further be divided into two winding methods: inductive (Figure 4a) and non-inductive (Figure 4b). The stacked type, on the other hand, involves stacking multiple layers of polymer films together and then inserting the stacked body into a shell (Figure 4c). Currently, wound-type film capacitors are more commonly used due to their ease of manufacturing.

Furthermore, film capacitors can be categorized into paper media and organic media based on their insulation materials. Paper dielectric capacitors are a type of wound capacitor that employs capacitor paper as the insulating medium and aluminum foil as the electrode. These capacitors consist of two or more layers of aluminum sheets interspersed with paper sheets. The paper sheets serve as the dielectric, whereas the aluminum sheets function as the capacitor electrodes, as shown in Figure 5a [31]. During the manufacturing process, paper and aluminum sheets are typically rolled into cylindrical shapes, with leads attached to both ends of the aluminum sheet, as depicted in Figure 5b. Subsequently, the entire cylindrical structure is coated with a protective layer of wax or plastic resin to safeguard it from exposure to moisture in the air.

Compared with traditional paper dielectric capacitors, the manufacturing process of metalized paper capacitors is more distinctive. It employs vacuum evaporation technology to deposit an ultra-thin and even layer of zinc or aluminum film onto the surface of the paper. Following this, the paper coated with this metal film is wound into a cylindrical structure, completing the overall fabrication of the metalized capacitor. Metalized paper capacitors feature a direct and thin coating of aluminum on paper, resulting in a thinner aluminum layer compared with traditional paper capacitors. This thinner layer contributes to a smaller capacitor size. Paper dielectric capacitors offer a diverse range of capacitance and operating voltage, along with a straightforward manufacturing process, low cost, and ease of metallization. Owing to these attributes, they are commonly utilized in high-voltage and high-current applications.

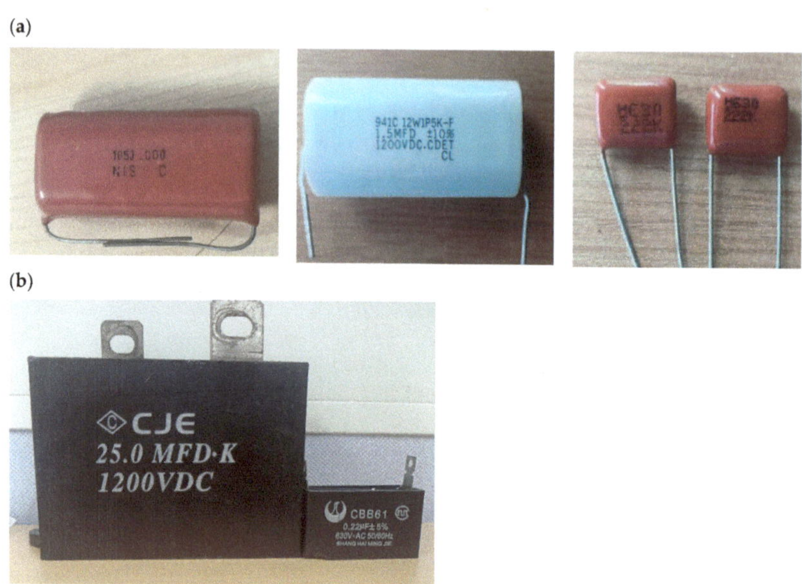

Figure 3. The digital images of film capacitors. (**a**) Wound-type film capacitors. (**b**) Stacked-type film capacitors. The capacitor on the left is used for DC circuits, while the one on the right is used for AC circuits.

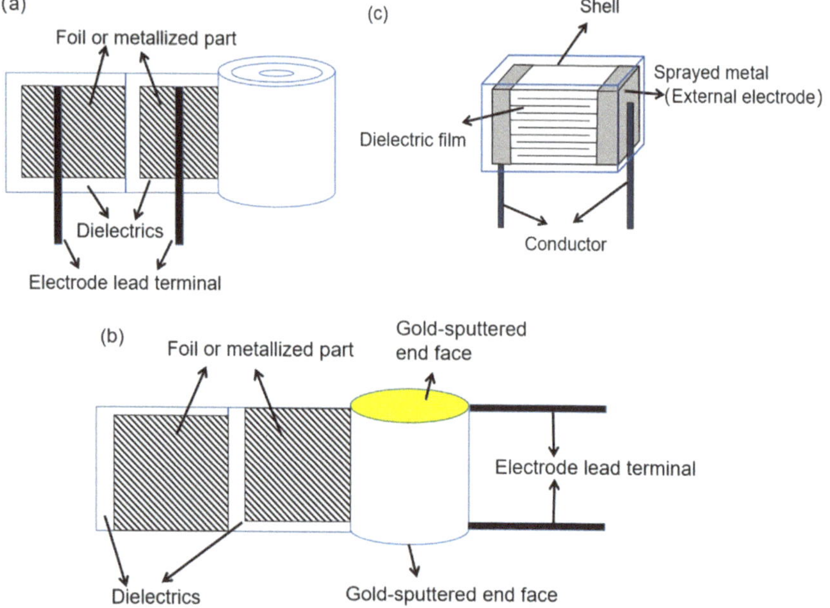

Figure 4. The structural diagram of film capacitors. (**a**) Inductive winding method. (**b**) Non-inductive winding method. (**c**) Stacked film capacitor.

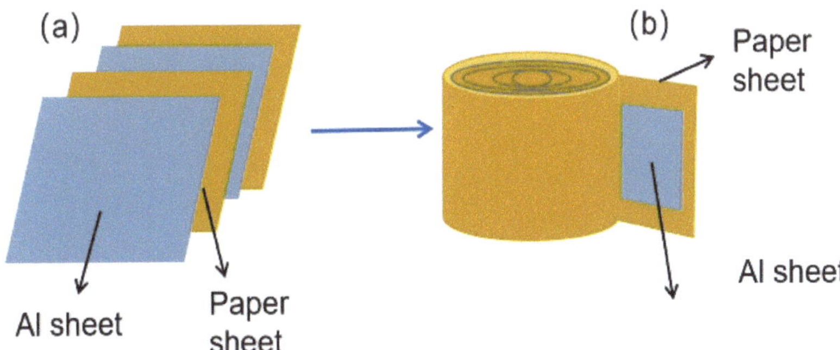

Figure 5. (**a**) Assembly of a paper sheet capacitor. (**b**) The process of rolling up a paper sheet capacitor into a cylinder. Reproduced from Ref. [31] with permission.

Organic dielectric capacitors primarily consist of synthetic organic thin films serving as dielectric materials and are typically constructed in a wound structure. Common dielectric materials employed in these capacitors include polypropylene (PP), polyethylene terephthalate (PET), polyethylene naphthalate (PEN), polyphenylene sulfide (PPS), and polycarbonate (PC) [28,32–34]. Ran et al. [35] improved the micro-morphology of PP films by utilizing an organic crystallization accelerator with good dispersion. The film capacitors produced using this method exhibit excellent breakdown strength. Ping et al. [36] prepared a double-layer polymer film by solution casting on the surface of PET film and coating it with boron nitride nanosheets. The resulting film capacitor exhibits excellent breakdown strength (736 MV m^{-1}) and high energy density (8.77 J cm^{-3}). Zou et al. [37] developed a high-performance thin-film capacitor through the controlled deposition of Si_3N_4 on PEN. This capacitor film possesses excellent mechanical properties, making it a promising candidate for power converters in electric vehicles. Zhang et al. [38] designed and prepared a P-3-P sandwich structured film with PC as the outer insulation layer and 3 wt% TiO_2– PC/polyvinylidene fluoride as the middle polarization layer using solution casting and hot-pressing processes. This film exhibits excellent charge-discharge characteristics, offering a promising possibility for the construction of high-energy storage film capacitors.

In comparison to inorganic dielectric capacitors, organic dielectric capacitors primarily utilize polymer materials as dielectrics, benefiting from the abundance of raw materials available. Additionally, the thickness of the films can be made exceptionally thin. Nevertheless, organic media are susceptible to aging and exhibit limited heat resistance, potentially compromising the capacitors' performance [39,40]. The characteristics of film capacitors vary significantly depending on the type of dielectric medium used, resulting in diverse application fields. Organic dielectric capacitors can be classified into two electrode types: metal foil electrodes (the foils are typically on the order of 6 µm in thickness) and metalized electrodes (the metalized layers are <100 nm thick) [41]. The organic dielectric capacitor of a metal foil electrode is made of two layers of plastic film or sheet. Each layer is interspersed with thin aluminum metal foil or sheet, serving as the electrode. Subsequently, the plastic sheets and aluminum sheets are rolled into a cylindrical jelly roll structure. To establish electrical connections, wire leads are attached to both ends of the aluminum sheets, typically through soldering or metal spraying techniques. The thickness of the plastic film determines the separation distance within the capacitor, while the operating area is dictated by the size of the electrodes. In metalized film organic dielectric capacitors, the aluminum sheets or foils are superseded by metal layers that are vacuum-deposited onto the thin film layer. These metal layers possess a thickness of approximately 1/100 of that of the metal foil, making them significantly thinner and more conducive to space conservation [31]. The most frequently utilized metal layer is composed of ultra-thin aluminum. The dielectric is typically formed by a plastic film layer made of synthetic materials, while the electrode is comprised of an aluminum layer [42]. The schematic diagram of the structure of organic dielectric capacitors is shown in Figure 6.

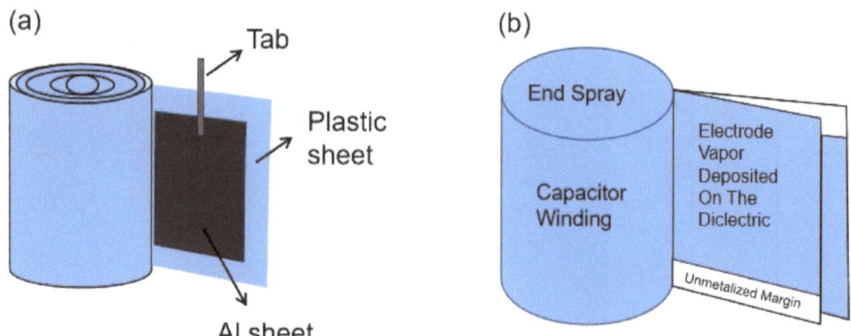

Figure 6. (**a**) Foil-type organic dielectric capacitor. (**b**) The capacitor with metalized electrodes, in which the electrodes are typically very thin aluminum layers deposited by vapor deposition onto an organic dielectric.

Finally, film capacitors can be categorized into two types: foil-type film capacitors and metalized film capacitors, depending on their distinct electrode formation techniques. The foil-type film capacitors represent the earliest incarnation of wound capacitors. Typically, they are crafted with clamping aluminum foil, which possesses exceptional ductility, within an insulating medium and then winding it, as shown in Figure 7. Conversely, the electrode of a metalized film capacitor eschews the use of metal foil. Instead, an ultra-thin metal film is deposited onto the capacitor through the process of vacuum evaporation, as shown in Figure 8. Compared with metal foil electrode film capacitors, metalized film capacitors have the obvious advantage of self-healing. Self-healing refers to the phenomenon that when a metalized capacitor experiences breakdown due to defects in the dielectric, an arc current is immediately generated at the breakdown point, and this current density is concentrated at the center of the breakdown point [43,44]. Due to the thin metallization film (less than 0.1 µm), the heat generated by this current is sufficient to melt and evaporate the metal near the breakdown point, forming a metal-free zone around the breakdown area, restoring insulation between the two electrodes of the capacitor, thus restoring normal operation of the capacitor. The self-healing properties of metalized film capacitors enable them to be applied in more complex situations.

(**a**) (**b**) (**c**)

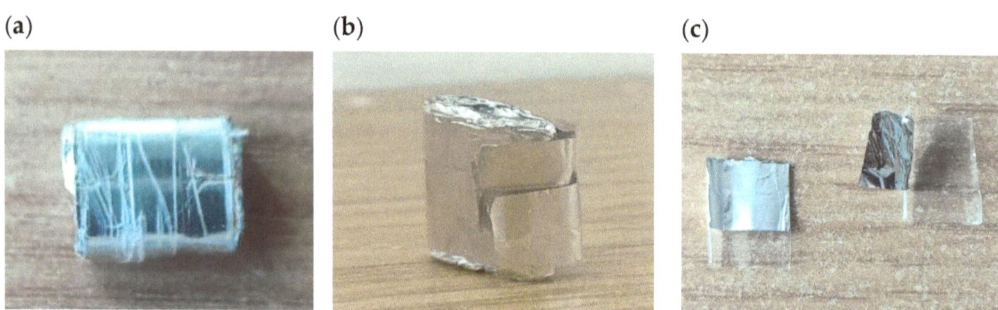

Figure 7. (**a**,**b**) Overall structural diagram of a disassembled foil-type film capacitor. (**c**) The aluminum foil is sandwiched between the plastic films.

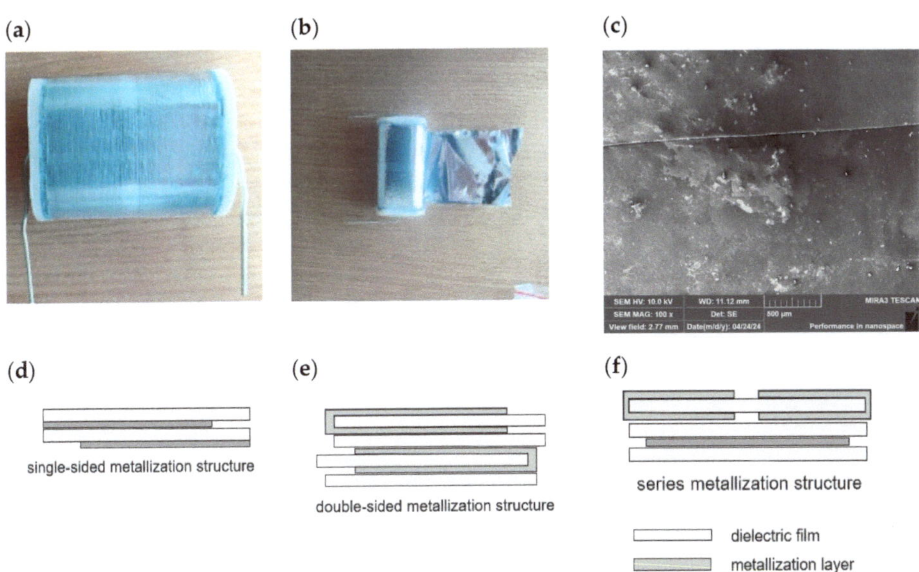

Figure 8. (**a,b**) Disassembled metalized film capacitor. (**c**) Placing the dismantled metalized film under a scanning electron microscope (SEM) reveals the deposition of aluminum on the film. (**d**) Single-sided metallization structure. (**e**) Double-sided metallization structure. (**f**) Series metallization structure.

2.2. Electrolytic Capacitor

Electrolytic capacitors are capacitors that exist in two forms: non-polar and polar. The anode of these capacitors typically comprises metal foil, such as aluminum or tantalum, with an oxide film, often aluminum oxide or tantalum pentoxide, serving as the dielectric and adhering closely to the anode. The cathode, on the other hand, consists of a combination of conductive materials, electrolytes (which can be either liquid or solid), and additional materials. The naming of electrolytic capacitors is derived from the electrolyte, which forms the principal component of the cathode. Notably, most electrolytic capacitors exhibit polarity, necessitating the application of voltage with the appropriate polarity. In the event of a reversed connection or incorrect polarity, the capacitor can undergo a short circuit, leading to a significant current flow that may result in permanent damage to the capacitor [31]. Electrolytic capacitors feature a thin dielectric layer, an extensive positive electrode area, and, consequently, a high capacitance per unit volume. This allows them to often boast higher capacitance values compared with other dielectric capacitors. However, they also exhibit a significant leakage current and a relatively short lifespan. When utilizing electrolytic capacitors, it is crucial to adhere to polarity requirements, ensuring that the positive and negative poles are connected correctly to avoid any mishaps.

In recent years, research on electrolytic capacitors has primarily concentrated on electrode materials and production processes. Bai et al. [45] utilized a protective atmosphere sintering process to produce sintered foils with added starch, evaluating the impact of starch addition within the range of 0–50 volume percent on the specific capacitance and anti-buckling performance of the sintered foils. This work unveiled the potential of starch addition in optimizing the properties of sintered foils, providing a valuable reference for developing an advanced powder metallurgy preparation process. Zeng et al. [46] employed additive manufacturing technology to produce anode foils, systematically investigating the influence of aluminum particle size and sintering temperature on the electrical properties of the anode foils. They also discussed the reaction mechanisms during sintering and the inherent relationship between the microstructure and electrical properties of the anode foils. Chen et al. [47] utilized metallic glass (MG) as a binder to adhere Ta powder at low temperatures (513 K), yielding MG-Ta composites. When applied in tantalum electrolytic capacitors, these composites exhibited a 57% increase in specific capacitance compared with pure Ta materials, accompanied by a 32% enhancement in mechanical properties.

2.2.1. Aluminum Electrolytic Capacitors

Aluminum electrolytic capacitors (AECs) offer a superior cost-to-energy ratio and volume efficiency compared with various other capacitor types [48]. As a result, they are frequently employed at the dc-link of power electronic converters (PECs) to serve as an energy buffer [49]. The physical structure and detailed structure of AEC are shown in Figure 9a. It comprises two aluminum electrodes, with a thin oxide film layer (known as alumina) serving as the dielectric [50]. Additionally, it includes a paper separator and an electrolyte, which is a blend of solvents and additives (shown in Figure 9b) [51].

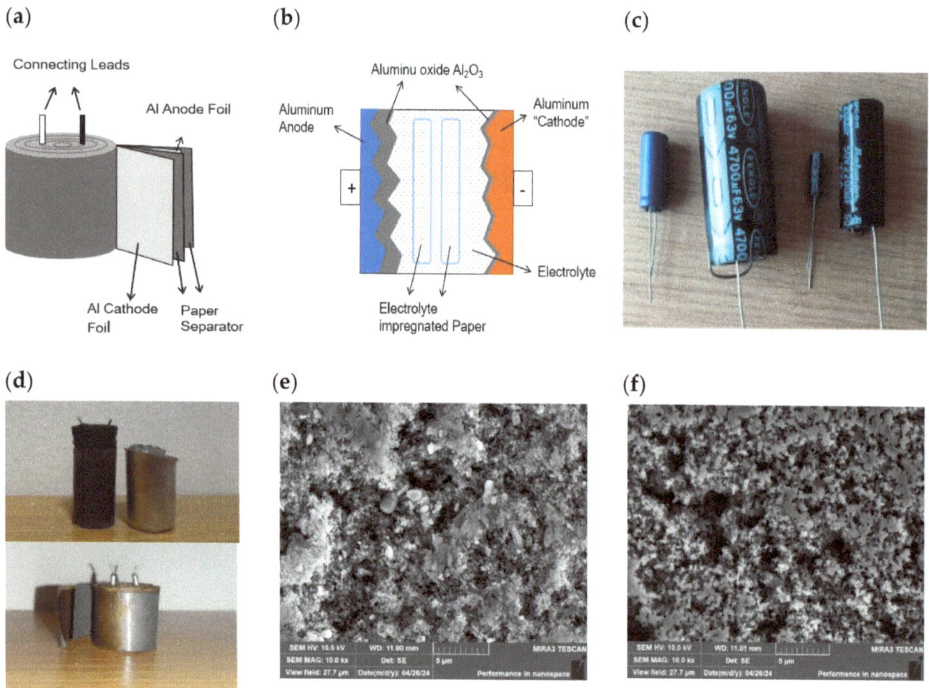

Figure 9. (**a**) Physical structure of ACE. Reproduced from Ref. [48] with permission. (**b**) A detailed structural diagram of the AEC, which consists of etched aluminum electrodes, alumina dielectrics, electrolyte carriers, and paper sheets impregnated with the electrolyte. Reproduced from Ref. [50] with permission. (**c**) Digital images of aluminum electrolytic capacitors. (**d**) The dismantled aluminum electrolytic capacitor. (**e**) The image of the anode aluminum foil of the aluminum electrolytic capacitor under a scanning electron microscope reveals that the anode aluminum foil is coated with aluminum oxide. (**f**) The image of the cathode aluminum foil of the aluminum electrolytic capacitor under a scanning electron microscope reveals that it is an etched aluminum foil.

In aluminum electrolytic capacitors, both the anode and cathode consist of pure aluminum foil. The anode foil is coated with a thin layer of aluminum oxide, electrically insulating in nature, serving as the dielectric. This oxide coating, along with the cathodes, is separated by electrolytic paper soaked in an electrolyte solution. This paper, possessing high porosity, can absorb and hold the maximum amount of electrolyte. While the cathode aluminum foil also possesses a naturally formed, very thin insulating oxide layer due to exposure to air, its thickness is significantly thinner compared with the oxide layer on the anode foil [52]. The electrolyte serves as the actual cathode of the capacitor when compared to the anode foil, while the cathode foil merely collects current from the electrolyte [48]. Figure 9c shows digital images of aluminum electrolytic capacitors, and the disassembled structure of one of the aluminum electrolytic capacitors is depicted in Figure 9d. After soaking the anode and cathode aluminum foils of the disassembled aluminum electrolytic capacitor in ionized water and allowing them to dry, they were observed under a scanning electron microscope. It can be observed that the anode aluminum foil is coated with aluminum oxide (as shown in Figure 9e), while the cathode aluminum foil is etched aluminum foil (as shown in Figure 9f).

Before forming a layer, the anode foil needs to undergo other processing. Initially, it is etched to enlarge its surface area, enabling better performance [53]. Subsequently, anodization is performed by applying a direct current voltage, resulting in the formation of a thin layer of aluminum oxide on the anode foil's surface. This oxide layer functions as the dielectric in AEC. Conversely, the thin layer of Al_2O_3 present on the cathode results from the natural oxidation of aluminum, which effectively mitigates corrosion.

AEC is a type of polarized capacitor that can only be subjected to a DC bias in one direction. Its electrochemical structure is shown in Figure 10 [48]. The dissociation of water molecules in the electrolyte produces proton (H^+) and hydroxyl (OH^-) ions, which can freely exist in the electrolyte. When a positive electric field is applied to the AEC terminal, staying within the rated voltage, the OH^- ions are drawn toward the aluminum anode foil. However, they encounter a dense dielectric barrier that prevents their passage. Conversely, if the applied negative voltage surpasses the potential hurdle posed by the aluminum oxide formed during passivation, the H^+ ions are drawn toward the anode foil. Given their significantly smaller size compared with OH^- ions, the H^+ ions can effortlessly penetrate the dielectric and arrive at the anode, resulting in the production of hydrogen gas. Meanwhile, the OH^- ions migrate toward the cathode and unite with aluminum (Al^{3+}) ions, leading to the formation of aluminum oxide on the cathode foil. In this case, the cathode capacitance will decrease, thereby reducing the total capacitance. If voltage is applied for a long time, it will damage the capacitor, so the polarity of ACE cannot be reversed.

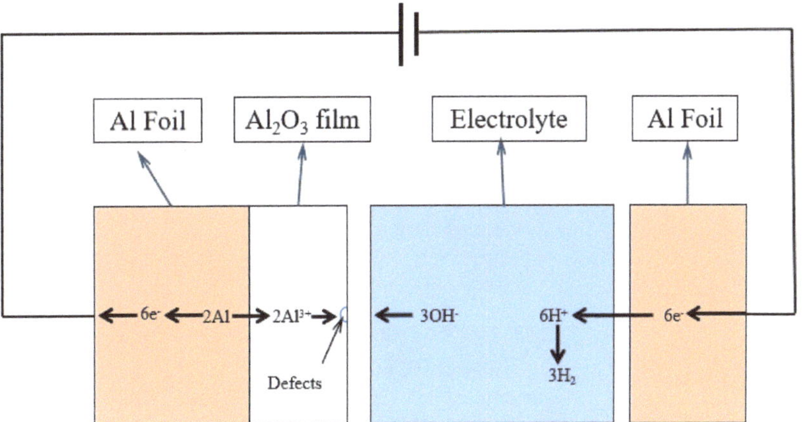

Figure 10. Electrochemical structure of AEC. Reproduced from Ref. [48] with permission.

The manufacturing process of aluminum electrolytic capacitors primarily comprises the following steps, as depicted in Figure 11 [50,54]:

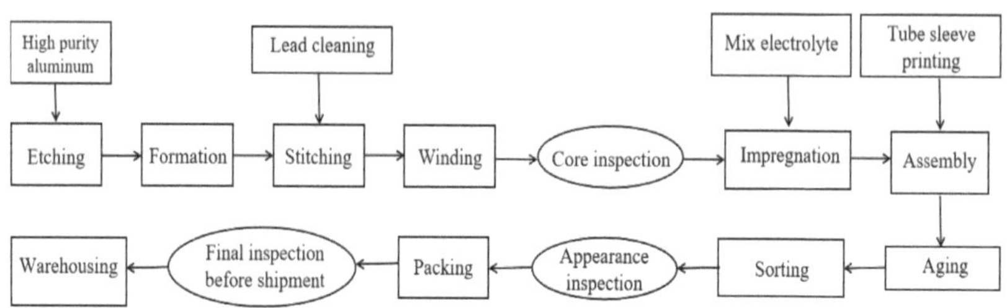

Figure 11. Manufacturing process diagram of aluminum electrolytic capacitors.

(1) Etching: High-purity aluminum foil undergoes an etching process through an electrochemical method in a chloride solution, utilizing either direct or alternating current. The anode and cathode foils are crafted from virtually pure aluminum foil. To enhance their effective surface area and minimize the size of the capacitor, anode foils ranging from 0.05 to 0.11 mm in thickness and cathode foils measuring 0.02 to 0.05 mm thick are continuously subjected to electrochemical etching in a chloride solution, utilizing either alternating current or direct current. Typically, AC electrolysis is employed for the production of low-voltage capacitors, whereas DC electrolysis is utilized for the fabrication of medium- and high-voltage capacitors.

(2) Formation: Through electrolysis, a continuous voltage exceeding the nominal value is applied, resulting in the formation of an aluminum oxide layer on the surface of the aluminum foil. The thickness of the alumina dielectric film can be controlled.

(3) Slitting: After etching and anodizing the aluminum foil roll, the foil is cut into a specified width according to the size of the capacitor shell.

(4) Winding: Secure the lead-out wires of the anode and cathode foils using rivets or welding and position the foils between isolation plates. Employ a winding machine to neatly wind them together, creating a capacitor core package.

(5) Impregnation: Soak the capacitor core with electrolyte to saturate the paper isolation layer and all parts of the corroded aluminum foil to ensure good contact between the oxide layer and the true cathode. This method requires the removal of gas from the core package and vacuum immersion of the electrolyte.

(6) Assembly: To prevent evaporation or moisture absorption of the electrolyte, which can lead to deterioration, it is imperative to insert the capacitor core into a metal casing and securely seal it. Furthermore, to safeguard against the potential for electrolytic capacitor explosion due to excessive gas pressure during faults, a pressure relief device must be integrated.

(7) Aging: Repair the oxide film that may be damaged during the manufacturing process, especially during cutting and assembly, by applying a DC voltage.

(8) Inspection: After sealing, inspect the product for capacitance, leakage current, appearance, and performance as required, and then proceed with packaging.

2.2.2. Tantalum Electrolytic Capacitor

After aluminum electrolytic capacitors gained widespread use, issues such as limited lifespan and inadequate high-temperature resistance became apparent, prompting the development of tantalum electrolytic capacitors. These capacitors, similar to other electrolytic types, consist of an anode, electrolyte, and cathode. The cathode can be either solid or liquid, but currently, the majority of tantalum electrolytic capacitors available on the market are of the solid variety.

Solid tantalum electrolytic capacitors are composed exclusively of stable, inorganic, and nonvolatile materials devoid of any water or other liquids. This composition endows them with numerous advantages, including compact size, excellent temperature characteristics, the elimination of sealing requirements, and an extended service life [55,56]. However, solid electrolytes have poor productivity and high costs, and the capacity achievement rate during use is generally poor [57].

Distinct from aluminum electrolytic capacitors, solid tantalum electrolytic capacitors employ tantalum powder sintered into porous tantalum blocks as the anode. The surface of these porous tantalum blocks is then oxidized to create an insulating medium composed of tantalum pentoxide [58]. The cathode, on the other hand, consists of manganese dioxide, which is in intimate contact with the tantalum pentoxide. The tantalum electrolytic capacitor is completed by drawing out the electrode. Notably, solid tantalum electrolytic capacitors are polarized capacitors, necessitating their use in a unipolar state; reverse polarity is strictly prohibited. The digital image and structure of this capacitor are illustrated in Figure 12.

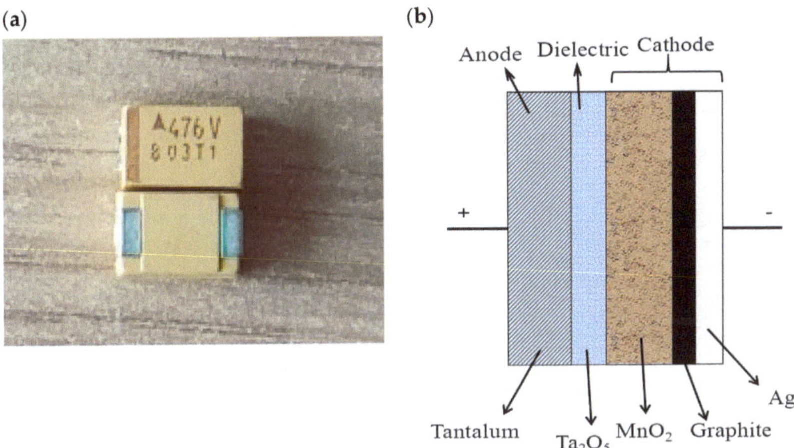

Figure 12. (**a**) Digital images of tantalum electrolytic capacitors. (**b**) Schematic diagram of tantalum electrolytic capacitor structure.

The production of solid tantalum electrolytic capacitors mainly involves the following steps (Figure 13) [31,59,60]:

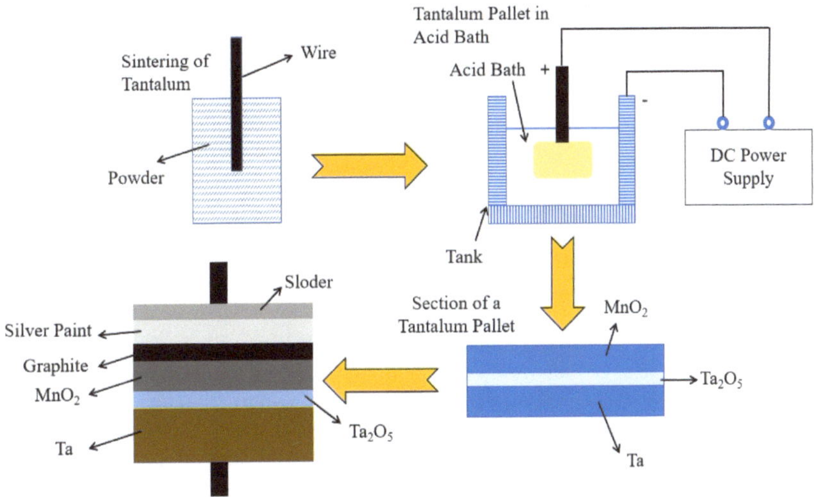

Figure 13. Fabrication process of solid electrolyte capacitor. Reproduced from Ref. [31] with permission.

(1) The tantalum metal is crushed into a fine powder and thoroughly mixed with organic solvents. This mixture is then pressed into a desired shape under pressure, with tantalum leads embedded within. Subsequently, the assembly is sintered in a vacuum high-temperature environment, transforming it into a sponge-like structure. This process creates a highly porous metal anode, which significantly enhances its capacitance value. At the same time, it is truly integrated with the lead wire.

(2) The sponge-like tantalum is submerged in a phosphoric acid solution for electrolysis. Through the process of oxidation, tantalum pentoxide is formed on its surface. This anode is then further coated with an insulating oxide layer, specifically tantalum pentoxide, serving as the dielectric layer. This comprehensive treatment process is referred to as anodizing.

(3) Liquid manganese nitrate is added to the tantalum blocks, followed by thermally decomposing them in an environment containing water vapor and a catalyst. This process results in the production of manganese dioxide. Due to the excellent adsorption properties of manganese nitrate, the generated manganese dioxide is able to be fully adsorbed into the numerous tiny pores within the sponge-like tantalum block. Alternatively, if a solid polymer with a lower melting point is utilized, it can be melted and directly placed into the small pores.

(4) Finally, silver powder and graphite are coated on the surface of manganese dioxide to reduce its equivalent resistance and enhance its conductivity. At the same time, external leads are added and packaged with epoxy resin.

In addition, there is another type of tantalum capacitor called a wet tantalum electrolytic capacitor, which uses a liquid electrolyte (usually sulfuric acid) instead of a solid electrolyte. The initial steps of anode production and dielectric deposition for wet tantalum electrolytic capacitors mirror those employed for solid tantalum capacitors. Subsequently, the dielectric-coated anode is submerged in a cage filled with the electrolyte solution. This cage, along with the electrolyte solution, collectively fulfills the role of the cathode in wet tantalum capacitors [31]. Wet tantalum electrolytic capacitors can be used at high temperatures and high ripple currents and are generally used in military and aerospace fields [61].

Although electrolytic capacitors share the self-healing ability with metalized thin film capacitors, the underlying mechanisms differ. In non-solid electrolytic capacitors, if the anodic oxide film sustains partial damage during operation or storage, the electrolyte serving as the cathode comes into play. Under the influence of the applied voltage, the non-solid electrolyte releases oxygen, which regenerates the oxide film at the damaged spot, thereby restoring its functionality. On the other hand, solid tantalum electrolytic capacitors may encounter cracks or metal impurities in the tantalum pentoxide film, resulting in an elevated leakage current. In this case, due to the high current density and temperature in the defective region, MnO_2 can locally transform into manganese oxides with high resistance, such as Mn_2O_3 and Mn_3O_4, thereby isolating the breakdown point and preventing capacitor failure [62]. This process effectively repairs the defect, exhibiting the self-healing property of these capacitors.

2.3. Ceramic Capacitors

In ceramic capacitors, ceramic materials serve as dielectrics, while conductive metals function as electrodes. These capacitors are extensively employed in electronic devices because of their abundant raw materials, uncomplicated design, affordable price, and vast range of electrical capacity. Ceramic capacitors come in diverse types, primarily categorized as Class I and Class II, based on the characteristics of the dielectric materials employed [63]. Class I ceramic capacitors, commonly referred to as high-frequency ceramic capacitors, exhibit low dielectric loss, high insulation resistance, and a linear variation in dielectric constant with temperature. These capacitors are ideal for resonant circuits, filters, and temperature compensation. On the other hand, Class II ceramic capacitors, also known as low-frequency ceramic capacitors or ferroelectric ceramic capacitors, utilize ferroelectric ceramics as their dielectric. They offer higher specific capacitance, non-linear capacitance changes with temperature, and increased losses. Therefore, they are commonly utilized for bypass or coupling applications in electronic devices [64]. In addition, according to their shape and structure, they can be divided into ceramic disk capacitors and multilayer ceramic capacitors (MLCCs).

Recently, research on ceramic capacitors has primarily focused on enhancing the performance of MLCC. Yong et al. [65] proposed a method of applying polydopamine (PDA) as a robust coating layer for MLCC. The barium titanate particles treated with PDA exhibited improved dispersion stability, preventing particle reagglomeration and contributing to the enhancement of the mechanical properties of MLCC. Lv et al. [66] developed a novel lead-free MLCC composed of $NaNbO_3-(Bi_{0.5}Na_{0.5})TiO_3-Bi(Mg_{0.5}Hf_{0.5})O_3$, which demonstrated exceptional performance, including a recoverable energy density of up to 12.65 J cm^{-3} and an energy efficiency of 88.5%.

2.3.1. Ceramic Disc Capacitors

Ceramic disc capacitors are constructed by applying silver contacts to both faces of a ceramic disc (shown in Figure 14a). Here, the ceramic disc serves as the dielectric material, while the silver coating on both sides functions as the electrodes of the capacitor. The leads, made of copper, are welded to the ceramic disc to establish electrical connections. Figure 14b shows actual images of ceramic disc capacitors. When its cross-section is placed under a scanning electron microscope, the observed images are shown in Figure 14c,d.

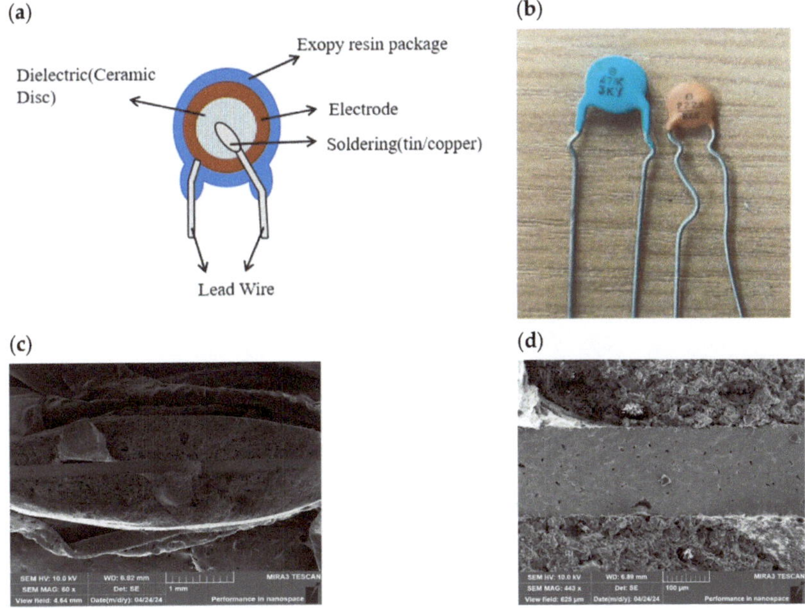

Figure 14. (**a**) Ceramic disc-type capacitor. (**b**) The actual images of ceramic disc capacitors. (**c**) The cross-section of the ceramic disc capacitor observed under a scanning electron microscope, which shows that the leads are bonded to the electrodes coated on both sides of the ceramic. (**d**) Enlarged cross-sectional view of a ceramic capacitor. In the middle is the ceramic dielectric, with electrodes coated on both sides and an outer layer of protective material.

Ceramic disc capacitors are extensively utilized in general electronic circuits due to their cost-effectiveness and ease of soldering. The capacitance of these capacitors is determined by the area of the ceramic disk or dielectric, as well as the spacing between the silver electrodes. For disc capacitors with lower capacitance, a single ceramic disc coated with silver contacts is sufficient, while for high capacitance ceramic disc capacitors, multilayer ceramic discs are required [64]. The preparation of ceramic discs is similar to traditional ceramic processing. Initially, the dielectric powder undergoes ball milling to disaggregate it. Subsequently, an organic binder is added to prepare the powder for the forming process. After spray drying, the powder is fed into an automatic granulator to shape it into a raw disk. Next, two firing processes are carried out: the first to burn out the binder at a temperature below 550 °C, removing organic components, and the second sintering process at over 1000 °C to consolidate the ceramic particles into a solid body. Then, electrodes are applied to both sides of the disc using a paste containing metal particles (such as silver) and glass frit, followed by heat treatment. During this treatment, the glass material melts, ensuring strong adhesion of the metal electrode to the ceramic surface at a relatively low processing temperature. Finally, after attaching the leads to the electrodes, an epoxy or phenolic coating is applied to safeguard the ceramics from environmental contaminants [64].

2.3.2. Multilayer Ceramic Capacitor

The multilayer ceramic capacitor (MLCC) stands as a pivotal passive surface mount component in contemporary electronic devices. Its genesis dates back to the 1960s when American companies pioneered its successful development. Subsequently, Japanese companies like Murata, TDK, and Sunpower spearheaded its rapid industrialization and refinement. These companies continue to uphold their global supremacy in the MLCC industry, exemplified by their proficiency in manufacturing MLCCs that boast high reliability, precision, integration, and frequency, along with intelligence, low power consumption, immense capacity, miniaturization, and cost-effectiveness.

The MLCC is fabricated by stacking multiple layers of ceramic material interspersed with conductive electrodes [67,68]. Each layer of ceramic material sandwiches the electrodes, serving as the dielectric for the capacitor. These multilayer ceramic media and electrodes are interconnected through the terminal's surface, creating a compact and efficient structure. In other words, the MLCC is constructed by alternately layering ceramic dielectric membranes (commonly rutile titanium dioxide or barium titanate) with printed electrodes (inner electrodes) in a staggered configuration. This assembly is then consolidated into a ceramic chip through a single high-temperature sintering process. Subsequently, a metal layer (outer electrode) is applied to both ends of the chip, completing the manufacturing process. Its structure is shown in Figure 15a. The MLCC boasts not only the benefits of standard ceramic dielectric capacitors but also exhibits a range of exceptional characteristics. These include compact size, substantial capacity, high mechanical strength, excellent moisture resistance, outstanding high-frequency performance, and remarkable reliability [69,70]. Given its versatility and superior performance, the MLCC finds widespread application in various electronic information fields, encompassing mobile phones, computers, the military industry, aerospace, and beyond [71,72]. The packaging forms of MLCC mainly include wire-bonded and surface-mount types. The wire-bonded type was previously known as the monolithic capacitor (shown in Figure 15b). Its name originated from the fact that the ceramic dielectric body coated with metal electrodes is sintered with the electrodes into a single unit, resembling a stone block, hence the name "monolithic capacitor." Removing the leads from the monolithic capacitor results in the surface-mount packaging form (shown in Figure 15c). The observed images of the surface-mount type MLCC placed under a scanning electron microscope are shown in Figure 15d–g.

The production of surface-mount MLCC encompasses multiple processes, including the formation of ceramic dielectric films, the fabrication of inner electrodes, the creation of capacitor chips, sintering them into ceramics, the construction of outer electrodes, performance testing, and packaging, among others [64,73,74]. The specific process diagram is shown in Figure 16. Firstly, the dielectric powder is mixed with solvents, dispersants, binders, and plasticizers to form a homogeneous suspension. This slurry is then transferred to a casting machine, where a scraper evenly applies it onto a tape. The tape, coated with the mixed slurry, is sent to a tape-casting machine where a doctor blade precisely scrapes the slurry onto it. The resulting wet sheet is subsequently dried, transforming into a flexible tape. Next, an electrode paste is applied to the tape in a designed pattern through screen printing. During the lamination process, the printed tapes are stacked layer by layer with precision alignment. After cutting and separation, each green MLCC undergoes high-temperature sintering to eliminate the adhesive and organic matter. This sintering process also solidifies the layers of dielectric tape and electrodes, creating a dense structure. Additionally, the sintered ceramic is chamfered to fully expose the inner electrodes.

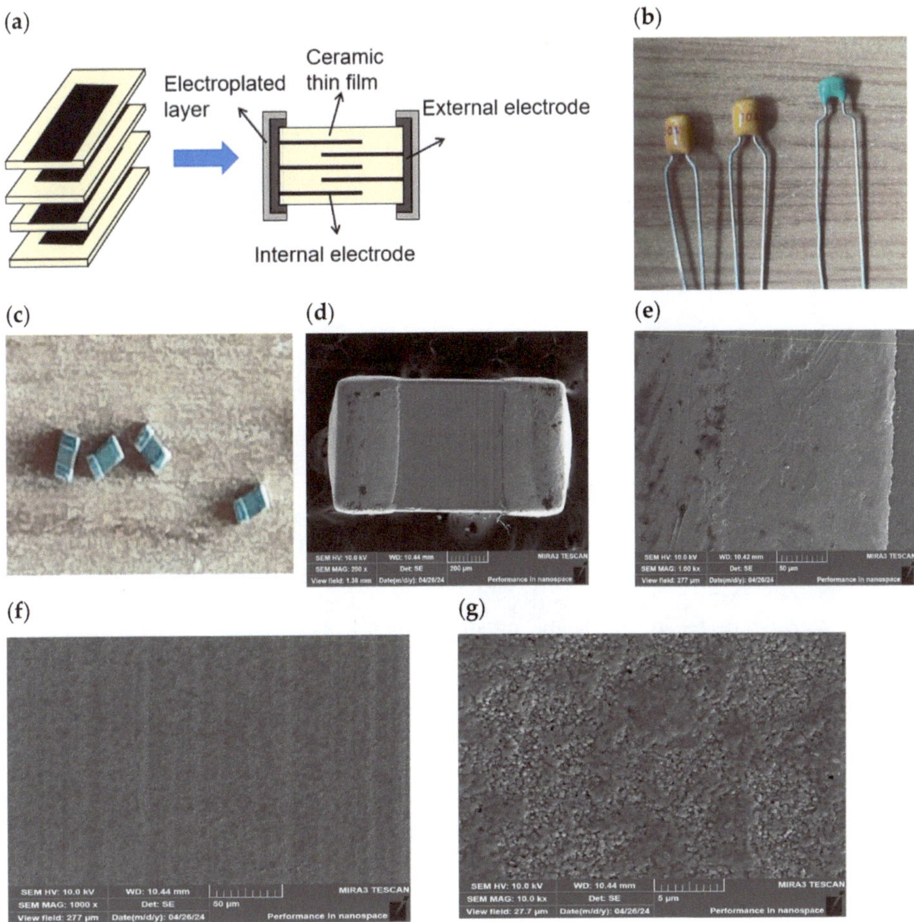

Figure 15. (**a**) Schematic diagram of MLCC structure. (**b**) The actual image of the monolithic ceramic capacitor. (**c**) The actual image of the surface-mount packaging form. (**d**) The overall structural diagram of the surface-mount type MLCC under a scanning electron microscope, which shows that MLCC is constructed by stacking ceramic dielectric sheets with printed electrodes in a staggered manner to form a layered structure. (**e**) The external electrode of MLCC, which is composed of an electroplated metal layer. (**f**) The internal electrode of MLCC, which is composed of metallic materials. (**g**) The dielectric of MLCC, which is composed of ceramic materials.

To form the outer electrode, electrode paste is applied to the exposed inner electrodes, connecting the inner electrodes on the same side. Notably, the surface-mount MLCC achieves electrical connection through three consecutive layers of termination electrodes [75]. The first layer, in contact with the inner electrode, is copper. Nickel serves as an intermediate layer, electroplated onto the copper, acting as a thermal barrier to safeguard the capacitor during welding. Finally, a layer of tin is applied to the nickel to enhance solderability. After the completion of outer electrode production, rigorous performance testing and packaging procedures are conducted to eliminate any defective products.

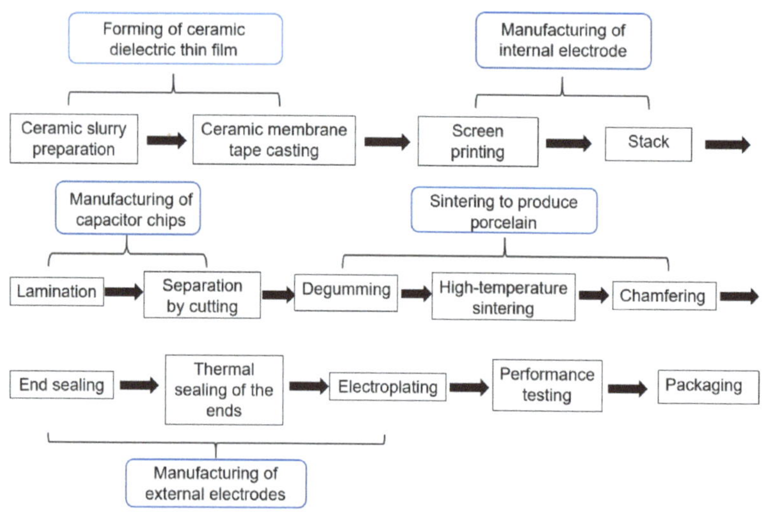

Figure 16. MLCC manufacturing process flowchart.

2.3.3. Others

In recent years, researchers have been paying attention to a new type of capacitor, namely, the multilayer polymer capacitor (MLPC), which has a similar design as MLCC and exhibits extremely high application value. Different from traditional solid capacitors, MLPC adopts a unique chip-type multilayer structure. Its production process is elaborately designed. Firstly, the aluminum foil is etched to form a porous oxide layer, and then a conductive polymer is deposited on the oxide layer as the cathode. After that, these treated aluminum foils are stacked into a multilayer structure, and then carbon paste is applied on them to form a carbon coating, followed by the application of silver paste to form a silver coating. Finally, the entire structure is completely sealed and fixed with epoxy resin, which can effectively protect its internal structure and prevent local overheating.

Compared with MLCC, MLPC has a higher capacity, which can significantly reduce the number of capacitors required in applications, thus reducing the overall cost. Moreover, MLPC exhibits excellent performance in high and low-temperature environments, overcoming the cracking issues that MLCC may encounter due to temperature or mechanical shock, thereby improving the reliability and stability of the product. Compared with solid tantalum capacitors, MLPC has a lower equivalent series resistance and higher safety. Currently, the conductive polymer of MLPC is mainly based on poly(3,4-ethylenedioxythiophene) (PEDOT), and the deposition technology often adopts in situ polymerization. The development of new conductive polymers and the optimization of polymer deposition technology have become research hotspots in recent years.

3. Electrochemical Capacitor

Electrochemical capacitors, commonly referred to as supercapacitors (SCs), possess remarkable charge and discharge efficiency, an outstanding cycle life, and exceptional power performance while being capable of operating across a broad temperature spectrum [76,77]. In comparison to batteries, supercapacitors exhibit a superior power density and the ability to rapidly store or discharge energy [78]. Nevertheless, their energy density is lower due to the constraints associated with electrode surface charge storage. When compared to traditional capacitors, they possess a lower power density but a higher energy density [79]. Supercapacitors can serve as rapid starting power sources for electric vehicles, as well as balancing power supplies for lifting equipment. Furthermore, they can be utilized as traction energy sources for hybrid electric vehicles, internal combustion engines, and trackless vehicles [80–82].

The history of supercapacitors can be traced back to 1853. In 1853, Helmholtz pioneered the exploration of electrical storage mechanisms within capacitors and introduced the concept of the double-layer model in the context of colloidal suspension research. In 1957, Becker filed the first patent for an electrochemical capacitor, which incorporated porous carbon electrodes immersed in an H_2SO_4 solution [83]. Advancing further, in 1971, Trasatti and his colleagues reported for the first time on the charge storage behavior of ruthenium oxide films in sulfuric acid, revealing the pseudocapacitance phenomenon in transition metal oxides. Subsequently, Shirakawa et al. garnered attention for the pseudocapacitive charge storage properties they developed in conductive polymer materials, thus propelling the interest in pseudocapacitors [84]. In 1978, NEC Corporation of Japan commercialized electrochemical capacitors, branding them "supercapacitors." A decade and a half later, in 1989, the U.S. Department of Energy initiated long-term research support for high-energy-density supercapacitors intended for use in electric drive systems as part of their electric and hybrid vehicle initiatives. Presently, leading global supercapacitor companies such as Maxwell (USA), Nesscap (South Korea), ELTON (Russia), and Nippon Chemicon (Japan) have developed and offered a diverse range of supercapacitors for commercial applications [85].

In recent years, researchers have proposed numerous approaches to improve the performance of supercapacitors. Zan et al. [86] employed a self-template biomimetic method to synthesize a novel mesoporous $Ni(OH)_2$ structure, which comprises house-of-cards-like cubic nanocages assembled from monolayer $Ni(OH)_2$ coupled with an exceptionally large interlayer spacing of approximately 1 nm. Tests have shown that it provides a specific capacity close to 100% of the theoretical value and exhibits an excellent cycling performance of over 10,000 cycles. Shwetha et al. [87] synthesized Co_3O_4 nanoparticles using a mixture of cobalt nitrate and ascorbic acid. Electrochemical data indicate that the Co_3O_4 nanoparticles exhibit good capacitive behavior, with the highest electrochemical performance observed when the cobalt nitrate/ascorbic acid ratio equals 1. Specifically, these nanoparticles deliver a specific capacitance of 166 F g^{-1} at a current density of 0.5 A g^{-1} and retain 90% of their capacitance after 5000 cycles. Luo et al. [88] reported a method for synthesizing heterogeneous Ni_3N-$Co_2N_{0.67}$/nitrogen-doped carbon (Ni_3N-$Co_2N_{0.67}$/NC) hollow nanoflowers by pyrolyzing a NiCo-TEOA (triethanolamine) complex precursor and employing urea as a nitrogen source. The assembled Ni_3N-$Co_2N_{0.67}$/NC//AC battery achieves a peak energy density of 32.4 W h kg^{-1} at a power density of 851.3 W kg^{-1}. Wang et al. [89] proposed an effective method for activating Ni-Co oxide nanosheet arrays (NiCoO NSAs) grown on carbon fiber cloths. The resulting ac-NiCoO NSA exhibits a high specific capacity (206.5 mAh g^{-1} at 0.5 A g^{-1}). The assembled capacitor demonstrates high energy density (45.4 Wh kg^{-1}), high power density (17.3 kW kg^{-1}), and ultra-long cycling stability, with a retention rate of 77.4% after 20,000 cycles (20 A g^{-1}).

Furthermore, in recent years, flexible supercapacitors, which exhibit both bendability and stretchability coupled with their high electrochemical performance retention, have garnered extensive and intensive research attention. Song et al. [90] have developed a self-wrinkled polyaniline (PANI)-based composite hydrogel (SPCH), featuring an electrolyte hydrogel and PANI composite hydrogel as its core and shell, respectively, through a stretching, low-temperature polymerization/release strategy. This SPCH exhibits remarkable stretchability (approximately 970%) and high fatigue resistance. Remarkably, upon cutting and reconnecting the edges, it can directly function as an intrinsically stretchable all-solid-state supercapacitor (A-SC), maintaining highly stable output with a capacitance retention rate of 92% even after 1000 stretching and releasing cycles at 100% strain. Huai et al. [91] synthesized cobalt-doped $NiMoO_4$ nanosheets via a hydrothermal method, which exhibited a specific capacitance of 906 C g^{-1} at a current density of 1 A g^{-1}. The assembled flexible supercapacitor delivered an energy density of 64 Wh kg^{-1} at a power density of 2880 W kg^{-1}. Notably, the device retained 78% of its initial capacitance after 10,000 cycles. Liu et al. [92] prepared graphene/MnO$_2$ composites by growing MnO$_2$ nanosheets on single-layer graphene via a water bath method and introducing two-dimensional black phosphorus during the pulping process. These composites were used to fabricate micro-supercapacitors that can be integrated with flexible film pressure sensors, holding promise for wearable electronic devices. Zhu et al. [93] provided an overview of solid-state flexible supercapacitors, reviewed the current research status of vanadium-based electrode materials in solid-state flexible SC, and proposed strategies to address the challenges associated with these materials. Wang et al. [94] introduced a fully biomass-based colloidal gel composed of a mononuclear anthraquinone derivative and porous lignin-based graphene oxide fabricated through a self-assembly process. This gel was successfully used to produce biomass-based flexible micro-supercapacitors via screen printing. Upon testing, these capacitors demonstrated significant areal capacitance (43.6 mF cm^{-2}), energy and power densities (6.1 μWh cm^{-2} and 50 μW cm^{-2}, respectively), and cyclic stability (>10,000 cycles).

In recent years, numerous review articles have outlined the research progress in supercapacitor electrode materials and electrolytes. Li et al. [95] comprehensively summarized the research advancements in nickel-based composites for supercapacitors, encompassing the properties of novel materials, preparation methods, and application potentials, offering insights into the research prospects and future trends of nickel-based composites. Troschke et al. [96] introduced the general chemical properties of Schiff bases and reviewed several nanomaterials and their carbonized derivatives obtained through Schiff-base formation, along with an outlook on the major obstacles and future prospects in this research field. Shi et al. [97] overviewed the research progress in redox electrolyte-enhanced carbon-based supercapacitors, analyzing the causes of self-discharge and corresponding suppression strategies from aspects such as separator modification, electrolyte formulation, and electrode design while outlining their development prospects. Wu et al. [98] reviewed the research progress of various spatially dimensional carbon materials in recent years, discussing their advantages and disadvantages as supercapacitor electrode materials, examining the key factors influencing their electrochemical performance, and proposing new development trends for carbon materials. Kong et al. [99] introduced the concept of superstructured carbons with customized functionality, comprehensively outlining their designs tailored to different energy storage mechanisms and prospectively pointing out potential challenges in their future development. Although these reviews have thoroughly explored supercapacitor electrode materials or electrolytes, a comprehensive overview of their structures and energy storage principles remains lacking. Thus, the present review focuses on elucidating the classification, fine structures, and unique energy storage mechanisms of supercapacitors, aiming to provide readers with a more complete and in-depth understanding of the framework.

The structure of a supercapacitor comprises four main components: two electrodes, an electrolyte, a separator, and current collectors. The function of current collectors is to collect the current generated by the active material in the capacitor and facilitate the formation of a larger current for external output. Typically, copper foil is used as the current collector for the negative electrode, while aluminum foil is employed for the positive electrode. Depending on the energy storage principle, SC can be categorized into three types, namely electrochemical double-layer capacitors (EDLCs), pseudocapacitors, and hybrid capacitors, as illustrated in Figure 17 [100,101]. Their respective energy storage mechanisms are based on non-Faradaic, Faradaic, and a blend of both processes [102]. In the non-Faradaic process, charges are distributed across the surface through physical means without the formation or breakdown of chemical bonds [103]. Conversely, the Faradaic process involves the transfer of charges between electrodes and electrolytes. Double-layer capacitors store energy through non-Faradaic reactions, commonly utilizing carbon-based materials with a high surface area and porosity as electrode materials. On the other hand, pseudocapacitors store energy through Faradaic reactions, typically employing transition metal oxides (like RuO_x) and conductive polymers (such as polyaniline) as electrode materials [104]. In hybrid capacitors, both Faradaic and non-Faradaic reactions occur for energy storage, combining the benefits of EDLC and pseudocapacitors [105].

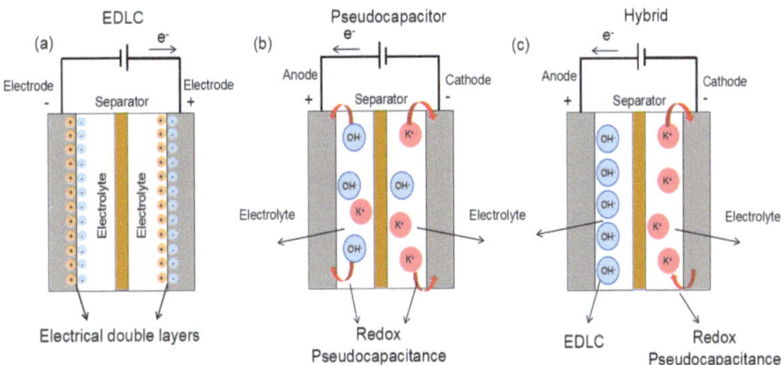

Figure 17. Types of supercapacitors. (**a**) EDLC. (**b**) Pseudocapacitors. (**c**) Hybrid supercapacitors. Reproduced from Ref. [106] with permission.

Supercapacitors and electrolytic capacitors seem superficially similar due to their shared electrolyte component. Both aim to enhance capacitance by increasing electrode area and reducing electrode distance. However, a deeper exploration reveals significant differences. The electrolyte in aluminum electrolytic capacitors is the actual cathode, while the electrolyte in supercapacitors is the dielectric, with porous activated carbon comprising the actual electrode. Traditional electrolytic capacitors utilize valve metal as the electrode and its oxide as the dielectric, resulting in polarity and intolerance to reverse voltage. Conversely, supercapacitors exhibit identical electrode structures, classifying them as non-polar capacitors.

3.1. Electrochemical Double-Layer Capacitors

The concept of the double layer was originally introduced and formulated by von Helmholtz in the 19th century, with subsequent modifications and enhancements made by Gouy, Chapman, and Stern [107]. EDLC comprises two carbon-based electrodes, a separator, and an electrolyte [108]. Charge storage is achieved by the formation of a double layer at the interface between the electrode and the electrolyte [109]. The physical image and the dismantled structural diagram of EDLC are shown in Figure 18. During the operation of an EDLC, charge accumulation occurs through a non-Faradaic process, meaning that there is no ion exchange between the electrode material and the electrolyte solution, and electrons do not transfer across the electrode interface [110]. When charging, electrons are driven by an external electric field to move from the positive electrode to the negative electrode through an external circuit, resulting in the formation of a layer of charge electrons on the surface lattice structure of the electrode material [110]. In electrolytes, anions are attracted toward the positive electrode, while cations migrate toward the negative electrode. These oppositely charged ions migrate toward the respective electrodes and accumulate on their surfaces, creating a double layer. Upon removal of the external electric field, the double layer persists, stabilizing the voltage due to the attractive forces between the opposing charges [80]. During discharge, the process reverses. The charged ions adsorbed on the electrode migrate in a directional manner, generating a current in the external circuit until the electrolyte returns to its electrically neutral state. This charge storage mechanism is fully reversible, and since it lacks chemical reactions, the electrode structure of the EDLC remains virtually unchanged. This attribute confers EDLC capacitors with a prolonged cycle life and exceptional power density [109,111].

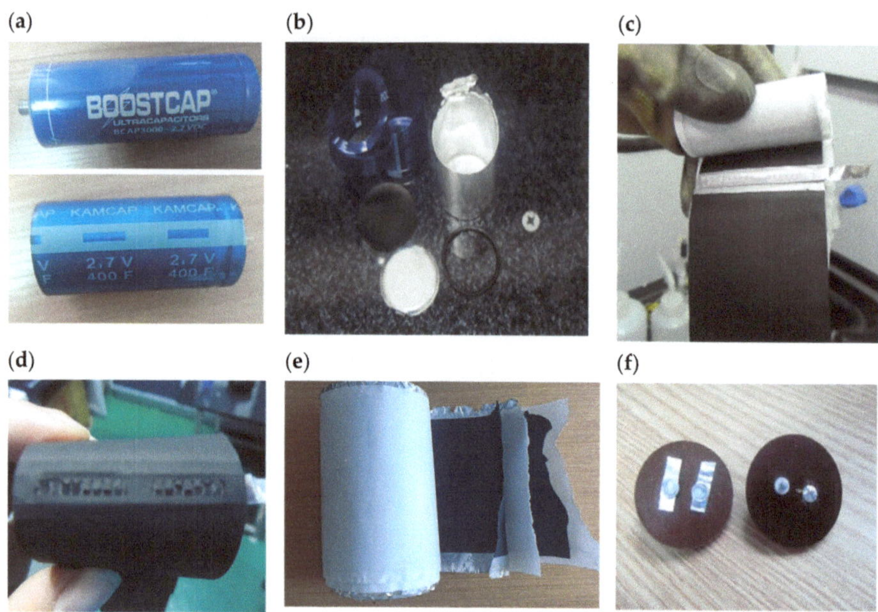

Figure 18. (**a**) The physical image of an EDLC. (**b**) The bottom gasket, insulating sheet, top sealing ring, outer aluminum case, and packaging plastic film of the EDLC. (**c**) Lead-out welding spot. (**d**) The back of the welding spot. (**e**) The overall structure of the EDLC. (**f**) Top rivet.

The electrode materials of EDLC commonly consist of carbon-based substances, encompassing activated carbon (AC) [112–114], carbon nanotubes (CNTs) [115], graphene [116,117], and various other carbon-based composite materials. These materials are deemed suitable for double-layer capacitors, attributed to their expansive specific surface area, robust thermal and electrochemical stability, and superior conductivity [107]. Currently, research in the field of EDLC is focused on two major directions: optimizing the performance of existing electrode materials and developing new materials. Fu et al. [118] obtained nitrogen and sulfur dual-doped activated carbon with a hierarchical pore structure by directly carbonizing/activating polymer-based monolithic materials. The EDLC assembled with this material exhibits favorable energy density and power density in a 6 mol·L^{-1} KOH aqueous electrolyte. Lv et al. [119] utilized dopamine and copper chloride precursors to form carbon flocculates embedded with ultrafine copper nanoparticles on carbon cloth through pyrolysis and electrochemical oxidation reactions. The results indicated that the obtained electrode possessed a large surface area of 55.5 m^2·g^{-1} and a high conductivity of 48.7 S·mm^{-1}. When the prepared material was applied in an EDLC, the EDLC exhibited outstanding performance with a power density as high as 179 mW·cm^{-3} and an energy density reaching 23 mWh·cm^{-3}, demonstrating promising application prospects. Yoo et al. [120] successfully prepared cost-effective carbon xerogels with large surface areas by substituting phenol for resorcinol and controlling the amount of catalyst, further confirming their excellent electrochemical performance as active materials for EDLC electrodes.

The performance of EDLC is also influenced by the compatibility between the pore size of the electrode material and the size of the electrolyte ions, as well as the ionic mobility. Consequently, the performance of EDLC can be altered by utilizing diverse electrolytes, including aqueous and organic electrolytes. Although aqueous electrolytes offer lower series resistance compared with organic electrolytes, their narrower potential window range restricts the energy density of EDLC [121]. During the production process, it is crucial to select an electrolyte solution that suitably adapts to the pore size of the electrode material.

3.2. Pseudocapacitors

In contrast to EDLC, a pseudocapacitor exhibits charge transfer at the electrode-electrolyte interface, enabling energy storage through a rapid and reversible Faradaic reaction. This mechanism enhances the energy density within the electrode [122]. Generally speaking, pseudocapacitance can be divided into three different types based on its reaction mechanism, including redox pseudocapacitance, under potential deposition (adsorption) pseudocapacitance, and intercalation pseudocapacitance [123]. The schematic diagrams of the three processes are shown in Figure 19.

3.2.1. Redox Pseudocapacitance

Redox pseudocapacitance (Figure 19a) is the most prevalent form of pseudocapacitance. Its working principle involves the electrochemical adsorption of active ions in the electrolyte onto or near the electrode surface when a potential is applied [124]. Simultaneously, the electrode material undergoes a rapid and reversible oxidation-reduction reaction, generating charges and facilitating Faraday currents to flow through the supercapacitor (Figure 19a) [125].

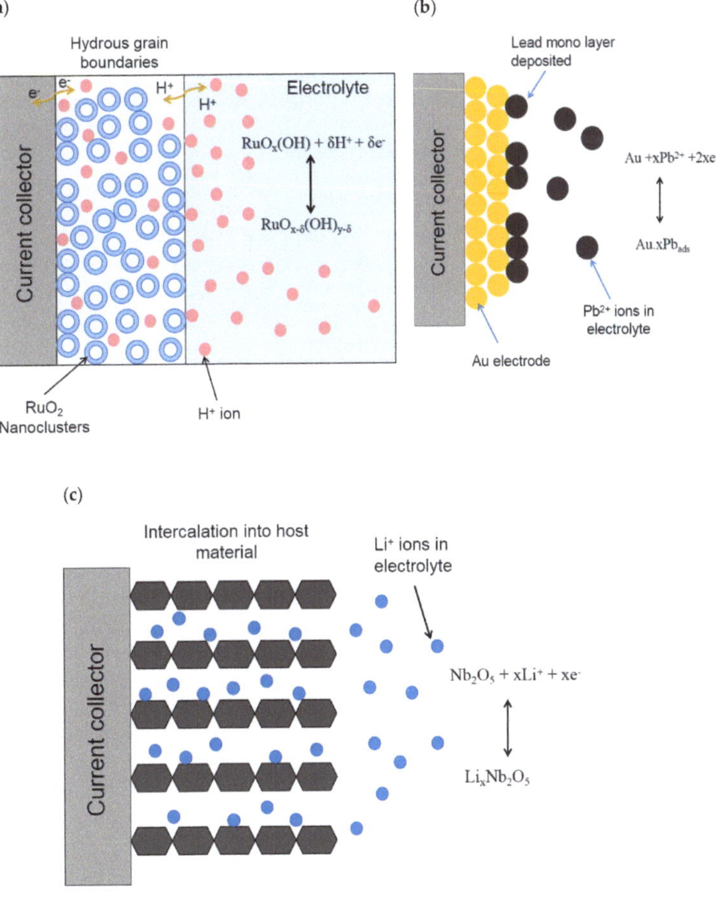

Figure 19. The schematic diagrams of pseudocapacitance. (**a**) A redox pseudocapacitance in RuO$_2$ schematics. (**b**) Underpotential deposition pseudocapacitance, Pb coating on Au electrode. (**c**) Intercalation pseudocapacitance, lithium ions embedded in the lattice of niobium oxide. Reproduced from Ref. [125] with permission.

3.2.2. Underpotential Deposition Pseudocapacitance

Underpotential deposition pseudocapacitance (Figure 19b), alternatively known as adsorption pseudocapacitance, occurs when an external potential is applied. This process involves the surface adsorption and reduction of metal ions, resulting in the deposition of a monolayer on the metal surface. This deposition causes a minor change in potential relative to its equilibrium potential. An example of underpotential deposition pseudocapacitance is the adsorption of lead onto the surface of an Au electrode (Figure 19b) [126].

3.2.3. Intercalation Pseudocapacitance

The principle of intercalated pseudocapacitors (Figure 19c) involves the embedding of ions from the electrolyte into the layered or tunnel-like structure of redox-active materials. This embedding process triggers rapid charge transfer, which occurrs without altering the crystal structure of the electrode material [125]. One example is the insertion of lithium ions into the lattice of niobium oxide (Figure 19c) [127].

Pseudocapacitors possess distinct electrochemical characteristics, allowing for high charge storage capacity and a boost in energy density through efficient charge transfer channels [128]. Typically, pseudocapacitors exhibit superior specific capacitance and energy density when compared to EDLCs [129]. However, redox reactions, particularly those occurring on the electrode surface, can alter the bulk phase or electrolyte composition of the electrode material. Consequently, over extended charging and discharging cycles, materials are prone to fatigue damage, resulting in a shorter cycling life [80]. While their lifespan is still longer than that of batteries, it falls short of that of double-layer capacitors.

During the early stages of pseudocapacitive electrode research, researchers primarily focused on transition metal oxides (TMOs) [130]. Nevertheless, their practical applications have been unsatisfactory due to low conductivity and relatively limited charge storage capacity, thus limiting their widespread use in pseudocapacitive electrodes. Currently, electrode materials for pseudocapacitors that are widely studied include carbon materials doped with heteroatoms (such as N, P, S) [131], transition metal oxides/hydroxides (like MnO_2, NiOOH) [132,133], two-dimensional transition metal carbides, and nitrides (MXene) [134]. Compared with pure carbon materials, carbon materials doped with heteroatoms possess characteristics such as low cost and excellent electrochemical performance. Commonly, N atoms, P atoms, and others are selected for doping. For instance, Cao et al. [135] prepared nitrogen-doped biomass-derived hierarchical porous carbon materials (HPC) using different nitrogen-containing compounds (NH_4Cl, $(NH_4)_2CO_3$, and urea) simultaneously as activators and dopants. When the three materials were used in pseudocapacitors, the pseudocapacitor based on HPC-urea exhibited the highest specific capacitance (300 $F \cdot g^{-1}$ at 1 $A \cdot g^{-1}$) and energy density (14.3 $Wh \cdot kg^{-1}$). As a transition metal oxide, MnO_2 boasts low cost and superior electrochemical performance, achieving a specific capacitance as high as 1100 $F \cdot g^{-1}$ within a potential window of 1.0 V. In recent years, MnO_2 has emerged as a research hotspot in the field of pseudocapacitive materials. The reversible transformation between MnO_2 and MnOONa in sodium sulfate electrolyte is the primary reason for the pseudocapacitive behavior exhibited by MnO_2 electrodes [136]. Currently, research on MnO_2 primarily focuses on enhancing its specific capacitance. For instance, Zhou et al. [137] successfully synthesized mesoporous manganese dioxide with a semicrystalline spinodal structure using mesoporous silica KIT-6 as a robust hard template. Within the potential range of -0.1 V to 0.55 V, this material exhibited stable and reversible electrochemical behavior, achieving an excellent capacitance performance of 220 $F \cdot g^{-1}$. Nayak et al. [138] obtained mesoporous manganese dioxide with an average pore diameter of 2–20 nm from potassium permanganate through a sonochemical method using triblock copolymers as soft templates. These materials possess a high specific capacitance of 265 $F \cdot g^{-1}$. MXene is a novel class of two-dimensional transition metal carbides, carbonitrides, or nitrides. Its interlayer structure is flexible, allowing various ions to be inserted between MXene sheets. Additionally, due to its abundant interlayer ion diffusion pathways and ion storage sites, combined with the conductive carbide/nitride core within the material, MXene has emerged as an important candidate for pseudocapacitive energy storage materials [139]. Cao et al. [140] have developed highly stretchable micro-pseudocapacitor electrodes composed of MXene nanosheets and in situ reconstructed silver nanoparticles (Ag-NP-MXene). These electrodes exhibit high energy density, a stable operating voltage of approximately 1 V, and rapid charging capabilities. Kim et al. [141] immersed carbon nanofibers (CNFs) into a colloidal solution of MXene (Ti_3C_2) to form a composite material through a dip-coating process. Subsequently, they constructed a three-electrode system using an aqueous solution of 1 M sodium sulfate as the electrolyte. Experimental results showed that this MXene-coated CNF exhibited remarkable performance, achieving a maximum specific capacitance of 514 $F \cdot g^{-1}$ at a current density of 0.5 $A \cdot g^{-1}$. Additionally, the energy density and power density reached 71.4 $Wh \cdot kg^{-1}$ at 0.5 $A \cdot g^{-1}$ and 2.3 $kW \cdot kg^{-1}$ at 5 $A \cdot g^{-1}$, respectively.

3.3. Hybrid Capacitors

As implied by its name, a hybrid capacitor is essentially a type of supercapacitor that consists of two electrode parts and a separator. The electrodes of a hybrid capacitor can be made from dissimilar materials, and the separator typically has a microporous structure. The diversity in hybrid capacitors is achieved through the combination of various redox and EDLC materials. The storage mechanism of hybrid supercapacitors integrates the principles of both EDLC and pseudocapacitors. This unique combination results in a significantly higher capacitance, often reaching levels two to three times greater than those of traditional capacitors, standalone EDLCs, or pseudocapacitors. Additionally, it exhibits a higher working potential, further enhancing its overall performance [142]. The electrodes of hybrid supercapacitors are usually asymmetric, with one electrode being a carbon electrode that charges and discharges based on the double-layer capacitance process and the other electrode being a pseudocapacitive electrode material or battery electrode material that undergoes Faradaic redox reactions during charging and discharging.

The positive electrode of a hybrid capacitor is a crucial component that supports its high-current discharge and high-power density capabilities. Carbon-based materials, known for their high electrical conductivity and large specific surface area, are commonly used as positive materials in hybrid capacitors. The reaction mechanism of carbon-based positive materials is primarily based on the electric double-layer principle, where energy is stored and released through the adsorption and desorption of ions from the electrolyte onto the surface of the carbon electrode. Common carbon-based positive materials include activated carbon and graphene. Activated carbon features a porous structure, a large specific surface area (approximately 1000–1500 $m^2 \cdot g^{-1}$), and strong adsorption capabilities [143]. However, its relatively low specific capacity (around 40–80 $mAh \cdot g^{-1}$) and electrical conductivity (approximately 1–5 $S \cdot cm^{-1}$) limit its applications to some extent. Currently, research on activated carbon materials focuses on optimizing their porous structure, morphology control, and surface modification to achieve higher specific capacitance. For example, Lu et al. [144] utilized corn stalks to prepare mesoporous activated carbon for use in high-performance supercapacitors. The activated carbon based on corn stalks exhibited a high specific capacitance of 188 $F \cdot g^{-1}$ at a current density of 1 $A \cdot g^{-1}$ in both organic and ionic liquid electrolytes. Piao et al. [145] reported an activated multi-hierarchical mesoporous carbon (MHPC). When tested in a three-electrode system using a 6 M KOH aqueous solution at a current density of 1 $A \cdot g^{-1}$, the assembled supercapacitor with the MHPC electrode achieved a specific capacitance of 318 $F \cdot g^{-1}$. Graphene is a two-dimensional monolayer material composed of carbon atoms arranged in a honeycomb lattice. It possesses a theoretical specific surface area of 2360 $m^2 \cdot g^{-1}$, a theoretical specific capacitance of 550 $F \cdot g^{-1}$, and high electrical conductivity (approximately 50–150 $S \cdot cm^{-1}$). However, irreversible stacking tends to occur between graphene sheets, reducing their surface area [113]. Currently, researchers have employed various methods to address the aforementioned issues. For instance, Gao et al. [146] synthesized a three-dimensional oriented graphene framework that resembles paper but possesses a directed surface, macropores, and interconnected parts through ordered assembly guided by hard templates. This framework exhibits a high specific surface area of up to 402.5 $m^2 \cdot g^{-1}$ and excellent mechanical flexibility, making it suitable for use as an electrode in hybrid capacitors. Yan et al. [147] prepared hollow graphene nanospheres through a combination of template separation, microwave heating, and graphitization of carbon layers. The synthesized graphene exhibits a specific surface area of 2794 $m^2 \cdot g^{-1}$, a capacitance exceeding 529 $F \cdot g^{-1}$ at a current density of 1 $A \cdot g^{-1}$, and a capacitance retention rate of 62.5% during continuous power supply.

For the negative electrode of hybrid capacitors, materials with rapid charge-discharge capabilities are often employed to compensate for the kinetic differences between the positive and negative electrodes. Transition metal oxides, with their high theoretical specific capacities, abundant sources, and low costs, are commonly used as negative electrode materials in hybrid capacitors. The reaction mechanism of transition metal oxide negative electrode materials is primarily based on Faraday redox reactions. When the hybrid capacitor is charged, ions in the electrolyte (such as sodium ions, lithium ions, etc.) diffuse to the surface of the transition metal oxide negative electrode material and undergo electrochemical reactions with it, storing charge. During this process, one or more redox pairs form on the surface of the transition metal oxide, which can reversibly convert during charging and discharging, thereby enabling the storage and release of charge. Common transition metal oxide materials include MnO and Co_3O_4. When MnO is used as a negative electrode material, its theoretical capacity ratio can reach 756 $mAh \cdot g^{-1}$, but pure MnO has poor electronic conductivity (approximately 10^{-8}–10^{-6} $S \cdot m^{-1}$), leading to easy capacity decay. Therefore, it is typically combined with highly conductive carbon materials. For example, Yang et al. [148] utilized graphene oxide (GO) and nanospherical K_xMnO_2 precursors to prepare a one-dimensional graphene nanoscroll-wrapped MnO nanoparticle (GNS@MnO) material through a simple liquid nitrogen quenching followed by atmospheric annealing process. The obtained material exhibited a high reversible capacity of 766 $mA \cdot h \cdot g^{-1}$ at a current density of 100 $mA \cdot g^{-1}$, high rate performance (437 $mA \cdot h \cdot g^{-1}$ at 5.0 $A \cdot g^{-1}$), and good cycling stability. Chen et al. [149] designed and synthesized one-dimensional graphene nanoscrolls wrapped with MnO nanoparticles featuring a unique nanocomposite structure. The constructed composite negative electrode material exhibits rapid ion and electron transport kinetics, as well as strong durability. When used as a negative electrode material, Co_3O_4 boasts a theoretical specific capacity of up to 890 $mA \cdot h \cdot g^{-1}$ and a low redox potential (<1 V). However, the volume expansion that occurs during charging and discharging can lead to the shedding of active materials and rapid capacity decay. To address this issue, researchers have taken various measures. For instance, Wang et al. [150] synthesized Co_3O_4 cryogels using a triblock polymer/ice crystal dual-template sol-gel method. These cryogels exhibited a specific capacitance of up to 742.3 F·g–1 and maintained 86.2% of their capacity after 2000 cycles. Liu et al. [151] successfully obtained uniformly sized and highly crystalline Co_3O_4 nanocubes with the assistance of mesoporous carbon nanorods. After heat treatment, mesoporous Co_3O_4 nanocubes were formed. Electrochemical tests revealed that the specific capacitance of the Co_3O_4 nanocube electrode was approximately 350 $F \cdot g^{-1}$ at a current density of 0.2 $A \cdot g^{-1}$.

Furthermore, the matching of electrode materials for hybrid capacitors is a crucial step in ensuring excellent capacitor performance. The compatibility of electrode materials is primarily related to their capacity, kinetics, and cycle life [152]. In the process of selecting compatible electrode materials, attention should be paid to balancing the energy density and power density of the electrode materials. It is important to choose negative and positive materials with similar physicochemical properties and consistent volume changes to enhance the stability of the cycling process. A common approach is to utilize a homologous strategy, where negative and positive materials derived from the same raw materials or undergoing similar processing procedures are selected to ensure similarity in their physicochemical properties, thereby improving material stability and compatibility. For instance, Xu et al. [153] employed a homologous strategy to prepare sulfur-doped channel carbon fiber negative electrode and active multi-channel carbon fiber positive electrode for potassium-ion capacitors. The resulting potassium-ion capacitors achieved high energy and power densities, as well as exceptional cycling stability.

Hybrid capacitors excel in their performance due to their extensive potential window, high specific capacitance, and minimal self-discharge rate. By seamlessly integrating Faradaic and non-Faradaic processes, these capacitors are able to store a significantly larger number of charges, leading to exceptional energy and power densities. When compared to double-layer capacitors and pseudocapacitors, hybrid capacitors offer superior capacitance. Furthermore, certain hybrid capacitors are capable of operating at high voltages, albeit with slightly reduced cycling performance [110]. The overall performance of hybrid supercapacitors hinges critically on the choice of electrode and electrolyte materials. These materials have a direct bearing on the performance characteristics of the hybrid supercapacitors. Consequently, selecting suitable electrode and electrolyte materials is paramount in enhancing the overall performance of hybrid supercapacitors.

Due to the exceptional energy density exhibited by lithium ions, the integration of supercapacitors with lithium-ion storage systems holds significant importance. As a cutting-edge electrochemical energy storage solution, lithium-ion capacitors (LICs) combine the lithium-ion intercalated electrode of lithium-ion batteries with the electrical double-layer electrode of supercapacitors, offering a unique blend of benefits [154,155]. They not only inherit the high energy density advantages of batteries but also incorporate the attributes of electric double-layer capacitors, such as high power density and prolonged cycle life, thereby significantly enhancing overall performance [156–159]. The concept of LIC dates back to 1987 when Yata and his team conducted groundbreaking research on the intercalation mechanism of lithium ions in polyacene semiconductor (PAS), which was successfully commercialized as a button cell in 1989. Subsequently, in 1992, the introduction of pre-lithiation technology in PAS capacitors significantly boosted the potential of these capacitors, leaping from 2.5 V to 3.3 V. As the 21st century dawned, LIC technology witnessed significant breakthroughs. In 2001, G. G. Amatucci and his research team at Telcordia Technologies in New Jersey designed and analyzed the first modern-day lithium-ion capacitor using lithium titanate (LTO) as the negative electrode in conjunction with an activated carbon positive electrode. This device demonstrated remarkable energy performance exceeding 20 Wh/kg within a potential window of 1.5 V to 3.0 V [160]. The following year, A. D. Pasquier and his research group introduced another innovative LIC featuring an LTO negative electrode paired with a poly(fluorophenylthiophene) positive electrode, achieving a maximum deliverable power of 12 kW/kg, further expanding the application prospects of LIC [161]. In 2005, activated carbon was utilized for the first time as the negative electrode material in LIC, combined with the $LiNi_{0.5}Mn_{1.5}O_4$ positive electrode [162]. Concurrently, Fuji Heavy Industries successfully commercialized pre-lithiated PAS-based LIC. By 2008, V. Khomenko and his research team had developed a high-energy-density LIC using commercial graphite and activated carbon, significantly advancing the commercialization of LIC technology [163]. Since then, researchers in the LIC field have relentlessly explored new materials and configurations, employing graphene and doped carbon and studying their symmetric and asymmetric configurations, driving the rise of LIC as potential hybrid energy storage devices for modern applications and ultimately achieving their commercialization [164].

LIC can be applied in scenarios such as railway transportation, automatic guided vehicles, and spacecrafts, which require maintenance-free, fast charging, high power, and high energy density [165,166]. These capacitors are constructed with multiple components, including a positive electrode (typically a capacitive one), a negative electrode (commonly a pre-lithiated battery negative electrode), an electrolyte, a separator, a current collector, a conductive agent, a binder, and metallic lithium foil [167]. Each of these components contributes to the overall functionality and performance of the capacitor. Figure 20a is a digital image of a lithium-ion capacitor manufactured by JM Energy Company. Figure 20b,c show the SEM images of its negative electrode and positive electrode.

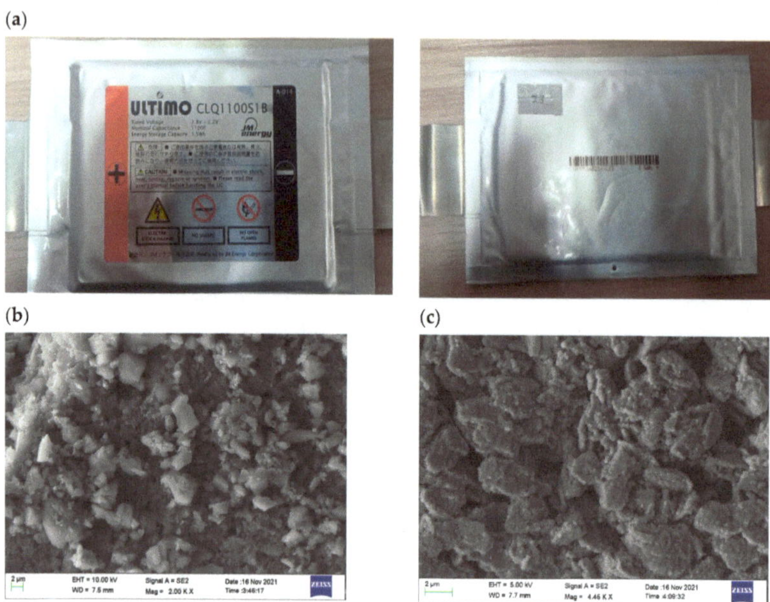

Figure 20. (**a**) Digital image of a soft-packaged lithium-ion capacitor manufactured by JM Energy Company. (**b**) An SEM image of the negative electrode. It shows that it has undergone pre-lithiation treatment. (**c**) An SEM image of the positive electrode, which is composed of carbon material.

The positive electrode-activated carbon material is primarily used for ion adsorption and desorption, typically prepared through carbonization and activation processes with raw materials such as coal and resin. Conversely, the negative electrode carbon material is primarily used for the insertion and extraction reactions of lithium ions, often prepared from materials like tar and asphalt through carbonization and graphitization processes. Electrolytes, the medium that facilitates ion transport between the electrodes, primarily comprise lithium salts dissolved in organic solvents. Moreover, the separator plays a crucial role as a key component, consisting of porous membranes crafted from polymers or cellulose. Its function is to avert direct contact between the positive and negative electrodes, thus safeguarding the cell's smooth and safe operation. On the other hand, the current collector, which is a thin metal foil measuring several micrometers, efficiently directs the current generated by the electrode, ensuring its effective utilization. Conductive agents, meanwhile, are tasked with establishing electronic pathways, often derived from raw materials like petroleum, through processes like cracking, carbonization, and deposition. The adhesive, a polymeric material typically synthesized artificially, plays a pivotal role in securely adhering the active material and conductive agent to the current collector. Lastly, metallic lithium foil fulfills a crucial pre-lithiation role in lithium-ion capacitor cells, with its preparation often involving the electrolysis and rolling of lithium salts. These components and materials collaborate seamlessly to guarantee the smooth operation and optimal performance of the capacitor.

The preparation flowchart of a single cell for lithium-ion capacitors is shown in Figure 21. The specific process involves mixing the positive and negative electrode materials, conductive agents, and binders and then coating them onto the current collector [168]. Subsequently, the electrodes are prepared by rolling, stripping, and slicing. These electrodes are then wound or stacked, welded, and encased. After drying, the electrolyte is injected to obtain a semi-finished cell. Finally, the semi-finished cell undergoes pre-lithiation, formation, aging, and capacity sorting to produce the final product [169]. There are various encapsulation types for lithium-ion capacitor cells. In application scenarios that require higher energy density and power density, the electrodes are assembled into cylindrical or rectangular cells with stacking or winding methods. Common encapsulation types include cylindrical, rectangular, and pouch-type designs. When lithium-ion capacitor cells are grouped together, they can be used in fields such as power grid frequency modulation, aerospace, and rail transportation.

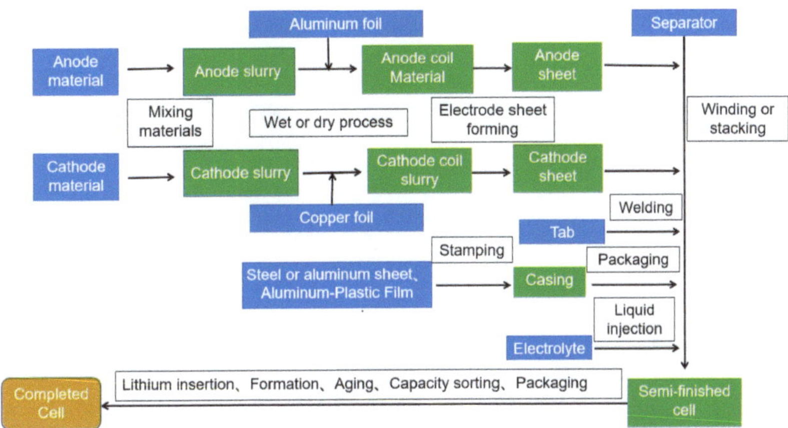

Figure 21. The preparation flowchart of a single cell for lithium-ion capacitors.

The energy storage mechanisms of the positive and negative electrodes in lithium-ion capacitors are different, and the currently common lithium-ion capacitor systems can be categorized into the following four types [170]:

(1)   The battery-type positive electrode and the capacitive-type negative electrode [171,172]. They operate through different mechanisms: electrochemical reactions occur at the positive electrode, while ion desorption processes take place at the negative electrode. During charging, lithium ions are desorbed from the positive electrode and released into the electrolyte, while simultaneously, lithium ions in the electrolyte are adsorbed onto the negative electrode, maintaining a constant ion concentration in the electrolyte. The electrolyte does not serve as an active component but merely functions as an ion carrier. However, due to the lower operating voltage, the energy density of this system is often significantly lower than other systems. Additionally, the high specific surface area of the carbon material in the negative electrode can easily lead to rapid growth of the solid electrolyte interface (SEI), resulting in poor cycling stability.

(2)   The capacitive-type positive electrode and the battery-type negative electrode [173–176]. Ion adsorption and desorption processes occur at the positive electrode, while electrochemical reactions take place at the negative electrode. During charging, anions in the electrolyte are adsorbed onto the positive electrode, while lithium ions are intercalated into the negative electrode material [177]. The ion concentration in the electrolyte decreases as the voltage increases. During discharge, the ions on the electrodes return to the electrolyte, gradually restoring the ion concentration in the electrolyte. In this system, the electrolyte serves not only as an ion carrier but also as an active component of the system. The energy density is limited by the electrode materials and the concentration of the electrolyte.

(3)   Capacitive-type positive electrode and pre-lithiated battery-type negative electrode [178,179]. The charging and discharging process of this system mainly consists of two parts: the consumption of electrolytes when the voltage is higher than the open-circuit voltage and the migration of lithium ions when the voltage is lower than the open-circuit voltage. When it is charged from the open-circuit voltage to the maximum voltage, anions in the electrolyte migrate to the positive electrode and are adsorbed onto its surface, while lithium ions are intercalated into the negative

electrode, resulting in a decrease in electrolyte ion concentration as the voltage increases. During discharge from the maximum voltage to the open-circuit voltage, anions are desorbed from the positive electrode, lithium ions are deintercalated from the negative electrode, and the electrolyte concentration returns to its initial state. When discharging from the open-circuit voltage to the minimum voltage, lithium ions stored in the negative electrode through pre-lithiation continue to be deintercalated, reducing the lithium–ion concentration in the negative electrode while gradually increasing the concentration of lithium ions adsorbed on the positive electrode. The ion concentration in the electrolyte remains basically unchanged. When charging from the minimum voltage to the open-circuit voltage, lithium ions are desorbed from the positive electrode into the electrolyte, and the lithium ions in the electrolyte are intercalated into the negative electrode. This system requires pre-intercalation of lithium into the negative electrode, which can eliminate the loss of lithium ions due to the formation of the SEI film during the initial cycle, thereby increasing the output voltage window and consequently enhancing the power density and energy density. This has become the most important category for commercialization at present.

(4) The battery-capacitor composite positive electrode and pre-lithiated battery-type negative electrode [180,181]. The introduction of battery-type materials into the positive electrode enhances the energy density of the system, but it comes with a tradeoff in the power density and cycle life of the device. Most of the energy in this system is provided by the battery materials, making it, strictly speaking, a battery-type capacitor.

4. Summary

To clarify the differences between dielectric capacitors, electric double-layer supercapacitors, and lithium-ion capacitors, this review first introduces the classification, energy storage advantages, and application prospects of capacitors, followed by a more specific introduction to specific types of capacitors. Regarding dielectric capacitors, this review provides a detailed introduction to the classification, advantages and disadvantages, structure, energy storage principles, and manufacturing processes of thin-film capacitors, electrolytic capacitors, and ceramic capacitors. For electrochemical capacitors, an overview of their classification, structure, and energy storage principles is given, followed by a further analysis of the differences between supercapacitors and electrolytic capacitors. Subsequently, the focus is on the structural composition, production process, and energy storage principles of lithium-ion capacitors.

Currently, research on film capacitors primarily focuses on metalized organic polymer capacitors, which exhibit high charge-discharge rates, high flexibility, and excellent self-healing capabilities, promising good application prospects in areas such as microwave communications, hybrid electric vehicles, and renewable energy. However, they face challenges in terms of relatively poor thermal stability and high production costs. In the future, more research should be conducted on multiphase/multicomponent high-temperature dielectric polymers, including blend polymers and multilayer polymers, to enhance their high-temperature performance. Additionally, advanced film technologies should be developed, and processing techniques optimized to reduce costs [182]. Electrolytic capacitors are known for their large capacitance and high volumetric efficiency, making them suitable for applications in electronic devices or as energy buffers. However, they suffer from drawbacks such as high equivalent series resistance (ESR) and relatively short service life. Therefore, future efforts should be directed toward the development of novel polymer-based electrolytic capacitors, accompanied by in-depth studies on their failure mechanisms and prediction methods to extend their service life. Research on ceramic capacitors primarily focuses on MLCC. These capacitors exhibit extremely low ESR and equivalent series inductance, coupled with high current-handling capabilities and outstanding high-temperature stability. As a result, they show immense potential for applications in electric vehicles, 5G base stations, clean energy generation, smart grids, and other fields. Future research in ceramic capacitors can focus on utilizing dielectric materials like antiferroelectric materials or barium titanate-based compounds. By optimizing their electrode structures or manufacturing processes, researchers aim to enhance the breakdown strength, dielectric stability, and energy density of ceramic capacitors, further expanding their capabilities and applications [183]. For supercapacitors, their high-power density and low energy density have made enhancing energy density a persistent research focus. Additionally, with the rapid development of flexible electronics technology, there has been a surge in the development of smart clothing and portable electronic devices, and flexible supercapacitors have demonstrated vast application prospects in areas such as flexible displays, wearable electronic devices, and portable biomedical monitoring devices. Therefore, in the future, significant efforts can be directed toward utilizing novel materials like metal-organic frameworks (MOFs), covalent organic frameworks (COFs), and hydrogen-bonded organic frameworks (HOFs) in supercapacitors to improve their chemical stability and energy density. Furthermore, the development of flexible supercapacitor electrode materials with good stability, excellent performance, and long service life is also a crucial direction for the future advancement of supercapacitors. For lithium-ion capacitors, future research should emphasize the exploration of new electrode materials like two-dimensional MXenes to enhance their energy density. Additionally, optimizing the pre-lithiation process of lithium-ion capacitors and improving the compatibility of pre-lithiation technology are also essential steps toward their further development.

Author Contributions: W.L.: Conceptualization, methodology, investigation, writing—original draft. X.S.: conceptualization, supervision, investigation, and writing—review and editing. X.Y.: investigation, methodology and resources. Y.G.: supervision, investigation, and writing—review and editing. X.Z.: supervision, investigation, and writing—review and editing. K.W.: methodology and resources. Y.M.: supervision, investigation, methodology, and writing—review and editing. All authors have read and agreed to the published version of the manuscript.

Funding: This research received no external funding.

Data Availability Statement: The original contributions presented in the study are included in the article; further inquiries can be directed to the corresponding author/s.

Acknowledgments: We are grateful to the Institute of Electrical Engineering, Chinese Academy of Sciences, for providing us with capacitors and experimental equipment.

Conflicts of Interest: The author Xinyu Yan was employed by the TBEA Sunoasis Co., Ltd. All authors declare that the research was conducted in the absence of any commercial or financial relationships that could be construed as a potential conflict of interest.

References

1. Adediji, Y.B.; Adeyinka, A.M.; Yahya, D.I.; Mbelu, O.V. A review of energy storage applications of lead-free $BaTiO_3$-based dielectric ceramic capacitors. *Energy Ecol. Environ.* **2023**, *8*, 401–419. [CrossRef]
2. Li, C.; Zhang, X.; Wang, K.; Su, F.Y.; Chen, C.M.; Liu, F.Y.; Wu, Z.S.; Ma, Y.W. Recent advances in carbon nanostructures prepared from carbon dioxide for high-performance supercapacitors. *J. Energy Chem.* **2021**, *54*, 352–367. [CrossRef]
3. Agajie, T.F.; Ali, A.; Fopah-Lele, A.; Amoussou, I.; Khan, B.; Velasco, C.L.R.; Tanyi, E. A Comprehensive Review on Techno-Economic Analysis and Optimal Sizing of Hybrid Renewable Energy Sources with Energy Storage Systems. *Energies* **2023**, *16*, 642. [CrossRef]
4. Deka, B.; Cho, K.-H. $BiFeO_3$-Based Relaxor Ferroelectrics for Energy Storage: Progress and Prospects. *Materials* **2021**, *14*, 7188. [CrossRef]
5. Sayed, E.T.; Olabi, A.G.; Alami, A.H.; Radwan, A.; Mdallal, A.; Rezk, A.; Abdelkareem, M.A. Renewable Energy and Energy Storage Systems. *Energies* **2023**, *16*, 1415. [CrossRef]
6. Xu, Y.; Wang, K.; Zhang, X.; Ma, Y.; Peng, Q.; Gong, Y.; Yi, S.; Guo, H.; Zhang, X.; Sun, X. Improved Li-Ion Conduction and (Electro) Chemical Stability at Garnet-Polymer Interface through Metal-Nitrogen Bonding. *Adv. Energy Mater.* **2023**, *13*, 2204377. [CrossRef]
7. Yang, L.; Kong, X.; Li, F.; Hao, H.; Cheng, Z.; Liu, H.; Li, J.-F.; Zhang, S. Perovskite lead-free dielectrics for energy storage applications. *Prog. Mater. Sci.* **2019**, *102*, 72–108. [CrossRef]
8. Navarro, G.; Torres, J.; Blanco, M.; Nájera, J.; Santos-Herran, M.; Lafoz, M. Present and Future of Supercapacitor Technology Applied to Powertrains, Renewable Generation and Grid Connection Applications. *Energies* **2021**, *14*, 3060. [CrossRef]

9. Liu, L.M.; Qu, J.L.; Gu, A.J.; Wang, B.H. Percolative polymer composites for dielectric capacitors: A brief history, materials, and multilayer interface design. *J. Mater. Chem. A* **2020**, *8*, 18515–18537. [CrossRef]

10. Conte, M. Supercapacitors Technical Requirements for New Applications. *Fuel Cells* **2010**, *10*, 806–818. [CrossRef]

11. Aravindan, V.; Gnanaraj, J.; Lee, Y.-S.; Madhavi, S. Insertion-Type Electrodes for Nonaqueous Li-Ion Capacitors. *Chem. Rev.* **2014**, *114*, 11619–11635. [CrossRef]

12. Fitzgerald, D. Improvements in electrical condensers or accumulators. *Br. Pat.* **1876**, *3466*, 1876.

13. McLean, D. Metallized paper for capacitors. *Proc. IRE* **1950**, *38*, 1010–1014. [CrossRef]

14. McLean, D.; Wehe, H. Miniature lacquer film capacitors. *Proc. IRE* **1954**, *42*, 1799–1805. [CrossRef]

15. Ho, J.; Jow, T.R.; Boggs, S. Historical introduction to capacitor technology. *IEEE Electr. Insul. Mag.* **2010**, *26*, 20–25. [CrossRef]

16. Simon, P.; Gogotsi, Y. Materials for electrochemical capacitors. *Nat. Mater.* **2008**, *7*, 845–854. [CrossRef]

17. Dong, J.-F.; Deng, X.-L.; Niu, Y.-J.; Pan, Z.-Z.; Wang, H. Research progress of polymer based dielectrics for high-temperature capacitor energy storage. *Acta Physica Sinica* **2020**, *69*, 217701. [CrossRef]

18. MacDougall, F.W.; Ennis, J.B.; Cooper, R.A.; Bates, J.; Seal, K. High energy density pulsed power capacitors. In Proceedings of the 14th IEEE International Pulsed Power Conference, Dallas, TX, USA, 15–18 June 2003; pp. 513–517.

19. Prateek; Thakur, V.K.; Gupta, R.K. Recent Progress on Ferroelectric Polymer-Based Nanocomposites for High Energy Density Capacitors: Synthesis, Dielectric Properties, and Future Aspects. *Chem. Rev.* **2016**, *116*, 4260–4317. [CrossRef]

20. Singh, M.; Apata, I.E.; Samant, S.; Wu, W.J.; Tawade, B.V.; Pradhan, N.; Raghavan, D.; Karim, A. Nanoscale Strategies to Enhance the Energy Storage Capacity of Polymeric Dielectric Capacitors: Review of Recent Advances. *Polym. Rev.* **2022**, *62*, 211–260. [CrossRef]

21. Dang, Z.-M. Polymer nanocomposites with high permittivity. In *Nanocrystalline Materials*; Elsevier: Amsterdam, The Netherlands, 2014; pp. 305–333.

22. Sherrill, S.A.; Banerjee, P.; Rubloff, G.W.; Lee, S.B. High to ultra-high power electrical energy storage. *Phys. Chem. Chem. Phys.* **2011**, *13*, 20714–20723. [CrossRef]

23. Sun, L.; Shi, Z.C.; Liang, L.; Wei, S.; Wang, H.L.; Dastan, D.; Sun, K.; Fan, R.H. Layer-structured BaTiO$_3$/P(VDF-HFP) composites with concurrently improved dielectric permittivity and breakdown strength toward capacitive energy-storage applications. *J. Mater. Chem. C* **2020**, *8*, 10257–10265. [CrossRef]

24. Fan, X.; Wang, J.; Yuan, H.; Zheng, Z.; Zhang, J.; Zhu, K. Multi-scale synergic optimization strategy for dielectric energy storage ceramics. *J. Adv. Ceram.* **2023**, *12*, 649–680. [CrossRef]

25. Yang, M.; Guo, M.; Xu, E.; Ren, W.; Wang, D.; Li, S.; Zhang, S.; Nan, C.-W.; Shen, Y. Polymer nanocomposite dielectrics for capacitive energy storage. *Nat. Nanotechnol.* **2024**, *19*, 588–603. [CrossRef]

26. Xiao, M.; Zhang, Z.; Du, B. Dielectric performance improvement of polypropylene film by hierarchical structure design for metallized film capacitors. *J. Phys. D Appl. Phys.* **2024**, *57*, 385501. [CrossRef]

27. Hu, T.-Y.; Ma, C.; Cheng, S.-d.; Hu, G.; Liu, M. Ultrahigh-temperature capacitors realized by controlling polarization behavior in relaxor ferroelectric. *Chem. Eng. J.* **2024**, *492*, 152365. [CrossRef]

28. Gnonhoue, O.G.; Velazquez-Salazar, A.; David, É.; Preda, I. Review of Technologies and Materials Used in High-Voltage Film Capacitors. *Polymers* **2021**, *13*, 766. [CrossRef]

29. Reuter, R.C., Jr.; Allen, J.J. Prediction of Mechanical States in Wound Capacitors. *J. Mech. Des.* **1991**, *113*, 387–392. [CrossRef]

30. Sarjeant, W.J.; Zirnheld, J.; MacDougall, F.W. Capacitors. *IEEE Trans. Plasma Sci.* **1998**, *26*, 1368–1392. [CrossRef]

31. Tahalyani, J.; Akhtar, M.J.; Cherusseri, J.; Kar, K.K. Characteristics of Capacitor: Fundamental Aspects. In *Handbook of Nanocomposite Supercapacitor Materials I: Characteristics*; Kar, K.K., Ed.; Springer International Publishing: Cham, Switzerland, 2020; pp. 1–51.

32. Fan, B.H.; Zhou, M.Y.; Zhang, C.; He, D.L.; Bai, J.B. Polymer-based materials for achieving high energy density film capacitors. *Prog. Polym. Sci.* **2019**, *97*, 101143. [CrossRef]

33. Xiong, J.; Wang, X.; Zhang, X.; Xie, Y.C.; Lu, J.Y.; Zhang, Z.C. How the biaxially stretching mode influence dielectric and energy storage properties of polypropylene films. *J. Appl. Polym. Sci.* **2021**, *138*, 50029. [CrossRef]

34. Streibl, M.; Karmazin, R.; Moos, R. Materials and applications of polymer films for power capacitors with special respect to nanocomposites. *IEEE Trans. Dielectr. Electr. Insul.* **2018**, *25*, 2429–2442. [CrossRef]

35. Ran, Z.Y.; Du, B.X.; Xiao, M.; Li, J. Crystallization Morphology-Dependent Breakdown Strength of Polypropylene Films for Converter Valve Capacitor. *IEEE Trans. Dielectr. Electr. Insul.* **2021**, *28*, 964–971. [CrossRef]

36. Ping, J.-B.; Feng, Q.-K.; Zhang, Y.-X.; Wang, X.-J.; Huang, L.; Zhong, S.-L.; Dang, Z.-M. A Bilayer High-Temperature Dielectric Film with Superior Breakdown Strength and Energy Storage Density. *Nano-Micro Lett.* **2023**, *15*, 154. [CrossRef]

37. Zou, C.; Zhang, Q.; Zhang, S.; Kushner, D.; Zhou, X.; Bernard, R.; Orchard, R.J., Jr. PEN/Si$_3$N$_4$ bilayer film for dc bus capacitors in power converters in hybrid electric vehicles. *J. Vac. Sci. Technol. B* **2011**, *29*, 061401. [CrossRef]

38. Zhang, C.; Yan, W.; Zhang, T.; Zhang, T.; Zhang, Y.; Zhang, Y.; Tang, C.; Chi, Q. Sandwich-Structured PC/PVDF-Based Energy Storage Dielectric with Inorganic Doping to Regulate Electric Field Distribution. *J. Phys. Chem. C* **2024**, *128*, 5717–5730. [CrossRef]

39. Reed, C.W.; Cichanowski, S.W. The Fundamentals of Aging in HV Polymer-Film Capacitors. *IEEE Trans. Dielectr. Electr. Insul.* **1994**, *1*, 904–922. [CrossRef]

40. Feng, M.N.; Chen, M.; Qiu, J.; He, M.; Huang, Y.M.; Lin, J. Improving dielectric properties of poly(arylene ether nitrile) composites by employing core-shell structured $BaTiO_3$@polydopamine and MoS_2@polydopamine interlinked with poly(ethylene imine) for high-temperature applications. *J. Alloys Compd.* **2021**, *856*, 158213. [CrossRef]

41. Valentine, N.; Azarian, M.H.; Pecht, M. Metallized film capacitors used for EMI filtering: A reliability review. *Microelectron. Reliab.* **2019**, *92*, 123–135. [CrossRef]

42. Ho, J.S.; Greenbaum, S.G. Polymer Capacitor Dielectrics for High Temperature Applications. *ACS Appl. Mater. Interfaces* **2018**, *10*, 29189–29218. [CrossRef]

43. Li, Q.; Cheng, S. Polymer nanocomposites for high-energy-density capacitor dielectrics: Fundamentals and recent progress. *IEEE Electr. Insul. Mag.* **2020**, *36*, 7–28. [CrossRef]

44. Rabuffi, M.; Picci, G. Status quo and future prospects for metallized polypropylene energy storage capacitors. *IEEE Trans. Plasma Sci.* **2002**, *30*, 1939–1942. [CrossRef]

45. Bai, G.; Chen, Z.; Liu, J.; Wang, F.; Zhang, Y. Microstructure Evolution and Performance Enhancement of Sintered Aluminum Foils for Aluminum Electrolytic Capacitors. *J. Electron. Mater.* **2024**, *53*, 2026–2039. [CrossRef]

46. Zeng, X.; Bian, J.; Liang, L.; Cao, Q.; Liu, L.; Chen, X.; Wang, Y.; Xie, X.; Xie, G. Preparation and characterization of anode foil for aluminum electrolytic capacitors by powder additive manufacturing. *Powder Technol.* **2023**, *426*, 118602. [CrossRef]

47. Chen, D.; Fu, J.; Huang, S.; Huang, J.; Yang, J.; Ren, S.; Ma, J. Design of La-based MG-Ta composite with high and tailorable properties for solid Ta electrolytic capacitor. *Mater. Des.* **2024**, *238*, 112743. [CrossRef]

48. Narale, S.B.; Verma, A.; Anand, S. Structure and Degradation of Aluminum Electrolytic Capacitors. In Proceedings of the 2019 National Power Electronics Conference (NPEC), Tiruchirappalli, India, 13–15 December 2019; pp. 1–6.

49. Bramoulle, M. Electrolytic or film capacitors? In Proceedings of the Conference Record of 1998 IEEE Industry Applications Conference. Thirty-Third IAS Annual Meeting (Cat. No.98CH36242), St. Loius, MI, USA, 12–15 October 1998; Volume 1132, pp. 1138–1141.

50. Torki, J.; Joubert, C.; Sari, A. Electrolytic capacitor: Properties and operation. *J. Energy Storage* **2023**, *58*, 106330. [CrossRef]

51. EPCOS, A. Aluminum Electrolytic Capacitors–General Technical Information. 2019. Available online: https://www.tdk-electronics.tdk.com/download/185386/e724fb43668a157bc547c65b0cff75f8/pdf-generaltechnicalinformation.pdf (accessed on 3 May 2024).

52. Chen, X.; Xi, L.; Zhang, Y.; Ma, H.; Huang, Y.; Chen, Y. Fractional techniques to characterize non-solid aluminum electrolytic capacitors for power electronic applications. *Nonlinear Dyn.* **2019**, *98*, 3125–3141. [CrossRef]

53. Both, J. The modern era of aluminum electrolytic capacitors. *IEEE Electr. Insul. Mag.* **2015**, *31*, 24–34. [CrossRef]

54. Film, F.C.M.P. 1. General Description of Aluminum Electrolytic Capacitors. 2015. Available online: https://www.nichicon.co.jp/english/products/pdf/aluminum.pdf (accessed on 3 May 2024).

55. McLean, D.A.; Power, F.S. Tantalum Solid Electrolytic Capacitors. *Proc. IRE* **1956**, *44*, 872–878. [CrossRef]

56. Freeman, Y.; Lessner, P.; Luzinov, I. Reliability and Failure Mode in Solid Tantalum Capacitors. *Ecs J. Solid State Sci. Technol.* **2021**, *10*, 045007. [CrossRef]

57. Nishino, A. Capacitors: Operating principles, current market and technical trends. *J. Power Sources* **1996**, *60*, 137–147. [CrossRef]

58. Romero, J.A.; Azarian, M.H.; Pecht, M. Life model for tantalum electrolytic capacitors with conductive polymers. *Microelectron. Reliab.* **2020**, *104*, 113550. [CrossRef]

59. Gill, J. Basic Tantalum Capacitor Technology. Available online: https://kyocera-avx.com/docs/techinfo/Tantalum-NiobiumCapacitors/bsctant.pdf (accessed on 3 May 2024).

60. Freeman, Y. Basic Technology. In *Tantalum and Niobium-Based Capacitors: Science, Technology, and Applications*; Freeman, Y., Ed.; Springer International Publishing: Cham, Switzerland, 2022; pp. 23–52.

61. Kaiser, C.J. *The Capacitor Handbook*; Springer Science & Business Media: Berlin/Heidelberg, Germany, 2012.

62. Teverovsky, A. Breakdown and Self-healing in Tantalum Capacitors. *IEEE Trans. Dielectr. Electr. Insul.* **2021**, *28*, 663–671. [CrossRef]

63. Jia, W.X.; Hou, Y.D.; Zheng, M.P.; Xu, Y.R.; Zhu, M.K.; Yang, K.Y.; Cheng, H.R.; Sun, S.Y.; Xing, J. Advances in lead-free high-temperature dielectric materials for ceramic capacitor applicationInspec keywordsOther keywords. *Iet Nanodielectr.* **2018**, *1*, 3–16. [CrossRef]

64. Randall, M.-J.P.C.A. A brief introduction to ceramic capacitors. *IEEE Electr. Insul. Mag.* **2010**, *26*, 44–50. [CrossRef]

65. Park, Y.; Park, J.J.; Park, K.S.; Hong, Y.M.; Lee, E.J.; Kim, S.O.; Lee, J.H. Enhancement of Mechanical Properties of Multilayer Ceramic Capacitors through a $BaTiO_3$/polydopamine Cover Layer. *Polymers* **2023**, *15*, 4014. [CrossRef]

66. Lv, Z.; Lu, T.; Liu, Z.; Hu, T.; Hong, Z.; Guo, S.; Xu, Z.; Song, Y.; Chen, Y.; Zhao, X.; et al. $NaNbO_3$-Based Multilayer Ceramic Capacitors with Ultrahigh Energy Storage Performance. *Adv. Energy Mater.* **2024**, *14*, 2304291. [CrossRef]

67. Van Trinh, H.; Talbot, J.B. Electrodeposition Method for Terminals of Multilayer Ceramic Capacitors. *J. Am. Ceram. Soc.* **2004**, *86*, 905–909. [CrossRef]

68. Wojewoda, L.E.; Hill, M.J.; Radhakrishnan, K.; Goyal, N. Use Condition Characterization of MLCCs. *IEEE Trans. Adv. Packag.* **2009**, *32*, 109–115. [CrossRef]

69. Gong, H.; Wang, X.; Tian, Z.; Zhang, H.; Li, L. Interfacial diffusion behavior in Ni-$BaTiO_3$ MLCCs with ultra-thin active layers. *Electron. Mater. Lett.* **2014**, *10*, 417–421. [CrossRef]

70. Sumithra, S.; Annapoorani, K.; Ellmore, A.; Vaidhyanathan, B. Microwave assisted processing of X8R nanocrystalline BaTiO$_3$ based ceramic capacitors and multilayer devices. *Open Ceram.* **2022**, *9*, 100214. [CrossRef]

71. Wang, Y.Q.; Ko, B.H.; Jeong, S.G.; Park, K.S.; Park, N.C.; Park, Y.P. Analysis of the influence of soldering parameters on multi-layer ceramic capacitor vibration. *Microsyst. Technol.-Micro-Nanosyst.-Inf. Storage Process. Syst.* **2015**, *21*, 2565–2571. [CrossRef]

72. Hong, K.; Lee, T.H.; Suh, J.M.; Yoon, S.H.; Jang, H.W. Perspectives and challenges in multilayer ceramic capacitors for next generation electronics. *J. Mater. Chem. C* **2019**, *7*, 9782–9802. [CrossRef]

73. Zhang, H.B.; Wei, T.; Zhang, Q.; Ma, W.G.; Fan, P.Y.; Salamon, D.; Zhang, S.T.; Nan, B.; Tan, H.; Ye, Z.G. A review on the development of lead-free ferroelectric energy-storage ceramics and multilayer capacitors. *J. Mater. Chem. C* **2020**, *8*, 16648–16667. [CrossRef]

74. Gurav, A.; Xu, X.; Freeman, Y.; Reed, E. KEMET Electronics: Breakthroughs in Capacitor Technology. In *Materials Research for Manufacturing: An Industrial Perspective of Turning Materials into New Products*; Madsen, L.D., Svedberg, E.B., Eds.; Springer International Publishing: Cham, Switzerland, 2016; pp. 93–129.

75. Laadjal, K.; Cardoso, A.J.M. Multilayer Ceramic Capacitors: An Overview of Failure Mechanisms, Perspectives, and Challenges. *Electronics* **2023**, *12*, 1297. [CrossRef]

76. Miller, J.R. Perspective on electrochemical capacitor energy storage. *Appl. Surf. Sci.* **2018**, *460*, 3–7. [CrossRef]

77. Zheng, L.X.; Xu, P.H.; Zhao, Y.J.; Peng, J.X.; Yang, P.J.; Shi, X.W.; Zheng, H.J. Unique core-shell Co$_2$(OH)$_2$CO$_3$@MOF nanoarrays with remarkably improved cycling life for high performance pseudocapacitors. *Electrochim. Acta* **2022**, *412*, 140142. [CrossRef]

78. Yi, S.; Wang, L.; Zhang, X.; Li, C.; Xu, Y.A.; Wang, K.; Sun, X.Z.; Ma, Y.W. Recent advances in MXene-based nanocomposites for supercapacitors. *Nanotechnology* **2023**, *34*, 432001. [CrossRef]

79. Zhao, J.; Burke, A.F. Review on supercapacitors: Technologies and performance evaluation. *J. Energy Chem.* **2021**, *59*, 276–291. [CrossRef]

80. Liu, Y.; Shearing, P.R.; He, G.; Brett, D.J.L. Supercapacitors: History, Theory, Emerging Technologies, and Applications. In *Advances in Sustainable Energy: Policy, Materials and Devices*; Gao, Y.-j., Song, W., Liu, J.L., Bashir, S., Eds.; Springer International Publishing: Cham, Switzerland, 2021; pp. 417–449.

81. Chen, Z.; Yu, D.; Xiong, W.; Liu, P.; Liu, Y.; Dai, L. Graphene-Based Nanowire Supercapacitors. *Langmuir* **2014**, *30*, 3567–3571. [CrossRef]

82. Snook, G.A.; Kao, P.; Best, A.S. Conducting-polymer-based supercapacitor devices and electrodes. *J. Power Sources* **2011**, *196*, 1–12. [CrossRef]

83. Becker, H. Low Voltage Electrolytic Capacitor. U. S. Patent 2,800,616A, 23 July 1957.

84. Shirakawa, H.; Louis, E.J.; MacDiarmid, A.G.; Chiang, C.K.; Heeger, A.J. Synthesis of electrically conducting organic polymers: Halogen derivatives of polyacetylene, (CH) x. *J. Chem. Soc. Chem. Commun.* **1977**, 578–580. [CrossRef]

85. Sun, J.; Luo, B.; Li, H. A Review on the Conventional Capacitors, Supercapacitors, and Emerging Hybrid Ion Capacitors: Past, Present, and Future. *Adv. Energy Sustain. Res.* **2022**, *3*, 2100191. [CrossRef]

86. Zan, G.; Li, S.; Chen, P.; Dong, K.; Wu, Q.; Wu, T. Mesoporous Cubic Nanocages Assembled by Coupled Monolayers With 100% Theoretical Capacity and Robust Cycling. *ACS Cent. Sci.* **2024**, *10*, 1283–1294. [CrossRef]

87. Shwetha, K.P.; Manjunatha, C.; Sudha Kamath, M.K.; Vinaykumar; Radhika, M.G.R.; Khosla, A. Morphology-controlled synthesis and structural features of ultrafine nanoparticles of Co$_3$O$_4$: An active electrode material for a supercapacitor. *Appl. Res.* **2022**, *1*, e202200031. [CrossRef]

88. Luo, Q.; Lu, C.; Liu, L.; Zhu, M. Triethanolamine assisted synthesis of bimetallic nickel cobalt nitride/nitrogen-doped carbon hollow nanoflowers for supercapacitor. *Microstructures* **2023**, *3*, 2023011.

89. Wang, T.; Wang, Y.; Lei, J.; Chen, K.-J.; Wang, H. Electrochemically induced surface reconstruction of Ni-Co oxide nanosheet arrays for hybrid supercapacitors. *Exploration* **2021**, *1*, 20210178. [CrossRef]

90. Song, H.; Wang, Y.; Fei, Q.; Nguyen, D.H.; Zhang, C.; Liu, T. Cryopolymerization-enabled self-wrinkled polyaniline-based hydrogels for highly stretchable all-in-one supercapacitors. *Exploration* **2022**, *2*, 20220006. [CrossRef]

91. Huai, X.; Liu, J.; Wu, X. Cobalt-doped NiMoO$_4$ nanosheet for high-performance flexible supercapacitor. *Chin. J. Struct. Chem.* **2023**, *42*, 100158. [CrossRef]

92. Liu, B.; Cao, Z.; Yang, Z.; Qi, W.; He, J.; Pan, P.; Li, H.; Zhang, P. Flexible micro-supercapacitors fabricated from MnO$_2$ nanosheet/graphene composites with black phosphorus additive. *Prog. Nat. Sci. Mater. Int.* **2022**, *32*, 10–19. [CrossRef]

93. Zhu, R.-J.; Liu, J.; Hua, C.; Pan, H.-Y.; Cao, Y.-J.; Li, M. Preparation of vanadium-based electrode materials and their research progress in solid-state flexible supercapacitors. *Rare Met.* **2024**, *43*, 431–454. [CrossRef]

94. Wang, T.; Hu, S.; Hu, Y.; Wu, D.; Wu, H.; Huang, J.; Wang, H.; Zhao, W.; Yu, W.; Wang, M.; et al. Biologically inspired anthraquinone redox centers and biomass graphene for renewable colloidal gels toward ultrahigh-performance flexible micro-supercapacitors. *J. Mater. Sci. Technol.* **2023**, *151*, 178–189. [CrossRef]

95. Li, J.; Dong, Z.; Chen, R.; Wu, Q.; Zan, G. Advanced nickel-based composite materials for supercapacitor electrodes. *Ionics* **2024**, *30*, 1833–1855. [CrossRef]

96. Troschke, E.; Oschatz, M.; Ilic, I.K. Schiff-bases for sustainable battery and supercapacitor electrodes. *Exploration* **2021**, *1*, 20210128. [CrossRef] [PubMed]

97. Shi, J.; Tian, X.; Yang, T.; Hu, S.; Liu, Z. Redox electrolyte-enhanced carbon-based supercapacitors: Recent advances and future perspectives. *Energy Mater. Devices* **2023**, *1*, 9370009. [CrossRef]

98. Wu, X.; Liu, R.; Zhao, J.; Fan, Z. Advanced carbon materials with different spatial dimensions for supercapacitors. *Nano Mater. Sci.* **2021**, *3*, 241–267. [CrossRef]

99. Kong, D.; Lv, W.; Liu, R.; He, Y.-B.; Wu, D.; Li, F.; Fu, R.; Yang, Q.-H.; Kang, F. Superstructured carbon materials: Design and energy applications. *Energy Mater. Devices* **2023**, *1*, 9370017. [CrossRef]

100. Wang, G.; Zhang, L.; Zhang, J. A review of electrode materials for electrochemical supercapacitors. *Chem. Soc. Rev.* **2012**, *41*, 797–828. [CrossRef] [PubMed]

101. Zhou, W.; Liu, Z.; Chen, W.; Sun, X.; Luo, M.; Zhang, X.; Li, C.; An, Y.; Song, S.; Wang, K. A Review on Thermal Behaviors and Thermal Management Systems for Supercapacitors. *Batteries* **2023**, *9*, 128. [CrossRef]

102. Shukla, A.K.; Banerjee, A.; Ravikumar, M.K.; Jalajakshi, A. Electrochemical capacitors: Technical challenges and prognosis for future markets. *Electrochim. Acta* **2012**, *84*, 165–173. [CrossRef]

103. Terasawa, N.; Asaka, K. High-Performance Hybrid (Electrostatic Double-Layer and Faradaic Capacitor-Based) Polymer Actuators Incorporating Nickel Oxide and Vapor-Grown Carbon Nanofibers. *Langmuir* **2014**, *30*, 14343–14351. [CrossRef]

104. Mohd Abdah, M.A.A.; Azman, N.H.N.; Kulandaivalu, S.; Sulaiman, Y. Review of the use of transition-metal-oxide and conducting polymer-based fibres for high-performance supercapacitors. *Mater. Des.* **2020**, *186*, 108199. [CrossRef]

105. Choi, H.; Yoon, H. Nanostructured electrode materials for electrochemical capacitor applications. *Nanomaterials* **2015**, *5*, 906–936. [CrossRef] [PubMed]

106. Pal, B.; Yang, S.; Ramesh, S.; Thangadurai, V.; Jose, R. Electrolyte selection for supercapacitive devices: A critical review. *Nanoscale Adv.* **2019**, *1*, 3807–3835. [CrossRef] [PubMed]

107. Zhang, L.L.; Zhao, X.S. Carbon-based materials as supercapacitor electrodes. *Chem. Soc. Rev.* **2009**, *38*, 2520–2531. [CrossRef] [PubMed]

108. Brza, M.A.; Aziz, S.B.; Anuar, H.; Ali, F.; Hamsan, M.H.; Kadir, M.F.Z.; Abdulwahid, R.T. Metal framework as a novel approach for the fabrication of electric double layer capacitor device with high energy density using plasticized Poly(vinyl alcohol): Ammonium thiocyanate based polymer electrolyte. *Arab. J. Chem.* **2020**, *13*, 7247–7263. [CrossRef]

109. Yao, F.; Pham, D.T.; Lee, Y.H. Carbon-Based Materials for Lithium-Ion Batteries, Electrochemical Capacitors, and Their Hybrid Devices. *ChemSusChem* **2015**, *8*, 2284–2311. [CrossRef]

110. Banerjee, S.; Sinha, P.; Verma, K.D.; Pal, T.; De, B.; Cherusseri, J.; Manna, P.K.; Kar, K.K. Capacitor to Supercapacitor. In *Handbook of Nanocomposite Supercapacitor Materials I: Characteristics*; Kar, K.K., Ed.; Springer International Publishing: Cham, Switzerland, 2020; pp. 53–89.

111. Forse, A.C.; Merlet, C.; Griffin, J.M.; Grey, C.P. New Perspectives on the Charging Mechanisms of Supercapacitors. *J. Am. Chem. Soc.* **2016**, *138*, 5731–5744. [CrossRef]

112. Frackowiak, E.; Béguin, F. Carbon materials for the electrochemical storage of energy in capacitors. *Carbon* **2001**, *39*, 937–950. [CrossRef]

113. Bizuneh, G.G.; Adam, A.M.M.; Ma, J. Progress on carbon for electrochemical capacitors. *Battery Energy* **2023**, *2*, 20220021. [CrossRef]

114. Li, C.; Zhang, X.; Lv, Z.S.; Wang, K.; Sun, X.Z.; Chen, X.D.; Ma, Y.W. Scalable combustion synthesis of graphene-welded activated carbon for high-performance supercapacitors. *Chem. Eng. J.* **2021**, *414*, 128781. [CrossRef]

115. Du, C.; Yeh, J.; Pan, N. High power density supercapacitors using locally aligned carbon nanotube electrodes. *Nanotechnology* **2005**, *16*, 350. [CrossRef]

116. Stoller, M.D.; Park, S.; Zhu, Y.; An, J.; Ruoff, R.S. Graphene-Based Ultracapacitors. *Nano Lett.* **2008**, *8*, 3498–3502. [CrossRef]

117. Yu, X.; Lu, B.; Xu, Z. Super Long-Life Supercapacitors Based on the Construction of Nanohoneycomb-Like Strongly Coupled CoMoO$_4$–3D Graphene Hybrid Electrodes. *Adv. Mater.* **2014**, *26*, 990. [CrossRef]

118. Fu, G.; Li, H.; Bai, Q.; Li, C.; Shen, Y.; Uyama, H. Dual-doping activated carbon with hierarchical pore structure derived from polymeric porous monolith for high performance EDLC. *Electrochim. Acta* **2021**, *375*, 137927. [CrossRef]

119. Lv, X.W.; Ji, S.; Wang, Z.H.; Wang, X.Y.; Wang, H.; Wang, R.F. Fabrication of highly-conductive porous capacitor electrodes by the insertion of Cu-nanoparticles into N-doped flocculated carbon catalysts. *J. Colloid Interface Sci.* **2022**, *610*, 106–115. [CrossRef]

120. Yoo, J.; Yang, I.; Kwon, D.; Jung, M.; Kim, M.-S.; Jung, J.C. Low-Cost Carbon Xerogels Derived from Phenol–Formaldehyde Resin for Organic Electric Double-Layer Capacitors. *Energy Technol.* **2021**, *9*, 2000918. [CrossRef]

121. Thirumurugan, A.; Ramakrishnan, K.; Ramadoss, A.; Thandapani, P.; Thangavelu, P.; Udayabhaskar, R.; Morel, M.J.; Dhanabalan, S.S.; Dineshbabu, N.; Ravichandran, K.; et al. Nanostructured Materials for Supercapacitors. In *Nanostructured Materials for Supercapacitors*; Thomas, S., Gueye, A.B., Gupta, R.K., Eds.; Springer International Publishing: Cham, Switzerland, 2022; pp. 1–26.

122. Ates, M.; Chebil, A.; Yoruk, O.; Dridi, C.; Turkyilmaz, M. Reliability of electrode materials for supercapacitors and batteries in energy storage applications: A review. *Ionics* **2021**, *28*, 27–52. [CrossRef]

123. Deng, T.; Zhang, W.; Arcelus, O.; Kim, J.-G.; Carrasco, J.; Yoo, S.J.; Zheng, W.; Wang, J.; Tian, H.; Zhang, H.; et al. Atomic-level energy storage mechanism of cobalt hydroxide electrode for pseudocapacitors. *Nat. Commun.* **2017**, *8*, 15194. [CrossRef] [PubMed]

124. Huang, M.; Li, F.; Dong, F.; Zhang, Y.X.; Zhang, L.L. MnO$_2$-based nanostructures for high-performance supercapacitors. *J. Mater. Chem. A* **2015**, *3*, 21380–21423. [CrossRef]

125. Waqas Hakim, M.; Fatima, S.; Rizwan, S.; Mahmood, A. Pseudo-capacitors: Introduction, Controlling Factors and Future. In *Nanostructured Materials for Supercapacitors*; Thomas, S., Gueye, A.B., Gupta, R.K., Eds.; Springer International Publishing: Cham, Switzerland, 2022; pp. 53–70.

126. Herrero, E.; Buller, L.J.; Abruña, H.D. Underpotential Deposition at Single Crystal Surfaces of Au, Pt, Ag and Other Materials. *Chem. Rev.* **2001**, *101*, 1897–1930. [CrossRef]

127. Brezesinski, K.; Wang, J.; Haetge, J.; Reitz, C.; Steinmueller, S.O.; Tolbert, S.H.; Smarsly, B.M.; Dunn, B.; Brezesinski, T. Pseudocapacitive Contributions to Charge Storage in Highly Ordered Mesoporous Group V Transition Metal Oxides with Iso-Oriented Layered Nanocrystalline Domains. *J. Am. Chem. Soc.* **2010**, *132*, 6982–6990. [CrossRef] [PubMed]

128. Yu, G.; Xie, X.; Pan, L.; Bao, Z.; Cui, Y. Hybrid nanostructured materials for high-performance electrochemical capacitors. *Nano Energy* **2013**, *2*, 213–234. [CrossRef]

129. Gao, D.; Luo, Z.L.; Liu, C.H.; Fan, S.S. A survey of hybrid energy devices based on supercapacitors. *Green Energy Environ.* **2023**, *8*, 972–988. [CrossRef]

130. Lokhande, C.D.; Dubal, D.P.; Joo, O.-S. Metal oxide thin film based supercapacitors. *Curr. Appl. Phys.* **2011**, *11*, 255–270. [CrossRef]

131. Yun, Y.S.; Kim, D.H.; Hong, S.J.; Park, M.H.; Park, Y.W.; Kim, B.H.; Jin, H.J.; Kang, K. Microporous carbon nanosheets with redox-active heteroatoms for pseudocapacitive charge storage. *Nanoscale* **2015**, *7*, 15051–15058. [CrossRef] [PubMed]

132. Sha, Z.; Zhou, Y.; Huang, F.; Yang, W.M.; Yu, Y.Y.; Zhang, J.; Wu, S.Y.; Brown, S.A.; Peng, S.H.; Han, Z.J.; et al. Carbon fibre electrodes for ultra long cycle life pseudocapacitors by engineering the nano-structure of vertical graphene and manganese dioxide. *Carbon* **2021**, *177*, 260–270. [CrossRef]

133. Xue, J.Y.; Zhang, F.M.; Ma, W.L.; Wang, M.M.; Cui, H.T. Hierarchically structured nanofelt-like β-NiOOH grown on nickel foam as electrode for high performance pseudocapacitor. *J. Nanopart. Res.* **2015**, *17*, 1–7. [CrossRef]

134. Kayali, E.; VahidMohammadi, A.; Orangi, J.; Beidaghi, M. Controlling the Dimensions of 2D MXenes for Ultrahigh-Rate Pseudocapacitive Energy Storage. *ACS Appl. Mater. Interfaces* **2018**, *10*, 25949–25954. [CrossRef] [PubMed]

135. Cao, L.; Li, H.; Xu, Z.; Zhang, H.; Ding, L.; Wang, S.; Zhang, G.; Hou, H.; Xu, W.; Yang, F.; et al. Comparison of the heteroatoms-doped biomass-derived carbon prepared by one-step nitrogen-containing activator for high performance supercapacitor. *Diam. Relat. Mater.* **2021**, *114*, 108316. [CrossRef]

136. Wang, Y.; Guo, J.; Wang, T.; Shao, J.; Wang, D.; Yang, Y.-W. Mesoporous Transition Metal Oxides for Supercapacitors. *Nanomaterials* **2015**, *5*, 1667–1689. [CrossRef]

137. Zhou, Q.; Li, X.; Li, Y.-G.; Tian, B.-Z.; Zhao, D.-Y.; Jiang, Z.-Y. Synthesis and Electrochemical Properties of Semicrystalline Gyroidal Mesoporous MnO$_2$. *Chin. J. Chem.* **2006**, *24*, 835–839. [CrossRef]

138. Nayak, P.K.; Munichandraiah, N. Rapid sonochemical synthesis of mesoporous MnO$_2$ for supercapacitor applications. *Mater. Sci. Eng. B* **2012**, *177*, 849–854. [CrossRef]

139. Fleischmann, S.; Mitchell, J.B.; Wang, R.; Zhan, C.; Jiang, D.-e.; Presser, V.; Augustyn, V. Pseudocapacitance: From fundamental understanding to high power energy storage materials. *Chem. Rev.* **2020**, *120*, 6738–6782. [CrossRef] [PubMed]

140. Cao, Z.; Zhu, Y.-B.; Chen, K.; Wang, Q.; Li, Y.; Xing, X.; Ru, J.; Meng, L.-G.; Shu, J.; Shpigel, N.; et al. Super-Stretchable and High-Energy Micro-Pseudocapacitors Based on MXene Embedded Ag Nanoparticles. *Adv. Mater.* **2024**, *36*, 2401271. [CrossRef] [PubMed]

141. Kim, S.K.; Kim, S.A.; Han, Y.S.; Jung, K.-H. Supercapacitor Performance of MXene-Coated Carbon Nanofiber Electrodes. *C* **2024**, *10*, 32. [CrossRef]

142. Muzaffar, A.; Ahamed, M.B.; Deshmukh, K.; Thirumalai, J. A review on recent advances in hybrid supercapacitors: Design, fabrication and applications. *Renew. Sustain. Energy Rev.* **2019**, *101*, 123–145. [CrossRef]

143. Kumar, N.; Kim, S.-B.; Lee, S.-Y.; Park, S.-J. Recent Advanced Supercapacitor: A Review of Storage Mechanisms, Electrode Materials, Modification, and Perspectives. *Nanomaterials* **2022**, *12*, 3708. [CrossRef]

144. Lu, Y.; Zhang, S.; Yin, J.; Bai, C.; Zhang, J.; Li, Y.; Yang, Y.; Ge, Z.; Zhang, M.; Wei, L.; et al. Mesoporous activated carbon materials with ultrahigh mesopore volume and effective specific surface area for high performance supercapacitors. *Carbon* **2017**, *124*, 64–71. [CrossRef]

145. Seong, K.-d.; Jin, X.; Kim, D.; Kim, J.M.; Ko, D.; Cho, Y.; Hwang, M.; Kim, J.-H.; Piao, Y. Ultrafast and scalable microwave-assisted synthesis of activated hierarchical porous carbon for high-performance supercapacitor electrodes. *J. Electroanal. Chem.* **2020**, *874*, 114464. [CrossRef]

146. Gao, Y.; Zhang, Y.; Zhang, Y.; Xie, L.; Li, X.; Su, F.; Wei, X.; Xu, Z.; Chen, C.; Cai, R. Three-dimensional paper-like graphene framework with highly oriented laminar structure as binder-free supercapacitor electrode. *J. Energy Chem.* **2016**, *25*, 49–54. [CrossRef]

147. Yan, Z.; Gao, Z.; Zhang, Z.; Dai, C.; Wei, W.; Shen, P.K. Graphene Nanosphere as Advanced Electrode Material to Promote High Performance Symmetrical Supercapacitor. *Small* **2021**, *17*, 2007915. [CrossRef]

148. Yang, B.; Chen, J.; Liu, B.; Ding, Y.; Tang, Y.; Yan, X. One dimensional graphene nanoscroll-wrapped MnO nanoparticles for high-performance lithium ion hybrid capacitors. *J. Mater. Chem. A* **2021**, *9*, 6352–6360. [CrossRef]

149. Chen, P.; Zhou, W.; Xiao, Z.; Li, S.; Chen, H.; Wang, Y.; Wang, Z.; Xi, W.; Xia, X.; Xie, S. In situ anchoring MnO nanoparticles on self-supported 3D interconnected graphene scroll framework: A fast kinetics boosted ultrahigh-rate anode for Li-ion capacitor. *Energy Storage Mater.* **2020**, *33*, 298–308. [CrossRef]

150. Wang, X.; Sumboja, A.; Khoo, E.; Yan, C.; Lee, P.S. Cryogel Synthesis of Hierarchical Interconnected Macro-/Mesoporous Co$_3$O$_4$ with Superb Electrochemical Energy Storage. *J. Phys. Chem. C* **2012**, *116*, 4930–4935. [CrossRef]

151. Liu, X.; Long, Q.; Jiang, C.; Zhan, B.; Li, C.; Liu, S.; Zhao, Q.; Huang, W.; Dong, X. Facile and green synthesis of mesoporous Co$_3$O$_4$ nanocubes and their applications for supercapacitors. *Nanoscale* **2013**, *5*, 6525–6529. [CrossRef]

152. Hu, T.; Li, J.; Wang, Y.; Chen, S.; Yu, T.; Cheng, H.-M.; Sun, Z.; Xu, Q.; Li, F. Coupling between cathode and anode in hybrid charge storage. *Joule* **2023**, *7*, 1176–1205. [CrossRef]

153. Xu, Y.; Ruan, J.; Pang, Y.; Sun, H.; Liang, C.; Li, H.; Yang, J.; Zheng, S. Homologous Strategy to Construct High-Performance Coupling Electrodes for Advanced Potassium-Ion Hybrid Capacitors. *Nano-Micro Lett.* **2020**, *13*, 14. [CrossRef]

154. Sun, X.; An, Y.; Zhang, X.; Wang, K.; Yuan, C.; Zhang, X.; Li, C.; Xu, Y.; Ma, Y. Unveil Overcharge Performances of Activated Carbon Cathode in Various Li-Ion Electrolytes. *Batteries* **2022**, *9*, 11. [CrossRef]

155. Li, C.; An, Y.B.; Wang, L.; Wang, K.; Sun, X.Z.; Zhang, H.T.; Zhang, X.; Ma, Y.W. Balancing microcrystalline domains in hard carbon with robust kinetics for a 46.7 Wh kg^{-1} practical lithium-ion capacitor. *Chem. Eng. J.* **2024**, *485*, 149880. [CrossRef]

156. Luo, J.; Zhang, W.; Yuan, H.; Jin, C.; Zhang, L.; Huang, H.; Liang, C.; Xia, Y.; Zhang, J.; Gan, Y.; et al. Pillared Structure Design of MXene with Ultralarge Interlayer Spacing for High-Performance Lithium-Ion Capacitors. *ACS Nano* **2017**, *11*, 2459–2469. [CrossRef] [PubMed]

157. Gao, Y.; Zhou, Y.S.; Qian, M.; Li, H.M.; Redepenning, J.; Fan, L.S.; He, X.N.; Xiong, W.; Huang, X.; Majhouri-Samani, M. High-performance flexible solid-state supercapacitors based on MnO_2-decorated nanocarbon electrodes. *RSC Adv.* **2013**, *3*, 20613–20618. [CrossRef]

158. Wei, W.; Wang, L.; Liang, C.; Liu, W.; Li, C.; An, Y.; Zhang, L.; Sun, X.; Wang, K.; Zhang, H.; et al. Interface engineering of $CoSe_2$/N-doped graphene heterostructure with ultrafast pseudocapacitive kinetics for high-performance lithium-ion capacitors. *Chem. Eng. J.* **2023**, *474*, 145788. [CrossRef]

159. Wang, L.; Zhang, X.; Xu, Y.N.; Li, C.; Liu, W.J.; Yi, S.; Wang, K.; Sun, X.Z.; Wu, Z.S.; Ma, Y.W. Tetrabutylammonium-Intercalated 1T-MoS_2 Nanosheets with Expanded Interlayer Spacing Vertically Coupled on 2D Delaminated MXene for High-Performance Lithium-Ion Capacitors. *Adv. Funct. Mater.* **2021**, *31*, 2104286. [CrossRef]

160. Amatucci, G.G.; Badway, F.; Du Pasquier, A.; Zheng, T. An Asymmetric Hybrid Nonaqueous Energy Storage Cell. *J. Electrochem. Soc.* **2001**, *148*, A930. [CrossRef]

161. Du Pasquier, A.; Laforgue, A.; Simon, P.; Amatucci, G.G.; Fauvarque, J.-F. A Nonaqueous Asymmetric Hybrid $Li_4Ti_5O_{12}$/Poly(fluorophenylthiophene) Energy Storage Device. *J. Electrochem. Soc.* **2002**, *149*, A302. [CrossRef]

162. Li, H.; Cheng, L.; Xia, Y. A Hybrid Electrochemical Supercapacitor Based on a 5 V Li-Ion Battery Cathode and Active Carbon. *Electrochem. Solid-State Lett.* **2005**, *8*, A433. [CrossRef]

163. Khomenko, V.; Raymundo-Piñero, E.; Béguin, F. High-energy density graphite/AC capacitor in organic electrolyte. *J. Power Sources* **2008**, *177*, 643–651. [CrossRef]

164. Bhattacharjee, U.; Bhowmik, S.; Ghosh, S.; Martha, S.K. A perspective on the evolution and journey of different types of lithium-ion capacitors: Mechanisms, energy-power balance, applicability, and commercialization. *Sustain. Energy Fuels* **2023**, *7*, 2321–2338. [CrossRef]

165. Song, S.; Zhang, X.; An, Y.B.; Hu, T.; Sun, C.K.; Wang, L.; Li, C.; Zhang, X.H.; Wang, K.; Xu, Z.C.J.; et al. Floating aging mechanism of lithium-ion capacitors: Impedance model and post-mortem analysis. *J. Power Sources* **2023**, *557*, 232597. [CrossRef]

166. Yi, S.; Wang, L.; Zhang, X.; Li, C.; Liu, W.J.; Wang, K.; Sun, X.Z.; Xu, Y.N.; Yang, Z.X.; Cao, Y.; et al. Cationic intermediates assisted self-assembly two-dimensional $Ti_3C_2T_x$/rGO hybrid nanoflakes for advanced lithium-ion capacitors. *Sci. Bull.* **2021**, *66*, 914–924. [CrossRef]

167. Wang, P.-L.; Sun, X.-Z.; An, Y.-B.; Zhang, X.; Yuan, C.-Z.; Zheng, S.-H.; Wang, K.; Ma, Y.-W. Additives to propylene carbonate-based electrolytes for lithium-ion capacitors. *Rare Met.* **2022**, *41*, 1304–1313. [CrossRef]

168. Liu, W.J.; An, Y.B.; Wang, L.; Hu, T.; Li, C.; Xu, Y.A.; Wang, K.; Sun, X.Z.; Zhang, H.T.; Zhang, X.; et al. Mechanically flexible reduced graphene oxide/carbon composite films for high-performance quasi-solid-state lithium-ion capacitors. *J. Energy Chem.* **2023**, *80*, 68–76. [CrossRef]

169. Zhou, W.; Liu, Z.; Chen, W.; Zhang, X.; Sun, X.Z.; Luo, M.J.; Zhang, X.H.; Li, C.; An, Y.B.; Song, S.; et al. Thermal characteristics of pouch lithium-ion battery capacitors based on activated carbon and $LiNi_{1/3}Co_{1/3}Mn_{1/3}O_2$. *J. Energy Storage* **2023**, *66*, 107474. [CrossRef]

170. Jin, L.; Shen, C.; Shellikeri, A.; Wu, Q.; Zheng, J.; Andrei, P.; Zhang, J.-G.; Zheng, J.P. Progress and perspectives on pre-lithiation technologies for lithium ion capacitors. *Energy Environ. Sci.* **2020**, *13*, 2341–2362. [CrossRef]

171. Karthikeyan, K.; Amaresh, S.; Aravindan, V.; Kim, H.; Kang, K.; Lee, Y. Unveiling organic–inorganic hybrids as a cathode material for high performance lithium-ion capacitors. *J. Mater. Chem. A* **2013**, *1*, 707–714. [CrossRef]

172. Krause, A.; Kossyrev, P.; Oljaca, M.; Passerini, S.; Winter, M.; Balducci, A. Electrochemical double layer capacitor and lithium-ion capacitor based on carbon black. *J. Power Sources* **2011**, *196*, 8836–8842. [CrossRef]

173. Wang, L.; Zhang, X.; Kong, Y.-Y.; Li, C.; An, Y.-B.; Sun, X.-Z.; Wang, K.; Ma, Y.-W. Metal–organic framework-derived $CoSe_2$@N-doped carbon nanocubes for high-performance lithium-ion capacitors. *Rare Met.* **2024**, *43*, 2150–2160. [CrossRef]

174. Ma, Y.; Wang, K.; Xu, Y.; Zhang, X.; Peng, Q.; Guo, Y.; Sun, X.; Zhang, X.; Wu, Z.S.; Ma, Y. Black Phosphorus Covalent Bonded by Metallic Antimony Toward High-Energy Lithium-Ion Capacitors. *Adv. Energy Mater.* **2024**, *14*, 2304408. [CrossRef]

175. Rajalekshmi, A.R.; Divya, M.L.; Lee, Y.S.; Aravindan, V. High-performance Li-ion capacitor via anion-intercalation process. *Battery Energy* **2022**, *1*, 20210005. [CrossRef]

176. Wang, L.; Zhang, X.; Li, C.; Xu, Y.N.; An, Y.B.; Liu, W.J.; Hu, T.; Yi, S.; Wang, K.; Sun, X.Z.; et al. Cation-deficient T-Nb_2O_5/graphene Hybrids synthesized via chemical oxidative etching of MXene for advanced lithium-ion capacitors. *Chem. Eng. J.* **2023**, *468*, 143507. [CrossRef]

177. Wang, L.; Zhang, X.; Li, C.; Sun, X.-Z.; Wang, K.; Su, F.-Y.; Liu, F.-Y.; Ma, Y.-W. Recent advances in transition metal chalcogenides for lithium-ion capacitors. *Rare Met.* **2022**, *41*, 2971–2984. [CrossRef]
178. Jin, L.; Gong, R.; Zhang, W.; Xiang, Y.; Zheng, J.; Xiang, Z.; Zhang, C.; Xia, Y.; Zheng, J.P. Toward high energy-density and long cycling-lifespan lithium ion capacitors: A 3D carbon modified low-potential Li_2TiSiO_5 anode coupled with a lignin-derived activated carbon cathode. *J. Mater. Chem. A* **2019**, *7*, 8234–8244. [CrossRef]
179. Cao, W.; Li, Y.; Fitch, B.; Shih, J.; Doung, T.; Zheng, J. Strategies to optimize lithium-ion supercapacitors achieving high-performance: Cathode configurations, lithium loadings on anode, and types of separator. *J. Power Sources* **2014**, *268*, 841–847. [CrossRef]
180. Arun, N.; Jain, A.; Aravindan, V.; Jayaraman, S.; Chui Ling, W.; Srinivasan, M.P.; Madhavi, S. Nanostructured spinel $LiNi_{0.5}Mn_{1.5}O_4$ as new insertion anode for advanced Li-ion capacitors with high power capability. *Nano Energy* **2015**, *12*, 69–75. [CrossRef]
181. Hagen, M.; Cao, W.J.; Shellikeri, A.; Adams, D.; Chen, X.J.; Brandt, W.; Yturriaga, S.R.; Wu, Q.; Read, J.A.; Jow, T.R.; et al. Improving the specific energy of Li-Ion capacitor laminate cell using hybrid activated Carbon/$LiNi_{0.5}Co_{0.2}Mn_{0.3}O_2$ as positive electrodes. *J. Power Sources* **2018**, *379*, 212–218. [CrossRef]
182. Feng, Q.-K.; Zhong, S.-L.; Pei, J.-Y.; Zhao, Y.; Zhang, D.-L.; Liu, D.-F.; Zhang, Y.-X.; Dang, Z.-M. Recent Progress and Future Prospects on All-Organic Polymer Dielectrics for Energy Storage Capacitors. *Chem. Rev.* **2022**, *122*, 3820–3878. [CrossRef]
183. Zhao, P.; Cai, Z.; Wu, L.; Zhu, C.; Li, L.; Wang, X. Perspectives and challenges for lead-free energy-storage multilayer ceramic capacitors. *J. Adv. Ceram.* **2021**, *10*, 1153–1193. [CrossRef]

Review

A Review on Design Parameters for the Full-Cell Lithium-Ion Batteries

Faizan Ghani [1], Kunsik An [2] and Dongjin Lee [1,*

[1] School of Mechanical and Aerospace Engineering, Konkuk University, 120 Neungdong-ro, Gwangjin-gu, Seoul 05029, Republic of Korea

[2] Department of Mechanical Engineering, Sejong University, 209 Neungdong-ro, Gwangjin-gu, Seoul 05006, Republic of Korea

* Correspondence: djlee@konkuk.ac.kr; Tel.: +82-2-450-0452

Abstract: The lithium-ion battery (LIB) is a promising energy storage system that has dominated the energy market due to its low cost, high specific capacity, and energy density, while still meeting the energy consumption requirements of current appliances. The simple design of LIBs in various formats—such as coin cells, pouch cells, cylindrical cells, etc.—along with the latest scientific findings, trends, data collection, and effective research methods, has been summarized previously. These papers addressed individual design parameters as well as provided a general overview of LIBs. They also included characterization techniques, selection of new electrodes and electrolytes, their properties, analysis of electrochemical reaction mechanisms, and reviews of recent research findings. Additionally, some articles on computer simulations and mathematical modeling have examined the design of full-cell LIBs for power grid and electric vehicle applications. To fully understand LIB operation, a simple and concise report on design parameters and modification strategies is essential. This literature aims to summarize the design parameters that are often overlooked in academic research for the development of full-cell LIBs.

Keywords: Li-Ion batteries; design parameters; electrodes; electrolytes; electrochemical reaction mechanism; physicochemical properties

Citation: Ghani, F.; An, K.; Lee, D. A Review on Design Parameters for the Full-Cell Lithium-Ion Batteries. *Batteries* **2024**, *10*, 340. https://doi.org/10.3390/batteries10100340

Academic Editor: Shaozhuan Huang

Received: 1 August 2024
Revised: 19 September 2024
Accepted: 23 September 2024
Published: 25 September 2024

1. Introduction

Environmental pollution has been minimized through campaigns aimed at reducing harmful and residual waste materials from energy storage techniques. According to 2020 reports from the China Energy Storage Alliance (CNESA) database, hydropower energy sources, primarily pumped hydro storage systems, account for 92.6% (171.03 GW) of the total energy storage capacity. In contrast, electrochemical energy storage systems, which produce electrical energy from chemical reactions, account for the remaining 5.2% (9.6 GW) of all energy technologies. Along with fuel cells and supercapacitors, batteries are the main electrochemical energy storage system, collectively accounting for 89% (8.5 GW) of the electrochemical energy capacity [1,2]. They store energy in the form of ions and electrons produced during the charge process and are the primary energy storage method for consumer products such as portable electronic gadgets, smartphones, tablets, laptops, and even electric vehicles. The first lithium-ion battery (LIB), invented by Exxon Corporation in the USA, was composed of a lithium metal anode, a TiS_2 cathode, and a liquid electrolyte composed of lithium salt ($LiClO_4$) and organic solvents of dimethoxyethane (glyme) and tetrahydrofuran (THF), exhibiting a discharge voltage of less than 2.5 V [3,4]. LIBs were designed to include a high-potential cathode material, a low-potential anode material, and an electrolyte with a sufficiently wide potential window to facilitate ion transport. Sony Corporation's LIBs exploited a graphite anode (specific capacity of 372 mAh/g), a $LiCoO_2$ cathode (specific capacity of 140 mAh/g), and an electrolyte containing $LiPF_6$ salt in an EC:DEC/DMC solvent, resulting in a working voltage of approximately 3.0 V [5,6].

Increasing energy consumption demands LIBs to have a high voltage window and efficient electrode materials to provide high energy density. Consequently, significant research efforts have been focused on improving energy density, power density, calendar life, thermal stability, and reducing maintenance costs for LIBs.

Numerous studies have been conducted to investigate the development and performance enhancement of LIBs thereby. Such enhancement includes the aging mechanisms, thermal stability responses, and heat generation during the charge/discharge processes of LIBs [7–9]. Most studies have summarized the performance enhancement of LIBs based on one to three experimental design parameters. For that, a design of experiment was suggested to perform the experiment effectively [10]. Many electrode materials have been specifically designed to maximize specific capacities and performance to meet practical consumer demands. These materials are categorized based on their nature, crystallography, and electrochemical reaction mechanisms. Investigations into their crystallography, mechanical, physicochemical, and electronic properties, as well as their impact on electrochemical performance, have been summarized in various reports [11–13]. Furthermore, different synthesis techniques, mechanisms, and modification routes have been explored to enhance the electrochemical responses of various electrode materials [14–16]. These studies highlighted the importance of understanding LIBs at the full cell level and emphasized the need for comprehensive exploration of the thermodynamics and reaction kinetics of LIBs [17,18]. The individual aspects correspond to electrochemical reaction mechanisms, design of electrode materials, various synthesis routes, their effect on morphology and particle size reduction, surface modification techniques, thermal response, and other parameters for the progress of LIBs [14,19–21]. Although most studies are experimental approaches, computational approaches have also been explored to develop mathematical models of thermodynamics and electrochemical reaction kinetics of LIBs, aiming at optimization of various parameters [17,22]. The computational optimization of full-cell design parameters has been aimed at maximizing specific energy density [23]. Lain et al. investigated the commercial lithium-ion cells of various production companies and reported the practical design strategies for high-power LIBs [24]. Similarly, many reports on the types of electrolytes and their modifications to enhance the charge transference and electrochemical potential window have been discussed and summarized [25,26].

Recently, design methodology was suggested in terms of energy density and cost of LIBs for electric vehicles using the computer simulation models [27]. Furthermore, recent material development progresses were reviewed for high-voltage LIBs [28]. Specifically, the design of the electrode material was comprehensively reviewed, but discussion was limited to organic materials [29]. While the existing literature covers these individual components of LIBs, there is a need for a comprehensive review that delves into experimental findings and explains key design parameters and factors affecting the efficiency and energy density of full-cell LIBs. This review aims to scrutinize the crucial design parameters necessary for achieving high energy density full-cell LIBs. Additionally, it summarizes the latest research results and strategies for designing various parameters and provides a detailed discussion on the critical factors influencing the performance of full-cell LIBs.

2. Working Principle of LIBs

The LIB generally consists of a positive electrode (cathode, e.g., $LiCoO_2$), a negative electrode (anode, e.g., graphite), an electrolyte (a mixture of lithium salts and various liquids depending on the type of LIBs), a separator, and two current collectors (Al and Cu) as shown in Figure 1. In LIBs, energy is stored and released through the movement of Li^+ ions between the anode and cathode. When LIB is charged, as depicted in Figure 1a, lithium ions migrate through the electrolyte from the cathode to the anode, where they are stored. LIBs are typically in a charged state with Li^+ ions stored within the crystal structure of the anode. During discharge, as demonstrated in Figure 1b, Li^+ ions are released from the anode, and the battery provides energy to an external load. Positively charged lithium ions move through the electrolyte from the anode to the cathode via the separator. It generates free electrons at the anode, leading to electric current at the load. Possessing porous structure, the separator blocks electron flow but allows ion movement within the electrolyte. The efficiency and lifespan of the battery depend on the quality of materials used and the management of ion transfer. The voltage of the battery is determined by the chemical potential difference between the cathode (μc) and the anode (μa), which is influenced by the electrochemical potential window of the electrolyte. The energy density of a battery, indicating how much energy it can store, is generally expressed in watt-hours per kilogram (Wh/kg). Power density, reflecting the rate at which energy is delivered, is expressed in watts per kilogram (W/kg). Energy density pertains to the total energy stored (specific capacity), while power density refers to the effective use of that energy to perform work [19,30].

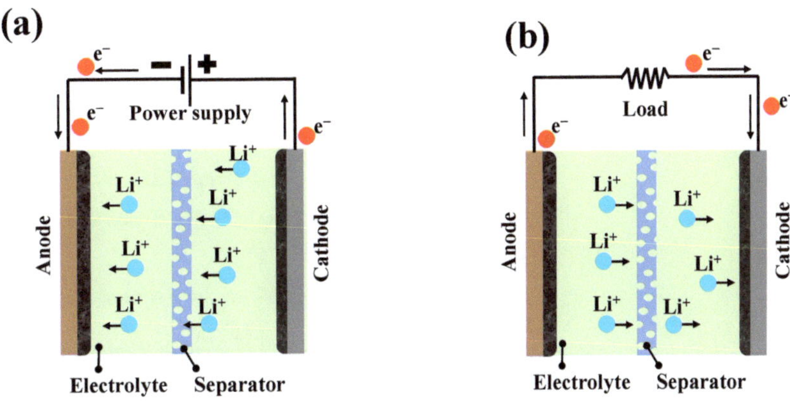

Figure 1. Working principle of the full-cell LIBs in case of (**a**) charging and (**b**) discharging.

3. Design Parameters for Full-Cell LIBs

The LIBs must meet the requirements of LIBs, such as voltage, current, and capacity, depending on the application. They also fit well in the space where the battery should be installed inside the final products, such as a laptop, smartphone, tablet, etc. The installed LIBs should be operated safely under use environments. To remain competitive in the market, finally, the LIBs should be fabricated and supplied at low cost. To meet such requirements, designing full-cell LIBs requires a comprehensive understanding of various design parameters suggested in this review. They include parameters such as form factor, material choices and types, the performance of main components, and productivity/cost as depicted in Figure 2. The form factor, such as geometry and dimension of the battery, ensures geometrical compatibility with electronic products. The choice and type of materials are most crucial for the design of full-cell LIBs, as they influence factors such as energy storage capacity, electrochemical reaction mechanisms, compatibility among components, performance of main components, and cost. Furthermore, the performance of the components of LIBs involves various factors that require thorough evaluation using advanced technological tools to understand their interdependencies. This performance depends on the compatibility of cell components and their effective interaction, the electrochemical potential window, mass loading (N/P ratio), reaction chemistry, intrinsic conductivity, and productivity. Finally, productivity directly influences the cost of full-cell LIBs, assuring market competitiveness. It is noted that these design parameters are closely related to each other, so that thorough consideration should be given when designing the LIBs.

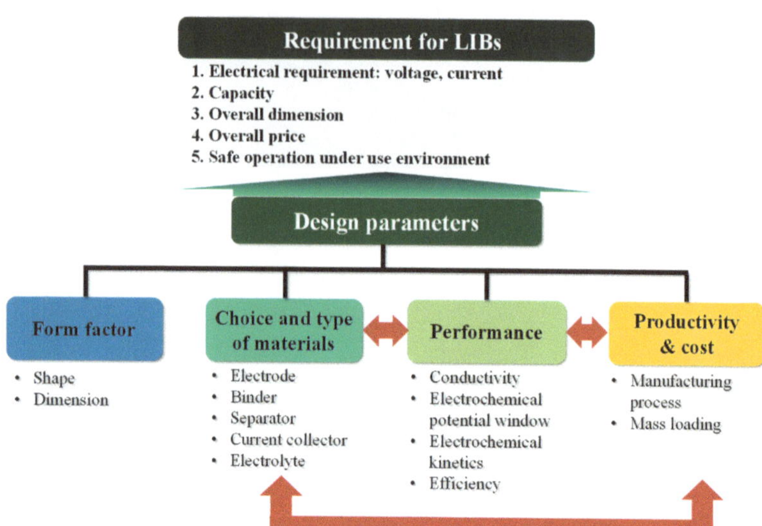

Figure 2. Design parameters for the full-cell LIBs discussed in this work.

3.1. Form Factor

The form factors of full-cell LIBs are essential design parameters, referring to attributes such as size, shape, volume, and weight. Size refers to the actual physical dimensions, including height, width, and thickness, while shape refers to the geometry of the battery, which can be coin-type, cylindrical, prismatic, or pouch-shaped, as shown in Figure 3. Volume indicates the space the battery occupies, and weight is the combined weight of various cell components, influenced by mass loading and the total number of cells. In full-cell battery design, therefore, the form factor is very elementary since it affects how the battery fits into devices, affects energy density, and influences overall performance. Particularly, the form factor is very important in applications like electric vehicles, mobile devices, and large-scale energy storage systems. The main cell types available on the market are summarized in Table 1. Limiting to coin-type cell configuration, CR2032, CR2016, or CR2025 can be chosen for single full-cell LIBs depending on the application. The full-cell configuration consists of the assembly of a casing (bottom and cap), a spacer, a wave-shaped O-ring, and a gasket to ensure a secure seal and prevent leakage during the charge/discharge process. Typically, these components are made of stainless steel, except for the gasket, which is made of polypropylene. The coin cell must be tightly sealed to prevent electrolyte leakage and gas emissions during operation.

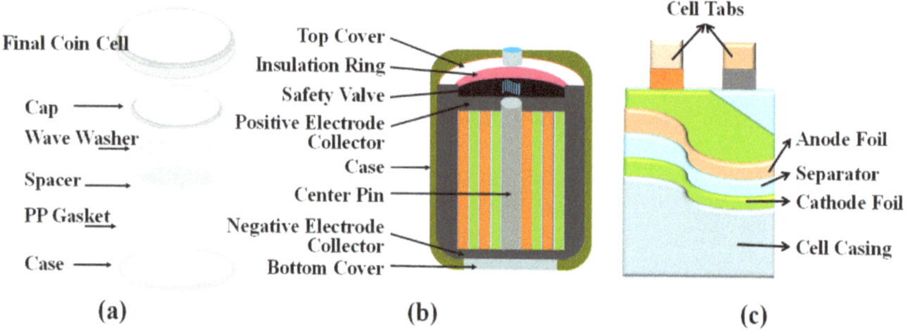

Figure 3. Various form factors of full-cell LIBs: (**a**) coin shape, (**b**) cylindrical shape, and (**c**) pouch-type.

Table 1. General properties of various form factors of LIBs.

Types of LIBs	Properties
Coin/Button Cell	<ul><li>Light weight, highly reliable, and cost effective</li><li>Wide operational window and long shelf life</li><li>High energy density and high cell voltage of ~3.0 V</li><li>Can be stacked for higher voltages with a high safety</li><li>Non-rechargeable (disposable/primary battery)</li><li>Delivers a low rate-capability and needs the assistance of a special holder</li><li>Operatable between temperature range of −30 °C and +60 °C</li><li>Applied in conventional calculators, camera, electronic watches, translators, etc.</li></ul>
Cylindrical Cell	<ul><li>First mass produced LIBs with a high mechanical strength</li><li>Suitable for automated manufacturing at a lower cost</li><li>Small and round enclosed cylinder can that prevents swelling and prevents undesired reactions from the accumulation of gases</li><li>Poor heat management (thermodynamically unstable)</li><li>The internal pressure from side reactions evenly distributed along the cell circumference to extend their shelf-life, safety, and stability</li><li>Applied in laptops, electric vehicles, e-bikes, medical devices, and satellites, and is crucial for space exploration</li></ul>
Pouch-Type Cell	<ul><li>Highly compact with small size cells (vulnerable to cause fires/explosions)</li><li>High efficiency and delivers highest power density</li><li>Relatively good safety performance and good ductility</li><li>High energy density and inexpensive to produce (cheaper cell than coin shaped and cylindrical cells)</li><li>Stable, safe to use, and with an enhanced cell weight</li><li>Comparatively degraded specific energy density</li><li>Gas-generated swelling</li><li>Standardized and knowledgeable automated production methodology increases the production rate</li><li>Applicable for high-end technology industry, portable applications such as mobile device, drones, and portable energy stations</li></ul>

3.2. Choice and Types of Materials for Main Components

Materials themselves are the most fundamental design factors that determine the electrochemical potential window, reaction chemistry (including reaction kinetics and mechanisms), and the types of batteries (e.g., aqueous, non-aqueous, polymeric, or solid-state). They also influence the cyclability, thermal stability, and overall performance of full-cell LIBs. Thus, most research was conducted to develop novel structures of materials for the main components of full-cell LIBs shown in Figure 4. Since the inception of LIBs, full-cell components have been thoroughly studied, and research continues to identify improvements with emphasis on the materials. It is well known that the materials are key factors in facilitating electrochemical reactions that generate high specific capacity and energy density within the electrochemical potential window. In detail, the main components of LIBs include electrodes (anode and cathode), binders (polymeric material), current collectors (metal foil of Cu or Al), separators (polyolefin thin sheet), and electrolytes (a mixture of salts and liquids). The properties of these components—including their electronic and crystal structures, chemical, electrical, and mechanical characteristics and intrinsic conductivities—play a crucial role in developing favorable reaction chemistries, enhancing thermal and mechanical stability, and improving the performance of full-cell LIBs.

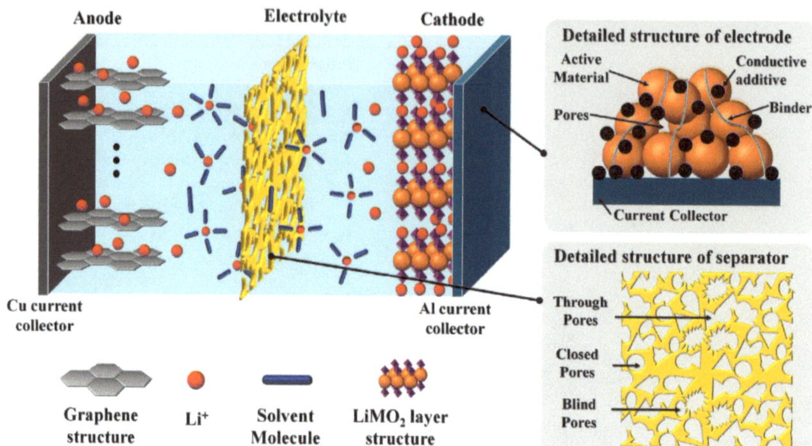

Figure 4. Detailed structure of main components of LIBs: each component has complex structures of several materials to enhance the performance.

3.2.1. Electrode

The electrode is one of the primary components of LIBs, determining the electrochemical potential window, specific capacity, energy and power density, overall performance, and electrochemical reaction mechanism. As a result, the design and modification of electrode materials is crucial for achieving high energy and power densities. Ideally, the electrode should exhibit high intrinsic conductivity, a wide potential window, excellent cycle and rate performance, low cost, and robust stability and safety to improve the overall performance of LIBs. Over the years, two major design strategies, intrinsic and extrinsic, have been explored to achieve these desired properties for LIBs in terms of many aspects as shown in Figure 5.

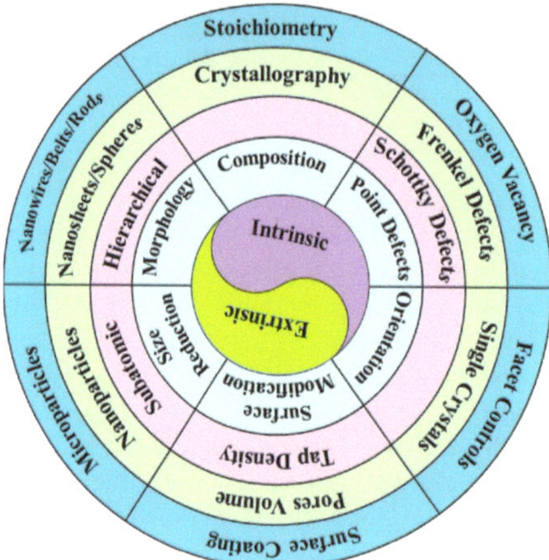

Figure 5. Two design strategies for electrode of LIBs: intrinsic vs. extrinsic.

The intrinsic design strategy primarily focuses on developing stoichiometric compositions, optimizing crystal defects, and controlling crystal orientation, which are discussed in detail.

- **Chemical compositions**: The chemical composition defines the crystal structure and governs key properties such as mechanical strength (adhesion/cohesion), stability (structural, chemical, and thermal), phase transformation, and intrinsic conductivity (electrical and ionic). It also specifies the amount of Li^+ ions that can be inserted or extracted from the crystal structure, directly impacting on the electrochemical properties of electrode materials [31–34]. Additionally, structural units, which represent the material's 'genes', provide insights into local chemical coordination and molecular chemistry, establishing the physical and chemical properties of the electrodes. Understanding the correlation between structural units and these physical/chemical properties offers critical evidence about charge-transfer characteristics, which are essential for intrinsic properties like structural/thermal stability, electronic/ionic conductivities, and Li^+ ion transport. These properties are crucial for enhancing the electrochemical performance of LIBs [35]. Therefore, it is essential to design and develop structurally tunable electrode materials that can accommodate additional Li^+ ions, improve intrinsic conductivities, expand the voltage window, enhance diffusion kinetics, and provide excellent electrochemical performance for LIBs.

- **Point defects**: Similarly, point defects such as Frenkel defects (where an atom migrates from its lattice site to an interstitial site, creating an interstitial defect), Schottky defects (which involve the simultaneous presence of cation and anion vacancies), and oxygen vacancies (absence of oxygen atoms or presence of hydroxyl ions within the crystal structure) play a significant role in defining the local structure of electrode materials. These defects can enhance intrinsic conductivity, improve thermal and structural stability, facilitate pseudo-capacitive kinetics, limit volume expansion, and boost the electrochemical performance of LIBs. Generally, electrode materials with symmetric compositions tend to act as semiconductors, while non-stoichiometric materials (doped or defect-induced) behave like metals, which helps alleviate structural, chemical, and thermal changes [36–39]. However, the effect of oxygen vacancies, compared to Frenkel and Schottky defects, has been insufficiently studied and further investigations are required for the development of innovative LIBs.

- **Crystal Orientation**: Crystal orientation influences specific facets, crystal structures, and surface energies, which in turn affect thermodynamics and reaction kinetics at the surface/interfaces. In batteries, supercapacitors, and fuel cells, physical and chemical interactions at the interfaces play an important role in promoting electrochemical energy storage activities [40]. Additionally, single crystals, which offer advantages such as a small specific surface area, excellent structural stability, high mechanical and thermal stability, superior reaction homogeneity, and good crystallinity, have been studied for their impact on crystal orientation. These studies aim to significantly enhance the electrochemical performance of electrode materials for LIBs, including safety, capacity retention, and cycle life. Electrode materials with low activation energy and substantial adsorption kinetics of crystal facets are promising for achieving high energy density and rate performance in LIBs [41–44]. The interest in exploring single crystal electrodes and their potential applications continues to grow, highlighting the need for advanced research methodologies to address future energy challenges.

On the other hand, the extrinsic design strategy fundamentally focuses on the effect of size reduction, morphological changes, and surface modifications of electrode materials.

- **Size reduction**: The particle size, size distribution, and shape of particles influence the contact area, diffusion resistance, diffusion path, energy density, and overall electrochemical performance of LIBs. Reducing particle size shortens the transportation length of Li^+ ions, decreases the Li^+ ion diffusion barrier, enhances ionic diffusion, increases the contact area among electrode active materials, current collectors, and electrolytes, and ensures the electroactivity of the electrode materials. However, smaller particle sizes also increase the surface area, which can promote electrochemical activity and lead to more side reactions, potentially causing thermal issues and internal short circuits in LIBs [21,45,46]. Particle size distribution affects the physical and chemical properties and overall surface energy activity of electrode materials. A broad size distribution results in high energy density but poor cell homogeneity due to particle size variance and surface energy differences. In contrast, a uniform size distribution, although challenging to produce, offers stable electroactivity by reducing stress strains during the charging process, thereby improving the cycle performance of LIBs [47,48]. Additionally, particle shape directly affects the effective surface area and mass flow properties, particularly the tap density, which influences the Li$^+$ ion diffusion channels and reaction kinetics, enhancing the cycle performance of LIBs. However, particles derived from single-crystal structures are expensive and difficult to manufacture and handle, requiring a highly regulated reaction environment [49–51].

- **Morphological change**: The shape and morphology of electrode materials affect various factors such as porosity, tap density, diffusion pathways, surface area, and interfacial contact area. These factors comprehensively lower the activation energy for electrochemical reactions, shorten the transportation length for Li$^+$ ions, enhance diffusivity and electroactivity, and improve specific capacity and rate capability, ultimately determining the electrochemical performance for energy storage applications [52,53]. Several morphologies, including nanosheets, nanowires/rods/belts/tubes, hierarchical nanostructures, microcubes, microspheres, and micro-flowers, have been developed depending on synthesis and calcination conditions. Nanowires/rods/belts/tubes and nanosheets, with improved compact density, provide unidirectional diffusion pathways for Li$^+$ ions. In contrast, microspheres/flowers, urchin-like structures, and 3D microspheres/microcubes with sizes around 5–10 μm increase electrode packing density, accommodate inactive components (binders and conductive additives used in slurry fabrication), offer extensive surface-active sites for electrolyte penetration, and promote Li$^+$ ion diffusion, resulting in high energy density for LIBs [54–56]. However, micron-sized particles can limit the rate performance and power density of LIBs by extending the diffusion pathways for Li$^+$ ions. Additionally, large cracks and deformations often appear between grain boundaries and at the electrode surface due to the accumulation of significant stresses during the charging process. These issues restrict electronic and ionic conductivity, leading to capacity fading, electrode detachment, and cell degradation [57,58].
- **Surface modification**: Surface modification is an accessible, cost-effective, and widely applied strategy and it is achieved through techniques such as surface coating, etching, and ion doping. These methods enhance ionic conductivity and create surface-active sites that facilitate electrolyte penetration, which is crucial for forming a solid electrolyte interface (SEI) layer. This layer helps buffer volume expansion and contraction, maintaining structural integrity and mitigating capacity fading during cycling [14,59,60]. Consequently, it is highly desirable to prepare electrodes with high voltage, high energy density, low cost, excellent intrinsic conductivity, and robust structural, chemical, and thermal stability. Additionally, electrodes should feature various morphologies with high surface area and porous characteristics. Surface modification techniques, including coating with carbonaceous materials or metal oxides, surface treatment (such as acid/base or metal oxide etching), and ion doping, are essential for enhancing electronic and ionic conductivity and developing coating layers. These modifications help alleviate volume changes, suppress microstrains in the crystal structure, and improve surface adsorption characteristics for additional Li+ ions, thus promoting the electrochemical performance of LIBs. Electrodes, whether designed intrinsically or extrinsically, are classified into various types based on the electrochemical reaction chemistry during the cycling process. Numerous reports detail the cathode and anode materials, synthesis methodologies, modifications, and investigations into electrochemical reaction mechanisms [24,61–63].

Tables 2 and 3 summarize the anode and cathode materials used in previous studies, respectively. Even though researchers do not use either of the two strategies, the above-mentioned electrode design factors were investigated to design better anode and cathode electrodes.

Table 2. Classifications, advantages, and disadvantages of various cathode materials for LIBs.

Class	Types	Advantages	Disadvantages	Refs
Layered Oxides	Co-based oxides	• High capacity • High energy density • Good cycle life • Good rate performance • High power density	• Highly toxic • High cost • Chemical/thermal instability • O_2 evolution	[64–66]
	Ni-rich oxides	• High capacity • Safety • Middle cost • Good cycle life • Chemical/structural stability • Good conductivity • Low toxicity	• Poor performance • Volume expansion • Structural/thermal instability • O_2 evolution • Li deficiency • Phase transition • Ni+4 existence	[67–70]
	Li-rich oxides	• High capacity • High energy density • Safety • Anions activity • Extra Li insertion	• Poor cycle and rate performance • High irreversible capacity • Voltage hysteresis • O_2 evolution • Phase transition	[71–74]
	V-based oxides	• Low cost • High capacity • Good rate performance • Structural stability • Thermal Stability • High voltage	• Poor cycle life • Poor conductivity • Irreversible phase formation • Poor reaction kinetics • Mass production	[75–77]
Spinal Oxides	LiM_2O_4 (M=Co, Mn, Ni)	• Low cost • High voltage • Good rate performance • Abundant Mn • Non-toxic/safety • Fast diffusion kinetics • Structural/Thermal stability	• Poor cycle life and energy density • Voltage hysteresis • Mn dissolution • Anions reduction • Jahn–Teller distortion • Restrict high energy density applications	[78–80]
Polyanionic	Phosphate oxides	• Low cost • Excellent cycle life • Good rate performance • High capacity • High safety • Structural/thermal stability • High voltage • Prevent O_2 evolution	• Poor conductivity • Low volumetric energy density • High processing cost • Low practical capacity • Capacity fading • Anti-site defects • Poor reaction kinetics	[81–84]
	Silicate oxide	• Low cost • High safety • High specific capacity • Availability • 2D Li-ion diffusion channel	• Poor conductivity • Poor rate performance • Structural instability • Sever capacity fading	[85,86]
	Borate oxides	• High theoretical specific capacity • Safety • High voltage	• Structural instability • Poor ionic conductivity • Poor cycle life	[87,88]
	Tavorites	• Good rate performance • Structural/Thermal stability • Strong O-P bond	• Low capacity retention • Low Li recovery • Low columbic efficiency	[89,90]

Table 3. Classifications, advantages, and disadvantages of various anode materials for LIBs.

Class	Types	Advantage	Disadvantage	Refs
Insertion/ Extraction	Carbonaceous	• Low cost, Abundance • High safety • Good cycle life • Good rate performance • Good working potential • Structural stability • High electronic conductivity	• Low capacity • High irreversible capacity • Low columbic efficiency • Dendrite formation • Internal short circuit	[19,61,91–94]
	Ti oxides	• Low cost • High safety/non-toxic • Abundant sources • High cycle performance • High rate performance • Chemical/structural stability	• Poor electronic conductivity • Poor ionic conductivity • Low energy density • Low capacity	[19,61,92–94]
Alloy/de-Alloy	Si, Ge, Sn, Sb, SnO, SiO, Zn, etc.	• Abundance/low Cost • Good safety/non-toxic • Excellent capacity • High energy density • Good rate capability • Good electrical conductivity • Good working potential • Large volumetric capacity	• Poor cyclability • Sluggish reaction kinetics • Large volume expansion • Pulverization • Large capacity fading • High irreversible capacity • Structural instability • Unstable SEI • Highly toxic	[19,61,92–94]
Conversion	Metal oxides	• Low cost/Availability • Eco-friendly/Good cycle • High capacity • High energy density • Rate performance	• Poor conductivity • Large potential hysteresis • Low C.E and unstable SEI formation • High irreversible capacity • Large capacity fading • Volume expansion	[12,16,61, 92–96]
	Chalcogenides	• Eco-friendly/low cost • Low working potential • Low polarization • High capacity • High energy density	• Poor conductivity • Sluggish reaction kinetics • Low potential hysteresis	[19,61,92–96]

3.2.2. Binders

LIB electrodes are typically fabricated by coating a slurry composed of active material, conductive additives, and non-conductive binder onto a current collector foil. The role of the binder in LIBs is as follows: The binder primarily maintains the structural integrity of the electrode materials by binding the active material particles (such as lithium cobalt oxide or graphite) and conductive additives (such as graphite, hard carbon, Super P, KS6, etc.) together. Binders also ensure good electrical contact within the electrode by maintaining a connection between the active material particles and the current collector. Finally, binders accommodate the stress and volume changes that occur in electrode materials during charge and discharge cycles.

Exothermic reactions might degrade the binder material, leading to thermal instability and potential internal short circuits in LIBs. Typically, being electrical insulators, binders should not hamper the movement of Li^+ ions to preserve the ionic conductivity, which is essential for charging efficiency and operation. The binder must absorb these changes without cracking or losing contact with the active material or current collector, which directly contributes to battery durability and cycle life. Therefore, it is essential to design binder materials with high mechanical strength, good adhesion properties, accessibility, low cost, recyclability, non-toxicity, and electrochemical stability [97–99]. Moreover, the binders must be chemically stable and inert in the battery's electrolyte to avoid degradation, which can compromise the electrode structure and lead to battery failure. Inhomogeneous distribution of conductive phases can limit electronic and ionic conductivities and affect the mechanical contact of the electrodes. Binders help create a uniform slurry, ensuring smooth coating onto the current collector. This uniformity is vital for consistent battery quality and performance. Furthermore, binders contribute to the battery's environmental stability, protecting electrodes from humidity, temperature fluctuations, and other external factors.

Table 4 summarizes various binder materials in terms of advantages and limitations. Various types of binder materials—aqueous, non-aqueous, solvent-based polymeric, and gel framework types—have been developed to provide improved contact and electrochemical stability during charging and discharging processes. They also prevented aggregation of active material particles and capacity fading. The overall thermal stability of LIBs encompasses the thermal stability of electrodes, separators, electrolytes, and the thermodynamic stability of the binder [100,101]. Therefore, binder materials with high electronic conductivity, mechanical strength, ductility, surface compatibility, self-healing properties, and effective ion transport capabilities significantly enhance the electrochemical performance, thermal and structural stability, and cycling stability of LIBs.

Table 4. Types and general properties of various binder used for LIBs.

Types	Materials	General Properties	Refs
Aqueous	Na based CMC, SBR, Chitosan, Alginate, etc.	• Low-cost and pollution free • Processability with no humid issues • Evaporate solvent fast • Average swelling tendency and Li+ ion conductivity • Strong chelation and adhesion properties	[102–104]
Non-aqueous	PVDF, SBR, NBR, CMC, PAN, CA, etc.	• Expensive, toxic, and humid sensitivity issues • Large swelling resulting peeling off the active mass from current collector during cycling process • Fast capacity fading for anode materials • Insufficient binding/chelation properties	[105,106]
Polymer	CMC, SBR, PVA, PVD, PVDF, SA, FPI, AR/CMC, Lignin, Sericin protein, etc.	• Excellent distribution of electrode components • Inhibited dissolution of transition metal ions • Good Li+ ion conductivities • Uniform passivation of high-voltage cathodes • Radical quenching ability • High cohesive/adhesion and mechanical properties • Scavenging HF and alleviating crystals of active materials	[100,107,108]
Hybrid	PAA-PAI, GO-StC, β-CDp, Natural polymer, WS-PS, etc.	• Good adhesion and mechanical properties • Strong chelation and binding properties • Negligible swelling and no peeling off the active mass from current collector • Cost effective, eco-friendly, abundant, and have low viscosity • Processability and uniformity of coating	[109,110]

3.2.3. Separators

The separator is also a crucial component of LIBs, as it isolates the electrodes, facilitates ionic diffusion pathways, prevents electronic conduction, and mitigates internal short circuits and thermal runaway. It plays a key role in defining the energy density, power density, and safety of LIBs. Separators must be chemically and thermally stable at elevated temperatures, electrically inactive, and resistant to degradation during battery operation [111]. Various types of separators have been explored and classified into categories such as (i) polyolefin films (e.g., PP, PE, PEI, PDA, PMMA, and glass fibers), (ii) thermally conductive materials (e.g., AlN/PP and BN/PP/C), (iii) carbon materials (e.g., cellulose-derived carbon, graphene, PDA/Gr-CMC, and Na-alginate), (iv) metals (e.g., Cu/PP, Si/PP, and Nb/PP), and (v) metal oxides (e.g., Al_2O_3/PP, SiO_2/PP, and ZrO_2/PP) [112]. The separator should facilitate ion transport through its pores, improve interfacial compatibility, and enhance the safety of LIBs. An inert surface of separators can weaken the electrode-separator interface, leading to longer diffusion paths, irregular Li plating, dendrite formation, and thermal runaway. Therefore, optimizing ionic and physical contact while minimizing voids is essential to improving interfacial compatibility, preventing dendrite formation, and reducing thermal shrinkage in separators for advanced battery systems [113]. Key designing parameters of the separator include thickness, porosity, mean pore size, tortuosity, permeability, wettability, thermal stability, and mechanical properties.

- **Thickness**: The thickness of a separator typically ranges from 20 to 50 μm, influencing the stability, mechanical properties, overall weight, and cell resistance of LIBs. For example, the commercially available Celgard 2400 separator has a thickness of 25 μm [114].
- **Porosity**: Porosity is a crucial factor in determining mass transport, as it ensures sufficient Li^+ ion conductivity and helps inhibit the formation of dendritic lithium. Common separators in the battery market typically exhibit around 40% porosity, which is defined as the ratio of the volume of pores to the apparent total volume of the pores. Porosity is typically measured by calculating the weight difference of the separator before and after soaking it in liquid, as shown below [115,116].

$$\text{Porosity}(\%) = \frac{(1 - \rho_m)}{\rho_P} \times 100 \tag{1}$$

$$\text{Porosity}(\%) = \frac{(W - W_o)}{\rho_L V_o} \times 100 \tag{2}$$

where ρ_m, ρ_P, W, W_o, ρ_L, and V_o represent the apparent density, separator material density, weight of void separator, weight of separator soaked in liquid, density of liquid, and geometric volume of separators, respectively.

- **Mean pore size**: The mean pore size is closely related to the size of Li^+ ions, active ionic species in the electrolyte, and active mass components. Pore size controls the flow of Li^+ ions, blocks lithium dendrites, and prevents short circuits. There exists a mean pore size of less than 1 μm for a commercially viable and safe separator to allow Li^+ ion transportation and block other active species. They can be classified into closed, blind, and through pores, as shown in Figure 4. Closed pores are fully enclosed without void spaces, while blind pores open to a void space on one side but are blocked on the other, trapping Li^+ ions and potentially leading to dendrite formation. Pores with open void spaces and high permeability allow effective Li^+ ion transport.
- **Geometric effect**: The geometric effect of pore morphology on the conductivity of Li^+ ions under certain pressure differences is known as tortuosity. It describes the morphological changes in the pores of the separator. Pores exhibit various morphologies, including interconnected, network-type hierarchical structures, circular shapes, and

other microstructures. Tortuosity is the ratio of the mean path length that ions must travel through the pores to the direct straight-line distance as follows:

$$\text{Tortuosity}(\tau) = \text{Sqrt}\left(\varepsilon \times \frac{R_s}{R_o}\right) \tag{3}$$

where ε is porosity, and R_s and R_o are resistivity of separators before and after soaking in liquid, respectively. Permeability is calculated using Darcy's law, which describes the rate of fluid flow through a porous surface as follows:

$$\text{Average Velocity}(\mu) = \frac{-\kappa}{\eta \nabla P} \tag{4}$$

where μ is average velocity, η and ∇P are viscosity and applied pressure gradient of the fluid, respectively, and κ is permeability of separators.

- **Wettability**: It is a key aspect of the separator that directly influences the capacity and cycle retention of LIBs. A separator must quickly absorb electrolytes and initiate uniform Li^+ ion transportation to prevent uneven Li^+ ion deposition on the electrodes. The wettability of the separator is measured through contact angle analysis and assesses its affinity for liquid electrolytes by determining the angle formed between the separator surface and the electrolyte droplet.

- **Thermal stability**: Thermal stability is a crucial factor for ensuring the safety of LIBs. The separator must remain thermodynamically stable and withstand rising heat flux during battery operation under extreme conditions. When the separator shrinks or wrinkles at high temperatures, it leads to poor interfacial contact with electrodes, resulting in significant energy loss. Excessive heat flow can trigger thermal runaway and internal short circuits in the LIBs. To prevent these issues, the thermal shrinkage of the separator must be kept below 5% after 60 min at 90 °C under vacuum and can be measured using the following equation:

$$\text{Thermal Shrinkage}(\%) = \left(\frac{A_i - A_f}{A_i}\right) \times 100 \tag{5}$$

where A_i and A_f indicate area of separators before and after heat treatment at a certain temperature, respectively [114–116]. Furthermore, the separator must possess a shutdown effect, where the pores are blocked once abnormal heat flow is detected. This feature prevents direct contact between electrodes, inhibiting thermal runaway and internal short circuits. In addition, the separator should be non-flammable, ideally flame retardant, because if thermal runaway occurs, a flammable separator could catch fire, potentially leading to a battery explosion [113,117].

- **Mechanical properties**: The mechanical properties (e.g., mechanical strength, strain percentages, compression percentages) play a crucial role in determining the stability of separators and LIBs. During cell assembly, the interaction between electrolytes and separators causes mechanical softening and swelling under compression. Battery operation induces volume expansion in the electrode's active materials, exerting pressures of up to ~5 MPa. Under such stresses, the elastic modulus of the separator decreases, reducing its tolerance, altering its microstructure, hindering ionic conductivity, and leading to swelling. This compromises stability, promotes dendritic Li formation, and increases the risk of internal short circuits in LIBs [118,119].

Therefore, separators with thermal stability, high porosity, and strong mechanical properties must be designed, taking into account the aforementioned critical parameters to achieve high-energy and high-safety LIBs.

3.2.4. Current Collector

Current collectors, though non-active materials in full-cell LIBs, play a crucial role in providing mechanical support for electrodes, enhancing electrical conductivity, corrosion resistance, and reducing contact resistance. These, in turn, improve the specific capacity, efficiency, cycle stability, and rate performance of LIBs. Comprising the second-largest weight percentage (15%) of the total weight of a full-cell LIB, excluding the case, efforts to reduce their thickness have been made to increase energy density. The essential design parameters for current collectors include electrochemical stability, density, mechanical strength, electrical conductivity, sustainability, and cost [120,121]. Furthermore, the choice of material, availability, recyclability, and cost are other important factors when designing current collectors for full-cell LIBs [122]. Each of these design parameters is discussed in detail below.

- **Electrochemical stability**: It is essential to keep the stable reduction/oxidation environment during the battery operation as cathode and anode require high and low electrochemical potentials in LIBs, respectively. Any undesired reactions between the current collector and the electrolyte at these extremes can destabilize the system, leading to capacity fading and a shortened cycle life. Therefore, selecting a current collector with excellent electrochemical stability is crucial for achieving LIBs [123].
- **Density**: Current collectors with low densities are advantageous for reducing weight and cost, which can enhance the energy density of LIBs. Furthermore, high mechanical strength is crucial for preserving the integrity of the electrode materials and ensuring strong bonding with the current collector, electrodes, and polymer binder materials.
- **Mechanical strength**: A current collector with high mechanical strength helps suppress volume expansion, prevent electrode pulverization and delamination, and maintain the integrity of active components, thereby enhancing cycle stability and prolonging the cycle life of LIBs [94].
- **Electrical conductivity**: The electrical conductivity of the current collector and the interfaces between the electrode and current collector is crucial for LIB performance, as electrons generated at the electrodes travel through the current collector to the external circuits. A current collector with high electrical conductivity improves energy efficiency and minimizes heat generation, thus reducing the loss of chemical/electrical energy as heat during battery operation [124].

The choice of materials for current collectors significantly impacts the electrochemical performance, electrode stability, and high-rate performance of LIBs. Various materials, such as copper (Cu), aluminum (Al), titanium (Ti), nickel (Ni), stainless steel, and carbonaceous materials, are used as current collectors. These materials can be processed into different forms, including foils, foams, meshes, etched surfaces, and coated layers. Chemical treatments like etching and coating can notably improve surface roughness, corrosion resistance, and contact resistance, thereby enhancing the overall performance of LIBs [125,126].

3.2.5. Electrolyte

Electrolytes, as non-electroactive components, should not participate in redox reactions or allow electron flow. Their primary role in a battery system is to facilitate the ionic transportation of species generated during redox reactions at the electrodes and to serve as the medium for charge transfer. They also facilitate chemical reactions and accommodate thermal and mechanical variations to prevent damage or explosions of the battery system. Ionic transport in electrolytes has been extensively studied, with computational models suggesting that it occurs through mechanisms like migration, diffusion, and convection. These processes contribute to the flux density of dissolved species within the electrolyte [127]. Therefore, the electrolyte must exhibit the following characteristics to ensure proper electrochemical reactions in LIB full cells:

- **Ionic conductivity**: The electrolyte must enable efficient ion transport, meaning it should be highly conductive to ions. High ionic conductivity within the electrolyte facilitates the rapid movement of ions between electrodes, promoting efficient charging and discharging of the LIB. This characteristic is crucial as it directly impacts the rate performance and power density of the LIB. Therefore, to achieve a high-rate and high-power density LIB, the electrolyte must exhibit excellent ionic conductivity.

- **Wide potential stability window:** The potential window of the electrolyte defines the range within which ions can effectively move between the cathode and anode. To ensure optimal performance, the electrolyte must support ionic movement from the highest occupied molecular orbital (HOMO) to the lowest unoccupied molecular orbital (LUMO) of the electrode materials. A broad potential window allows for a wider range of electrochemical reactions, enhancing the specific capacity, cycle life, and overall electrochemical performance of the LIB. Moreover, the potential window must remain stable during cycling to preserve the reaction chemistry, prevent thermal runaway, and maintain the structural and operational stability of the full-cell LIBs. Any fluctuation in this window can compromise the cell's performance and safety. Therefore, a stable and wide potential window is essential for achieving high-performance, long-lasting, and safe LIBs.
- **Chemically inert:** Electrolytes must be electrochemically inert and should not participate in the electrochemical reaction of the full-cell LIBs.
- **Low cost:** The electrolyte should be cost-effective and have easy accessibility.
- **Reducible:** The electrolyte must undergo reduction during the electrochemical reaction so that Li^+ ions can transfer under the migration, diffusion, and convection phenomena.
- **Environment-friendly:** The electrolyte materials should be non-toxic and environment-friendly and should not cause any harm to human beings, animals, or the environment.
- **Electron insulator:** The electrolyte must block electron flow while allowing uninterrupted ionic transport. In other words, it should act as an electrical insulator, preventing electron involvement in any reactions during the electrochemical process.
- **High fluidity and low vapor pressure:**

The electrolytes are composed of metal salts and various solvents. Conducting salts must be fully dissolved to enhance ionic mobility and remain inert to the solvent and other battery components to prevent degradation. These salts should also be non-toxic, non-corrosive, and thermally stable to ensure safety during battery operation. Common salts used in LIBs include lithium perchlorate ($LiClO_4$), lithium hexafluorophosphate ($LiPF_6$), lithium bis(trifluoromethanesulfonyl)imide (LiTFSI), lithium triflate (LiTf), lithium hexafluoroarsenate ($LiAsF_6$), and lithium tetrafluoroborate ($LiBF_4$) [128]. Among these, $LiPF_6$ is the most widely used commercial salt for LIBs. However, LiTFSI has been proposed for high-energy density LIBs, despite its potential to corrode current collectors at high voltages [129].

Electrolytes are categorized into various types based on the nature of solvents used, and their general properties are outlined in Table 5. The types of electrolytes include:

(a) Aqueous electrolytes;
(b) Non-aqueous electrolytes;
(c) Ionic liquids;
(d) Polymer electrolytes (gel polymer, solid polymer);
(e) Hybrid electrolytes.

Furthermore, a brief comparison of their inherent properties is summarized in Table 6. Aqueous electrolytes have undergone modifications through several approaches, such as increasing salt concentration, incorporating additives, adjusting interfaces between electrodes, electrolytes, and current collectors, and developing new types such as gelled (hydrogel) and hybrid-solvent electrolytes. These modifications and their effects on different properties are detailed in Table 7. Aqueous electrolytes, with high wettability and improved interfacial kinetics, contribute to stabilizing the SEI layers, thus enhancing reversibility and providing thermodynamic stability to LIBs. Consequently, optimizing the potential window, increasing ionic conductivity, lowering the freezing point, and reducing costs are crucial for the development of promising hybrid electrolytes for LIBs.

Table 5. Types, advantages and disadvantages of various electrolytes used for LIBs.

Types	Advantages	Disadvantages	Refs
Aqueous	• Non-flammable • Non-toxic • High fluidity • High dielectric constant • High ionic conductivity • Low cost • Safe operation • Long cycle life	• Narrow potential window • Low energy density • Large overpotential • Poor mechanical stability	[130,131]
Non-aqueous	• Combination of cyclic and linear solvents enhance ionic conductivity and lower viscosity. • Form stable SEI layer at the anode surface. • Wide temperature range • Wide potential window (~6.0 V) • High energy density batteries	• Poor ionic conductivity and low salts solubility • High flammability and toxic reaction products • Instability at high temperature and high current density • Humid sensitive • Thermal instability and high cost • Extreme purification and handling	[107,132,133]
Ionic Liquids	• Environment-friendly • High chemical/thermal stability • Tunable molecular structure • High oxidation potential (~6.0 V) • Zero vapor pressure • Flame retardant • Non-volatile	• High viscosity • Poor ionic conductivity • Poor wettability • Poor mechanical stability • Large scale productivity	[134–136]
Polymers (GPEs, SPEs)	• Low electrolyte leakage • High flexibility • Low flammability • High stability • No liquid solvents involvement • Light weight • Ease of processing • Strong adhesion properties • Large scale productivity	• High viscosity • Poor ionic conductivity • Poor wettability • Li+ ion transference number • Moderate potential window (4.5 V~5.0 V)	[106,107,137,138]
Hybrids	• Mechanical robustness • Excellent processability • Reduced interface resistance	• Poor ionic conductivity • Poor cation transference number • Medium potential window	[100,108,109]

Table 6. Summary of the inherent properties of various types of electrolytes used for LIBs.

Types	Aqueous	Non-Aqueous	Ionic Liquids	Polymer (Gels, Hybrid)
Mechanical strength	Poor	Good	Good	Medium
Ionic conductivity	$>10^{-3}$ Scm^{-1}	$>10^{-3}$ Scm^{-1}	10^{-3}~10^2 Scm^{-1}	$>10^{-4}$ Scm^{-1}
Thermal stability	Poor	Medium	Good	Medium
Electrochemical stability	Poor	Good	Good	Poor
Safety	Poor	Medium	Medium	Medium
Interfacial properties	Good	Medium	Good	Medium

Table 7. Modification strategies and general properties of aqueous electrolytes used for LIBs.

Electrolytes	Strategies	General Properties	Refs
Aqueous	Enriching salt concentration	Cut anions at anode surface Enhance anion/cation interaction Break hydrogen bonding to reduce O_2 solubility and H_2 evolution Develop eutectic system for better ORR kinetics to improve the potential window (1.23 to ~3.0 V) Improve the ionic conductivity ($\approx 10^2$) Affect the rate performance and overpotential Suppresses OER at the cathode surface Prevent corrosion/dendrite formation	[130,131,139]
	Incorporating additive	Modify interfaces (electrodes/current collectors/electrolyte/separator) Change solvation sheath to widen potential window for high temperature and freezing temperature Suppresses free radicals, reactive anions/cations to control side reactions	[139]
	Tuning interfaces \ (electrodes/current collectors/electrolyte)	Affects interfaces stability and reactivity To achieve thermodynamics (chemical, thermal) and kinetic (charge/mass transportation activity) stability Solidifies water (lower fluidity)	[140,141]
	Addition of decoupling gel/polymer material	Develop anti-freezing function at low temperature Stabilize/widen the potential window and working temperature range Lowers the production cost	[142,143]
	Solvent-hybrid electrolyte	Efficiently reduce the cost and environmental problems Improve interfacial chemistry Enhance performance of LIBs	[144]

The non-aqueous electrolytes, typically comprising lithium salts and organic solvents (such as carbonates, acetals, ethers, esters, sulfones, sulfites, and sulfoxides), are designed to maintain desired conductivity, viscosity, and compatibility with battery components. Key parameters include ionic conductivity, viscosity (ideally < 2 mPa·s), dielectric constant, wettability, working temperature range, flash point, and environmental impact [145]. Carbonates are commonly used due to their excellent electrochemical properties and ability to support a high-quality SEI layer, crucial for preventing dendrite formation and ensuring battery safety. For instance, ethylene carbonate (EC) is frequently employed in electrolytes like 1 M $LiPF_6$ in EC:DMC:DEC (1:1:1) and 1.2 M $LiPF_6$ in EC:EMC (3:7), which showed electrochemical stability windows of 4.5 V and 4.9 V, respectively [146,147]. Physicochemical and electrochemical properties of different organic solvents (carbonates, acetals, ethers, esters, sulfones, sulfites, and sulfoxides) are summarized in Table 8. Selecting solvents with high dielectric constants, low viscosity, and high ionic conductivity is essential for enhancing the electrolyte's oxidation potential and promoting electrocatalytic reactions, thereby improving the performance of high energy density LIBs [148].

Table 8. Physicochemical and electrochemical data of some organic solvents for non-aqueous electrolytes.

Solvents	Names	m.p	b.p	η (20 °C)	ε_r	μ	ρ (V_m)	κ	E_{ox} vs. Li^+/Li	Ref
Carbonates	EC	36.4	248	1.90	89.8	4.61	1.32 (66.71)	8.3	6.7	[149]
	PC	−48.8	242	2.53	64.9	4.81	1.20 (85.08)	5.6	6.0	[150]
	DMC	4.6	96	0.59	3.1	0.76	1.06 (84.98)	6.0	5.5	[148]
	DEC	−74.3	126	0.75	2.8	0.96	0.97 (121.78)	2.4	5.2	[133]

Table 8. *Cont.*

Solvents	Names	m.p	b.p	η (20 °C)	ε_r	μ	ρ (V_m)	κ	E_{ox} vs. Li$^+$/Li	Ref
Esters	EA	−84	77	0.45	6.0	1.83	0.90 (97.90)	11.5	5.4	[151]
	MA	−98.2	57	0.37	6.7	1.70	0.93 (79.66)	14.8	5.2	[152]
	MB	−84	102	0.60	5.5	1.71	0.90 (113.48)	4.2	4.6	[153]
Ethers/Acetals	EPE	−126.7	63	0.31	-	1.16	0.73 (120.75)	4.5	5.5	[154]
	DEE	−74	121	0.56	5.1	1.76	0.84 (140.69)	5.8	4.5	[155]
	THF	−109	66	0.46	7.4	1.70	0.88 (81.94)	9.1	3.5	[156]
Sulfur Compounds	DMSO	18.5	189	1.90	46.6	3.90	1.09 (71.68)	8.6	4.1	[157]
	DMS	−141	126	0.87	22.5	2.90	1.20 (91.78)	13.6	4.2	[158]
	DES	−112	156	0.83	15.6	2.96	1.08 (127.94)	10.2	2.9	[159]

Abbreviations: m.p = melting point, b.p = boiling point, η = dynamic viscosity in mPa, $\bar{\varepsilon}_r$ = dielectric constant, μ = dipole moment, ρ = density (g·cm^{-3}), V_m = molar volume (cm^3·mol^{-1}), κ = ionic conductivity of 1 M LiPF6 (m·Scm^{-1}), E_{ox} = oxidation potential. Chemical Name: EC = ethylene carbonate, PC = polycarbonate, DMC = dimethyl carbonate, DEC = diethyl carbonate, EA = ethyl acetate, MA = methyl acetate, MB = methyl butyrate, EPE = 3-(1,1,2,2-tetrafluoroethoxy)-1,1,2,2-tetrafluoropropane, DEE = 1,2-diethoxyethane, THF = tetrahydrofuran, DMSO = dimethyl sulfoxide, DMS = dimethyl sulfide, and DES = diethyl sulfite.

Ionic liquids (ILs) offer benefits such as flame retardancy, non-volatility, and thermal stability, but their high viscosity can reduce ionic conductivity and affect the rate-performance of lithium-ion batteries (LIBs). To address this, anti-freezing agents like acetonitrile or ethylene glycol are added to enhance dielectric constants, miscibility, and oxidation stability at lower temperatures [160]. Despite their advantages, ILs face challenges such as high viscosity and poor wettability with battery components, which limit their practical use. ILs are considered "green solvents" and are used to mitigate issues like volatility and low oxidation potential seen in traditional organic solvents [146,148,161]. They are composed of ionic bonds between cations and anions and can dissolve a variety of substances. Adding low-viscosity compounds can improve their ionic conductivity [134,136]. ILs are categorized into aprotic, protic, and zwitterionic types based on their chemical composition. Aprotic ILs, which lack free protons, are used as co-solvents, while protic ILs, which contain free protons, offer better ionic conductivity and are used in fuel cells and batteries [162]. However, ILs generally have high melting points and strong ionic bonds, leading to high viscosity and reduced ionic conductivity, which can hinder LIB performance [135]. To overcome these limitations, strategies such as blending ILs with organic solvents and modifying the cation side chains are being explored. These solutions, while improving performance, often increase costs and require extensive purification, limiting their commercialization in energy storage applications.

Polymer electrolytes (PEs) are emerging as a solution to the limitations of liquid electrolytes, such as leakage, poor volume suppression, and safety issues. PEs offer benefits including low leakage, high flexibility, low flammability, stability between electrolyte and electrodes, no liquid solvents, lightweight, easy processing, and strong adhesion. Also known as solid polymer electrolytes (SPEs), they consist of lithium metal salts dissolved in a polymer matrix. These can be classified into organic electrolytes with polymer composites and inorganic electrolytes with materials like ceramics and perovskites. While organic electrolytes often suffer from poor ionic conductivity, inorganic electrolytes face issues with mechanical stability and large-scale production. High ionic conductivity, large Li$_+$ ion transference number, wide electrochemical stability window, high mechanical stability, low electrolyte/electrode impedance, electrical insulation, low cost, ease of manufacturing, sustainability, and environmental compatibility are the significant design parameters for polymer electrolytes. [163–165] Ionic conductivity can be improved by choosing suitable polymers (linear, branched, or crosslinked), adding new salts, and incorporating fillers. Common polymers used in PEs include PEO, poly (vinylidene fluoride; PVDF), poly (ethylene carbonate; PEC), poly (acrylonitrile; PAN), and poly (ethylene glycol; PEG) [166,167]. The size of the anionic component in lithium salts affects ionic conductivity; larger anions increase ionic contact distance and dissociation, thus enhancing conductivity. The dissociation constants of commonly used salts can be explained in the following order: $LiCF_3SO_3 < LiBF_4 < LiClO_4 < LiPF_6 < LiAsF_6 < LiN(CF_3SO_2)_2$ [161,168]. Fillers, both active (e.g., ionic liquids and lithium salts) and passive (e.g., carbonaceous compounds and ceramics), can improve the ionic conductivity, thermal stability, and mechanical properties of SPEs [137,169,170]. However, poor compatibility with separators and electrodes, along with a moderate potential window (4.5 V~5.0 V), limits their application in high voltage LIBs, necessitating further research for high-potential window polymer electrolytes.

Hybrid electrolytes (HEs) are an advanced research focus, combining polymer and inorganic/organic electrolytes to achieve mechanical robustness, high ionic conductivity, and reduced interfacial resistance. Despite these advantages, they generally exhibit lower ionic conductivity, a lower cation transference number, and a medium potential window. The properties of HEs, such as melting point, glass transition temperature, and crystallinity, are influenced by the molecular groups attached to the polymer chain, affecting their overall performance. HEs are classified into solid/liquid, solid/solid, or liquid/solid/liquid types, incorporating inorganic or organic compounds as fillers to modify the polymer matrix. The liquid electrolytes are organic or aqueous, whereas the solid electrolytes are polymers and LISICON [171–173]. While they combine characteristics of both polymers and other types of electrolytes, further research is needed to improve ion transport, interfacial properties, and potential window. New polymer matrices and fillers need to be designed and tested to enhance high-performance hybrid electrolytes [174,175]. For effective design, hybrid electrolytes must have high ionic conductivity, a wide electrochemical stability window, versatility across temperatures, good interfacial compatibility, high cation transference number, mechanical strength, low viscosity, high dielectric constants, and safety features such as non-flammability, non-toxicity, and eco-friendliness. These qualities are essential for developing high-power, high energy density LIBs.

The liquid electrolyte offers high ionic conductivity, good wettability, and superior interfacial contact, contributing to excellent electrochemical performance. However, its volatility and flammability pose safety risks, leading to irreversible active material loss during charge–discharge cycles. In contrast, solid-state electrolytes prevent the transfer of intermediate products, reducing the shuttle effect and lithium dendrite growth. They provide low flammability, high thermal stability, and no leakage, enhancing safety and battery lifespan. Solid-state electrolytes also minimize material usage, resulting in reduced mass and volume per capacity unit and enable the use of higher-capacity electrode materials, boosting energy density and safety [176]. Common solid-state electrolytes include oxide (e.g., garnet-type $Li_7La_3Zr_2O_{12}$) [177–180], sulfide (e.g., $Li_2S\text{-}SiS_2$) [181–183], and polymer electrolytes (e.g., PVDF) [184–186]. Oxide electrolytes achieve ionic conductivity of 10^{-4} Scm^{-1} at room temperature, while sulfide electrolytes can reach up to 10^{-3} Scm^{-1} due to their wider ion transport channels. Polymer electrolytes facilitate ion transport through Li$^+$ complexation, reducing reactivity with electrodes and enhancing safety. Xiong et al. reported that the mechanical failure of solid-state electrolytes in lithium metal batteries, driven by lithium anode growth, is linked to interfacial and internal defects [187,188]. A modified electro-chemo-mechanical model reveals how these defects influence stress transmission and damage propagation, providing insights for designing safer, high energy density solid-state batteries.

3.3. Design Parameters Directly Affecting Performance

Performance is a crucial metric for assessing the energy storage capability of LIBs, specifically their ability to endure electrochemical reactions over time under severe conditions. It encompasses a correlation among all design parameters, material selections, reaction kinetics, and thermodynamics. Reaction kinetics, in particular, explores the relationship between physicochemical and electrochemical reactions. The physicochemical reaction includes both physical properties such as surface area, porosity, density, wettability, conductivity, and thickness and chemical properties such as chemical potentials, reactions, and thermodynamics. They together clarify the reaction kinetics at the interfaces among electrodes, electrolytes, and current collectors, as well as the transportation of ions within the electrolyte (refer to Figure 6). Electrochemical reaction kinetics, on the other hand, focus on the ion/charge transportation mechanism during the chemical reactions within the electrolyte throughout the charge and discharge processes. The relationship between various physicochemical and electrochemical properties defines key critical design parameters, such as conductivity, electrochemical potential window, reaction chemistry, and thermodynamics of LIBs. Optimizing these design parameters is essential for achieving optimal performance that meets industrial energy consumption demands.

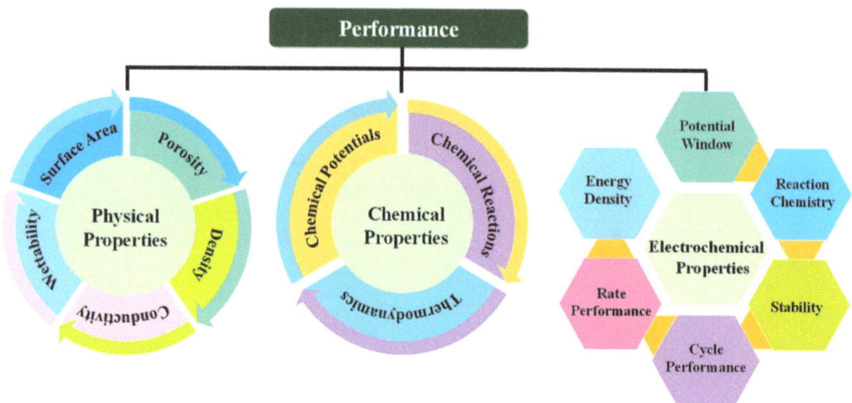

Figure 6. Substantial properties influencing the overall performance of full-cell LIBs.

3.3.1. Conductivity

Both electronic and ionic conductivities are intrinsic performances that play a crucial role in determining the flow of electrons and ions within the electrodes. They can be extrinsically modified to improve the capacity, cyclability, and rate performance of LIBs. The electrical conductivity depends mainly on the electronic structure of electrode materials. The electronic structure has multiple electrons in the valence band with the minimum energy band gap to increase electrical conductivity. Materials such as metalloids, post-transition metals, and some reactive nonmetals exhibit higher electrical conductivity compared to transition metals and their oxides. Several methods have been employed to enhance electrical conductivity, such as introducing point defects in the crystal structure, doping with other elements, forming composites, surface etching, and coating materials with metals, metal oxides, or carbonaceous compounds [189–193]. For instance, doping metal elements into electrode materials can increase electrical conductivity by up to 10^3 times, while coating with carbonaceous or metal oxide layers can reduce the energy band gap and improve conductivity. Enhancing electrical conductivity lowers charge transfer resistance and boosts the specific capacity and energy density of LIBs. A recent study demonstrated that developing an eco-friendly carbon composite made from microalgae combined with an SnO_2 anode led to increased reversible capacity, improved cycle stability, and mitigation of volume expansion issues in LIBs [60]. Additionally, coating CN-LTO using a chemical reflux technique increased the electrical conductivity of pristine LTO from 1.58×10^{-9} to 4.29×10^{-6} S/cm [194].

Ionic conductivity, which governs ion transport mechanisms, is another factor influencing the rate performance and power density of LIBs. It can be improved by modifying the morphology, particle size, and physicochemical properties of the electrode materials, electrolyte, separators, and binders, which in turn affect the adsorption and diffusion behavior at electrode surfaces. In particular, nanomaterials significantly reduce the diffusion length of Li^+ ions and enhance their diffusivity. The residence time (τ) is the duration for which Li^+ ions remain within the electrode, while diffusion length (L) describes the length of the diffusion pathway through the crystal structure of the electrode. They are related through the diffusivity (D) as follows:

$$\tau = \frac{L^2}{D} \tag{6}$$

Moreover, ionic conductivity is related to porosity and tortuosity of electrodes and separators. Porous structures in electrodes and separators provide extensive diffusion pathways, facilitating faster diffusion of Li^+ ions, reducing charging time, and enabling rapid charging in LIBs. Tortuosity, the complexity and geometry of the surfaces, offers multiple diffusion channels that enhance the movement of Li^+ ions. Voids and pores, formed by particle interactions and distances, play a crucial role in electrolyte penetration, thereby increasing the diffusion of Li^+ ions. Moreover, the surface area of the electrode materials significantly affects the adsorption kinetics of Li^+ ions, as well as the behavior of redox couples and free radicals generated during electrochemical reactions. A high surface area allows for more active sites for Li^+ ion adsorption and promotes better electrolyte penetration, facilitating improved Li^+ ion diffusion during electrochemical reactions, which is essential for achieving high specific capacity and energy density.

The cationic transference number, affecting the rate performance and power density of LIBs, is another critical factor, particularly in selecting appropriate electrolytes to enhance conductivity. Therefore, it is necessary to identify electrode materials with high porosity, appropriate tortuosity, many voids, and a substantial specific surface area to optimize Li^+ ion diffusion and reduce the charging times of LIBs. Furthermore, it is advisable to use electrolytes with high conductivity and an optimized cationic transference number to improve rate performance, ensure electrochemical potential stability, and enhance specific capacity, cycle life, and safety. In addition, separators must be porous to maintain open pores and prevent the formation of lithium dendrites, which can cause internal short circuits and thermal runaway. Consequently, the careful design and optimization of both electrical and ionic conductivity are vital for enhancing the overall performance of full-cell LIBs.

3.3.2. Electrochemical Potential Window

An electrochemical potential window is a design parameter that determines the range within which various electrochemical reactions can occur, directly impacting the performance of a full-cell LIB. It is influenced by the standard redox potentials of electrodes, electrolytes, separators, binders, and current collectors versus the standard hydrogen potential (SHE) [195]. The difference in potentials between cathode and anode in the absence of applied current is called electromotive force (EMF) or open-circuit voltage (OCV) of the LIBs and is determined by the nature of electrode materials. The voltage potential V_{OC} (ϕ) is determined by the Nernst equation given below:

$$V_{OC}\,(\phi) = -\Delta F_F = -\left(\frac{\mu_C - \mu_A}{F}\right) \tag{7}$$

where F is the Faraday constant, and μ_C and μ_A are the chemical potentials of cathode and anode materials in full-cell LIBs, respectively [196,197]. The energy band gap between the highest occupied molecular orbital (HOMO) and the lowest unoccupied molecular orbital (LUMO) of the electrolyte determines the voltage potential. The energy band gap between the highest occupied molecular orbital (HOMO) and the lowest unoccupied molecular orbital (LUMO) of the electrolyte determines the voltage potential of a full-cell LIB. For optimal performance, the chemical potential of the cathode material should be above the HOMO, while that of the anode should be below the LUMO of the electrolyte, as shown in Figure 7 [198]. This arrangement ensures a stable electrochemical environment, minimizing undesirable reactions and improving the overall voltage stability and efficiency of the LIBs.

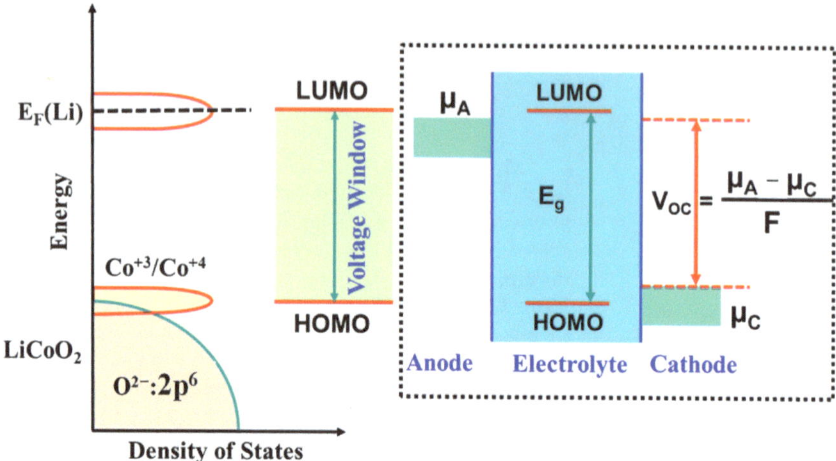

Figure 7. Energy phases of electrochemical potentials of Li metal and LiCoO$_2$, and their respective HOMO and LUMO energy positions in electrolytes. (Dotted Box) Relative energy of electrolyte, and HOMO and LUMO positions of electrolytes and electrodes.

In addition, electronegativity, the ability of an atom or a functional group to attract electrons, plays a crucial role in determining the electrochemical potential of electrode materials. The higher the electronegativity of an atom or functional group, the greater its oxidation potential. Obviously, fluorine and oxygen atoms are commonly used in the development of cathode materials for LIBs. Conversely, atoms or functional groups with lower electronegativities possess low reduction potentials, making them suitable for constructing anode materials. A functional group with high oxidation potential and low reduction potential creates electrodes with a high potential difference, leading to high energy density [199]. The potential difference between electrodes is influenced by the nature of polyanionic groups and the percentage of ionic bonding between cations and anions. Replacing oxygen atoms in polyanionic groups with fluorine increases the ionic bond characteristics, resulting in a larger potential difference. For example, transition metal phosphates, silicates, and sulfates exhibit higher potential differences compared to oxygen-based compounds [200].

Another crucial factor affecting the potential difference is the occupation of different lattice sites by various atoms, each corresponding to a distinct energy level and crystal structure. The insertion and extraction of Li+ ions and corresponding electrons from the 3D orbitals of transition metals alter the crystal structure or cause phase transformations. The energy required for these processes correlates with the electrochemical potential of the LIBs [201]. Lattice energy, the enthalpy change (ΔH) involved in the Li+ ion intercalation process, affects the Gibbs free energy (ΔG), which ultimately determines the electrochemical potential of the battery [202,203]. Crystal defects, such as the substitution of foreign atoms or ions in place of inherent lattice atoms, cause lattice distortion, altering the lattice energy and influencing both the electrochemical potential and thermal properties of electrode materials [204]. The electronic structure, which defines the number of electrons inserted or extracted during oxidation and reduction reactions, also correlates with the electrochemical potential. Changes in crystal and electronic structures can cause phase transformations, contributing to changes in enthalpy and Gibbs free energy (ΔG) [201]. Thus, it is essential to consider the electronegativity of atoms, the nature of chemical bonds (describing polyanionic groups), crystal structures, lattice energy, crystal defects, and electronic structures of both electrodes and electrolytes to design a stable and wide electrochemical potential window for full-cell LIBs.

3.3.3. Electrochemical Reaction Kinetics

The electrochemical reaction mechanism is a key determinant of the redox behavior of electrodes and their overall performance in LIBs. This mechanism is influenced by several factors, including the nature of the electrode materials, their crystal and electronic structures, the properties of the electrolyte, and the electrochemical potential window. Based on the characteristics of lithiation and de-lithiation processes, electrochemical reaction mechanisms are generally categorized into intercalation/de-intercalation, alloy/de-alloy, conversion, and mixed-type reactions. Each type of reaction has its own fundamental issues and theoretical specific capacity range, as illustrated in Figure 8. These mechanisms play a pivotal role in dictating the efficiency, stability, and specific capacity of the electrodes, with each presenting its own set of challenges and advantages. Understanding and optimizing these mechanisms is crucial for the development of high-performance LIBs.

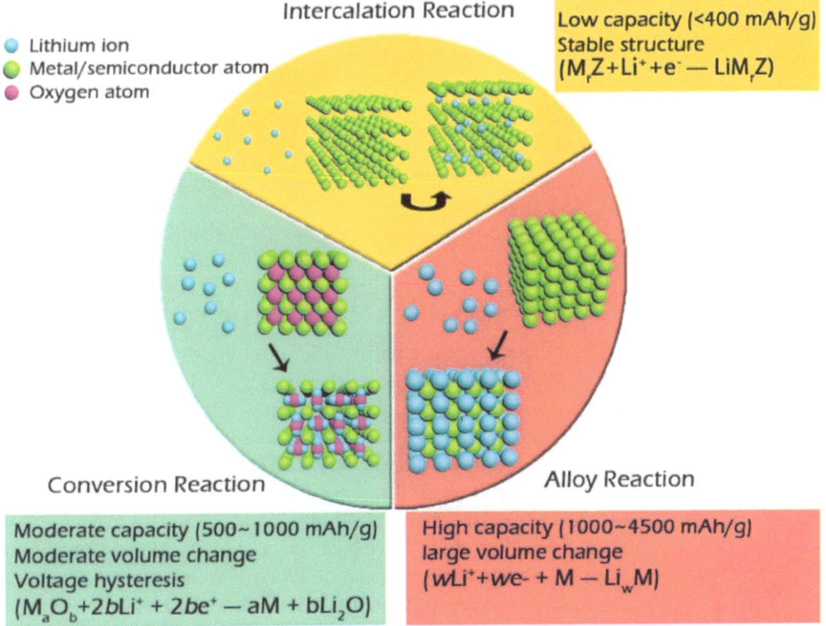

Figure 8. Illustration of fundamental electrochemical reaction mechanisms for LIBs (Reprinted/adapted with permission from Ref. [205] published by Royal Society of Chemistry, 2015).

The movement of Li$^+$ ions between cathode and anode, known as intercalation/de-intercalation mechanism, is the foundational electrochemical reaction mechanism in LIBs. It is explained as follows:

$$LiMO_2 \rightarrow Li_{1-x}MO_2 + xLi^+ + xe^- \tag{8}$$

Nevertheless, some electrode materials undergo phase transition during this process, which can hinder the electrochemical process and restrict the electrochemical activity of LIBs. For example, V_2O_5 cathode material undergoes various phase transformations during lithiation, as expressed by Refs [76,205].

$$V_2O_5 + 0.5Li^+ + 0.5e^- \rightarrow \varepsilon\text{-}Li_{0.5}V_2O_5 \tag{9}$$

$$\varepsilon\text{-}Li_{0.5}V_2O_5 + 0.5Li^+ + 0.5e^- \rightarrow \delta\text{-}LiV_2O_5 \tag{10}$$

$$\delta\text{-Li}_{0.5}\text{V}_2\text{O}_5 + \text{Li}^+ + e^- \rightarrow \gamma\text{-Li}_2\text{V}_2\text{O}_5 \tag{11}$$

The intercalation/de-intercalation process maintains the system's thermal stability by preserving structural integrity and limiting the evolution of gases such as H_2 and O_2, making it suitable for long-life LIBs. Anode materials like carbon compounds and titanium oxides are preferred for LIBs because they follow this mechanism. For cathode materials, oxides of vanadium (e.g., V_2O_5, VO_2, LiV_3O_8), spinel oxides (e.g., LiM_2O_4 where M = Mn, Co, Ni), polyanionic oxides (e.g., $LiMPO_4$ where M = Fe, Mn, Co, Ni), and certain layered oxides (e.g., $LiMO_2$ where M = Co, Mn, Ni, including NCM variants such as 811, 622, 333) are prominent choices. However, electrodes using the intercalation/de-intercalation mechanism often suffer from poor intrinsic conductivity, limited specific capacity, and potential hysteresis. To address these issues, the alloy/de-alloy reaction mechanism has been developed. Metal electrodes such as Si, Sb, Zn, and Sn, which follow the alloy/de-alloy reaction, offer high specific capacities, high-rate performance, and high energy density. With the exception of Ge and Sb, these metals are characterized by high electrical conductivity, abundance, low cost, and non-toxicity. They form distinct lithium alloys until the lower cut-off voltage of LIBs. Therefore, it is crucial to investigate the distinct phases of lithium alloys and the number of moles of Li^+ ions involved in the alloy/de-alloy reaction [19]. For example, SnO_2 nanoparticles encapsulated in a mesoporous carbon composite (SnO_2@MPC) anode form lithium alloys with distinct phases under several steps as follows:

$$SnO_2 + 4Li^+ + 4e^- \rightarrow 4Li_2O + Sn \ (711 \ \text{mAhg}^{-1}) \tag{12}$$

$$Sn + xLi^+ + xe^- \longleftrightarrow Li_xSn \ (x < 4.4) \ (783 \ \text{mAhg}^{-1}) \tag{13}$$

The alloy/de-alloy reaction in LIBs often results in an unstable SEI layer during charge and discharge cycles. This reaction is characterized by significant volume expansion and contraction—up to ~300% for Si anodes—far exceeding the ~150% expansion seen with intercalation/de-intercalation mechanisms [60]. These large volume changes lead to severe issues such as substantial capacity fading, poor rate performance, sluggish reaction kinetics, and structural instability [48,49]. In contrast, the conversion reaction involves transforming the electrode material into its constituent or derivative compounds. For example, the formation of Li_2O has been observed with N-doped reduced graphene oxide (rGO) wrapped around Mn_2O_3 nanorods, along with the conversion of MnO and Mn [206]. It addresses major issues such as poor specific capacity and environmental concerns, offering excellent rate performance and high energy density for LIBs. This improvement is demonstrated by the following general electrochemical reaction:

$$M_aX_b + (b.\ n)\ Li^+ + (b.\ n)\ e^- \rightarrow aM + bLi_nX \tag{14}$$

where 'X' is any of chalcogenides such as oxides, sulfides, nitrides, carbides, etc. Common materials used as anode materials for LIBs in the conversion-type reaction mechanism include 3D-transition metal oxides such as CuO, NiO, and M'_2O_3/M'_3O_4 (M' = V, Mn, Fe, Co, and Ni), as well as binary oxides like ferrites, manganites, and cobaltites ($M'M''_2O_4$, where M'/M'' = Ni, Zn, Fe, and Co). These materials typically exhibit potential stability windows around ~3.0 V vs. Li/Li+, deliver high specific capacities (~700 mAhg^{-1}), and are suitable for high energy density LIBs. However, they encounter significant challenges, including large potential hysteresis, capacity fading, poor intrinsic conductivities, and irreversible capacity loss [207]. To address these issues, a new electrochemical reaction mechanism known as the mixed reaction mechanism has been developed. This mechanism involves lithium insertion/extraction combined with alloy/de-alloy or conversion reactions. Vanadium-based binary oxides such as MV_2O_4/M_2VO_4, MV_2O_6, MV_2O_7, and $M_3V_2O_8$ (where M = Ni, Zn, Co, Mn, and Fe) are used in this mixed reaction mechanism. For instance, Zn-based vanadium metal oxides undergo alloy/de-alloy reactions upon lithiation, leading to the formation of a vanadium-based matrix that follows the insertion/extraction reaction mechanism, while Zn metal/metal oxides participate in the alloy/de-alloy reaction. For example, ZnV_2O_4 converts into ZnO and a lithiated vanadium oxide matrix (Li_yVO_2). During the charge/discharge process of LIBs, these components separately and spontaneously undergo alloy/de-alloy (Li_xZn) and insertion/extraction reactions as follows:

$$ZnV_2O_4 + (2x + y)\,Li^+ + (2x + y)\,e^- \rightarrow 2Li_xVO_2 + Li_yZn \tag{15}$$

$$ZnO + (y + 2)\,Li^+ + (y + 2)\,e^- \rightarrow Li_yZn + Li_2O \tag{16}$$

$$Li_xVO_2 \leftrightarrow LiVO_2 + (x - 1)\,Li^+ + (x - 1)\,e^- \tag{17}$$

$$Li_yZn \leftrightarrow Zn + y\,Li^+ + y\,e^- \tag{18}$$

$$Zn + Li_2O \leftrightarrow ZnO + 2\,Li^+ + 2\,e^- \quad (1 \leq x \leq 2, y \leq 1) \tag{19}$$

Transition metals based on vanadium oxides typically follow conversion and insertion/extraction reaction mechanisms. For instance, $Ni_3V_2O_8$ hollow microspheres convert into NiO/Ni metal and a lithiated vanadium oxide matrix ($LixV_2O_5$), delivering a high specific capacity [208]. The excellent rate performance of these materials for LIBs is attributed to the in situ formation of metal/metal oxide layers during the mixed electrochemical reaction mechanism. This process enhances thermal stability, minimizes irreversible capacity loss and volume changes, and improves conductivity due to the electroactively lithiated vanadium oxide matrix. Consequently, intercalation and mixed electrochemical reactions are considered the most desirable mechanisms for the development of high energy density LIBs.

3.3.4. Efficiency

The efficiency, the ratio between output energy to input energy for a full-cell LIBs, measures the battery's ability to deliver a specific amount of energy for applications such as smartphones, laptops, and tablets. It is described in terms of coulombic efficiency and energy efficiency as follows:

$$\text{Coulombic Efficiency}(\%) = \frac{\text{Total Discharge Capacity}}{\text{Toal Charge Capacity}} \times 100 \tag{20}$$

$$\text{Energy Efficiency}(\%) = \frac{\text{Discharge Energy Density}}{\text{Charge Energy Density}} \times 100 \tag{21}$$

Coulombic efficiency refers to the proportion of Li^+ ions effectively cycled within a full-cell LIB, comparing the amount of Li^+ ions extracted from the cathode to those inserted into the anode during cycling. It represents the ratio of the obtained specific capacity to the amount of Li^+ ions cycled. Ideally, if the same amount of Li^+ ions is extracted and inserted, coulombic efficiency would be 100%. However, practical observations show that some Li^+ ions are not fully recovered due to factors like SEI layer formation and electrode material characteristics. For instance, $LiCoO_2$ might deliver a certain number of Li atoms to reach a fully charged state ($Li_{0.5}CoO_2$) but only recover a reduced amount ($0.9\times$ number of Li atoms), leading to a coulombic efficiency of around 90%.

Energy efficiency, on the other hand, pertains to the overpotential of the LIB. Each electrode has a different redox potential, and Li atoms require varying amounts of energy for insertion or extraction. If some Li atoms are extracted at 2.0 V while others require 4.0 V during charging, the energy used to insert or extract these atoms differs, indicating large overpotentials that cannot be recovered. Thus, energy efficiency is determined by the discharge/charge curves of the LIBs, where each point on the curve represents the energy required for the insertion or extraction of Li atoms. The height of these profiles, alongside the length of the x-axis, is crucial for assessing the energy efficiency of the LIB. Both the nature of the electrodes, including their physicochemical properties, particle sizes, and electrochemical reaction kinetics, play significant roles in defining the efficiency of full-cell LIBs.

3.4. Productivity and Cost of Full-Cell LIBs

The cost and weight of a full-cell LIB are influenced by the selection of materials, the fabrication process, and the desired dimensions, which are determined based on the intended application. We primarily focused on coin-type full cells with CR2032 dimensions for simplicity. The weight of a coin cell is determined by several components: current collectors (aluminum and copper foils), reference electrodes (lithium foil), the amount of electrolyte, the mass loading of electrode materials (both anode and cathode), and additional winding components such as wave springs, gaskets, casings, spacers, and separators. The mass loading and weight of the electrodes need to be controlled during fabrication. The active mass of the electrodes is determined by the proper mixing of carbon additives and polymer binders with the active material to produce the final electrodes. Variations in the mixing ratios (e.g., 60:20:20, 70:20:10, and 80:5:5) affect the final weight of the electrodes in the coin cell. The total weight of the coin cell is composed of 25.5% cathode material, 14.5% anode material, 6.9% aluminum current collector, 8.1% copper current collector, 11.2% electrolyte, 3.6% separator, and 30.2% battery casing, as depicted in Figure 9 [209]. Likewise, the cost of a full coin cell is dependent on the cost of each cell component, including spacers, gaskets, wave springs, casings, production line equipment (e.g., mixing machines, roller presses, electrode and electrolyte materials, and electrode cutters), lithium foil, current collectors, and the coin cell assembly environment, such as an argon-filled glove box. Therefore, optimizing the performance of the full coin cell LIBs is essential for determining its overall cost and achieving high energy density LIBs.

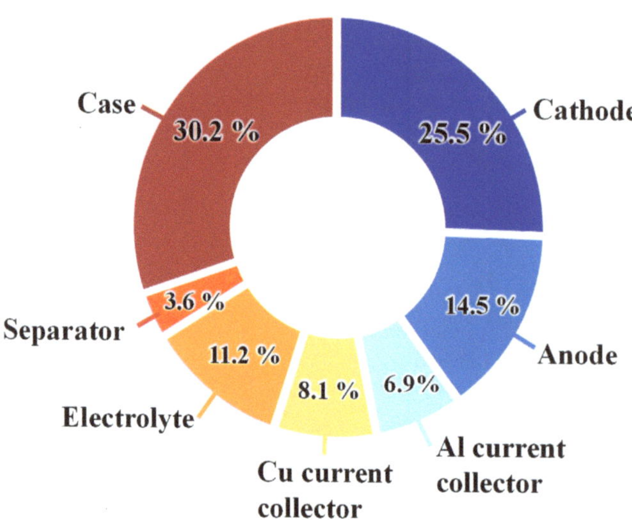

Figure 9. Weight percentage of main components of full-cell LIBs (re-drawn from the data from Ref. [209]).

3.4.1. Cell Fabrication Processes

The cell manufacturing process plays a crucial role in determining various parameters such as electrode weight, thickness, porosity, tortuosity, mixing ratio, slurry rheology, and granule characteristics, all of which impact the potential window and performance of LIBs [210]. The manufacturing of electrodes typically involves a two-step process: slurry preparation and film formation, as shown in Figure 10 [211]. This process encompasses the entire sequence from active material mixing to slurry coating, solvent evaporation (drying), and further processing, such as calendaring, cutting, notching, etc. [104].

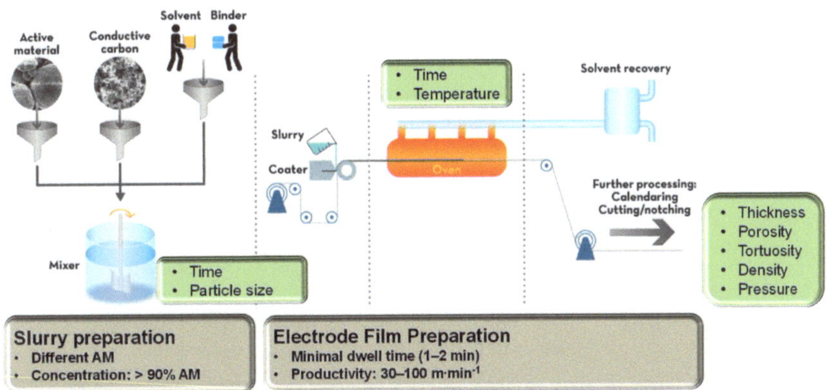

Figure 10. The manufacturing process of electrodes for full-cell LIBs along with main steps with most influential parameters. Adapted from Ref. [104] and modified.

Slurry preparation involves mixing active materials, carbon additives, and binders. Key factors include mixing time, particle size, chemical characteristics, and the choice of solvent. The mixing ratio of active materials with carbon additives and binders, along with the uniformity of granules and slurry rheology, directly affects the quality of the electrode film. A homogeneous slurry with minimal agglomeration and sedimentation ensures stable particle–solvent interactions, impacting viscosity and stability. At the laboratory level, proper particle dispersion and uniform mixing can be achieved through manual grinding with a mortar and pestle, ball milling, ultrasonic grinding, or magnetic stirring. In industrial sectors, methods such as ball milling, planetary mixing, high-speed mixers, hydrodynamic shear mixing, and homogenizers are used to achieve optimal particle dispersion and mixing [212]. The choice of solvent is crucial for determining the dispersibility and wettability of the slurry used in electrode manufacturing. The contact angle describes how well the solid particles interact with the solvent in the slurry. Key factors such as mixing time and the appropriate amount of solvent are vital for achieving uniformity and the desired rheological properties of the slurry. Extended mixing may improve the uniformity of the slurry's rheology but can lead to solvent evaporation. For instance, mixing carbon additives into the binder enhances mechanical and cohesive properties, forming an electronic network within the binder. However, this can make it challenging to achieve uniform mixing with active materials, potentially suppressing volume expansion, providing thermal/mechanical stability, and increasing irreversible capacity loss. On the other hand, when carbon additives are uniformly mixed with active materials, they cover the surfaces of active particles, enhancing the electron conductive network, improving conductivity, and potentially increasing specific capacity and energy density, although it may lead to slurry detachment from the current collector during the charging process.

The film preparation process involves spreading the slurry over the current collector, followed by drying to evaporate the solvent, and calendaring to optimize the thickness, porosity, tortuosity, and density of the film. The electrode thickness is controlled by the slurry spreading process, which can be performed using methods such as doctor blade coating in academic sectors or electrostatic deposition, roll coating, slot-die coating, and screen printing in industrial sectors [213,214]. The solvent evaporation is managed by drying the slurry at specific temperatures, times, and under vacuum conditions to ensure good adhesion and mechanical strength without cracking [215]. The calendaring process reduces electrode thickness and porosity, optimizing electrode density and affecting the wettability of the electrolyte and the overall performance of the full-cell LIBs [216,217].

In cell assembly, electrodes and separators are cut or slit to specific dimensions, stacked and wound together with the required amount of electrolyte before being packed using a punching machine. The dimensions of the LIBs are determined during the cutting/slitting process, and electrolyte filling creates the necessary environment for the electrochemical reactions. Proper stacking and winding are essential to prevent gas leakage (O_2/H_2), avoid component disintegration, and ensure mechanical stability. This delicate assembly process is typically conducted in an argon-filled glove box to maintain a dirt- and moisture-free environment. Finally, cell aging and inspection involve two stages: calendar aging and cycle aging. Calendar aging assesses capacity loss over time, while cycle aging measures capacity fading due to factors like voltage range, operating temperature, and charging rate [210,218]. The efficiency of the full-cell LIB is analyzed by charging and discharging cells under constant current (CC) and constant voltage (CV) phases for a specified number of cycles. Specific capacity, energy density, power density, efficiency, and charge/discharge times are determined, with specific C-rates correlating to the inspection time. The test scheme must specify the working voltage window, C-rate, weight, and thickness of electrodes to accurately determine the lifespan of the LIBs.

3.4.2. Mass Loading of Active Material

The specific capacity, energy density, overall cell packaging weight, and cell performance of LIBs are determined by the N/P ratio, which is defined as the ratio of the active mass loading amounts of negative (anode) to positive (cathode) electrode materials:

$$\text{N/P ratio} = \frac{\text{Mass } of \text{ anode material}}{\text{Mass } of \text{ cathode material}} \tag{22}$$

Higher or lower mass loading leads to poor specific capacity and hinder charge transportation. The mass loading amount is crucial for balancing the capacity of a full-cell LIB and is adjusted through the thickness of the electrode materials during film fabrication.

The anode materials such as Li metal (3860 mAhg^{-1}), graphite (372 mAhg^{-1}), silicon (4200 mAhg^{-1}), and SnO (782 mAhg^{-1}) deliver higher theoretical specific capacities than those of cathode materials, such as NCM811 (220 mAhg^{-1}), LiCoO$_2$ (140 mAhg^{-1}), LiMn$_2$O$_4$ (148 mAhg^{-1}), LiFePO$_4$ (170 mAhg^{-1}), and LiV$_3$O$_8$ (280 mAhg^{-1}) [219–222]. To achieve higher capacity and energy density in LIBs, the specific capacities should be considered based on nearly twice the mass loading amount of cathode materials. An anode-free configuration (0 N/P ratio) indicates no extra lithium is involved, which helps extend the life of LIBs. Thus, the recommended N/P ratio for full-cell configurations typically ranges between 1 and 1.2 [223]. The N/P ratio can be adjusted by varying the density of the anode materials. Increasing the N/P ratio generally reduces initial coulombic efficiency and increases the electrochemical potential of the anode material at the end of charge. This, in turn, can degrade the specific capacity and cycle life of the LIB. Research on different N/P ratios (1.10, 1.20, and 1.30) at room temperature (25 °C) and a C-rate of 0.85C shows that cells with an N/P ratio higher than 1.10 suppress Li plating, while an N/P ratio of 1.20 enhances cycle life [224]. The impact of different N/P ratios (1.02, 1.06, 1.10, and 1.14) on the electrochemical performance of LiFePO$_4$ batteries at various temperatures (0 °C, 45 °C) indicates that higher N/P ratios (1.10 and 1.14) provide better capacity retention compared to lower ratios [225]. Additionally, studies with different current collectors (Al and C) and mass loadings (1, 4 mg cm^{-2} and 1, 4, 8 mg/cm^{-2}) for LiAl$_{0.1}$Mn$_{1.9}$O$_4$ cathodes demonstrate that although high mass loading reduces specific capacities, it significantly improves capacity retention. Carbon-based current collectors showed higher specific capacity and retention compared to aluminum-based ones [226]. The tap density of materials affects LIB performance as it regulates the total mass per unit volume. The N/P ratio is influenced by the amount of active mass loading, which determines the utility of electrode materials during cycling. The diameter and thickness of materials also play a role; for example, a CR2032-sized coin cell LIB typically has an anode diameter of 10–12 mm and a cathode diameter of 12–14 mm. Using a smaller diameter for the anode plate helps adjust the utility of anode and cathode materials and minimize waste during cycling. Optimizing the N/P ratio is crucial for fabricating high energy density LIBs in a full coin cell configuration.

4. Summary and Outlook

LIBs are prominent energy storage devices to meet the growing energy demands of the modern era. They offer high specific capacity, energy density, thermal stability, and long calendar life compared to other types of batteries. LIBs are used in a diverse range of applications, from powering household appliances to supporting electric vehicles. Effective LIB design must accommodate a significant number of Li^+ ions while maintaining structural integrity, thermal and mechanical stability, and an optimal balance between energy and power density. This requires careful consideration of electrochemical reaction mechanisms. The design of LIBs involves numerous parameters that collectively impact their overall performance. This review aimed to detail key design parameters, their modification strategies, and their effects on the electrochemical performance of LIBs and reached the following summary.

- The full-cell configuration of LIBs includes electrodes (cathodes, anodes), current collectors, a separator, and an electrolyte. The cathode functions as the positive electrode with a high oxidation potential, facilitating the delivery of Li^+ ions to the battery system. On the other hand, the anode acts as the negative electrode with a low reduction potential, accepting incoming Li^+ ions. Current collectors are typically metal foils, metal oxides, or carbon fibers. Commonly used commercial current collectors include copper and aluminum foils. PP sheets, glass fibers, and sodium alginates are commonly used separators that prevent the flow of electrons while allowing the conduction of Li^+ ions within the electrolyte. The electrolyte manages the transportation of Li^+ ions and supports the chemical reactions. The electrolyte is a mixture of lithium salts ($LiClO_4$, $LiPF_6$, LiTFSI, LiTf, $LiAsF_6$, $LiBF_4$) and solvents (aqueous solutions, organic solvents, ionic liquids, polymers, and gels). A commercially used electrolyte is 1.0 M $LiPF_6$ in a solvent mixture of ethylene carbonate (EC) and diethyl carbonate (DEC) at a 1:1 volume ratio, or in EC and dimethyl carbonate (DMC) at a 3:7 volume ratio.
- The design of full-cell LIBs involves several critical factors, including form factors (such as length, width, height, shape, and volume), material selection, performance, and productivity/cost aspects. Obviously, the form factors must be carefully considered to meet the requirements of LIBs. Material selection is most fundamental and crucial since it defines the electrochemical reaction mechanism, performance, and cost of LIBs. Designing electrode materials requires careful consideration of both intrinsic and extrinsic approaches. Binders should be designed with robust adhesion and cohesion properties, optimal binder selection, excellent distribution, free radical quenching capabilities, strong chelation, and electrochemical compatibility. Current collectors are evaluated based on electrochemical stability, density, mechanical strength, electrical conductivity, sustainability, and cost. Separators should be designed with attention to thickness, porosity, mean pore size (typically less than 1 μm), pore morphology, wettability, thermal stability, and mechanical properties. The development of electrolytes involves considering characteristics such as ionic conductivity, wide potential stability window, temperature tolerance, mechanical and thermal stability, chemical stability, and the ability to support the reaction kinetics of LIBs.
- Performance in full-cell LIBs is determined by several factors: conductivity, electrochemical reaction mechanisms, voltage window, efficiency, and thermodynamics. Both electrical and ionic conductivities significantly impact the specific capacity, energy density, power density, and cycle stability of LIBs. Electrical conductivity is governed by the electronic structure of the electrode materials, whereas ionic conductivity is

influenced by the crystal structure, physicochemical properties, morphology, and particle size of the electrode materials, as well as the porosity and geometry of the separator. These factors directly or indirectly influence Li$^+$ ion diffusion, which in turn affects the rate performance and power density of LIBs. Thus, a thorough investigation is necessary to evaluate the performance of LIBs. The potential window indicates the range of electrochemical reactions that can occur within HOMO and LUMO of the electrolyte. It is influenced by factors such as the electronegativity of atoms, the nature of chemical bonds, lattice energy, crystal defects, and the crystal and electronic structures of both electrodes and electrolytes.

- Productivity is determined by factors such as the electrode fabrication process, mass loading amount, and processability, all of which impact the cost and weight of the final product. Key factors influencing the final electrode's properties include process parameters that affect the compact density, thickness, mass loading amount, and porosity of the electrode. The mass loading amount of the cathode and anode, determined by the thickness and mixing ratio of the electroactive materials, directly influences the specific capacity, energy, power density, and overall performance of the LIBs. Thus, each step in the fabrication process should be carefully managed to meet the requirements of full-cell LIBs. The final cost of full-cell LIBs is influenced by the costs of materials, cell components, and manufacturing processes, necessitating the optimization of cost analysis for each component to minimize the overall expense.

Consequently, numerous parameters have been specifically designed, modified, and optimized to enhance the efficiency of full-cell LIBs. Critical parameters include the form factor (shapes and dimensions) of the battery, choice of materials for the main component, and factors affecting performance such as the electrochemical potential window, electrochemical reaction chemistry, conductivity, efficiency, and thermodynamics. The last factor to be considered is productivity and cost of LIBs. It is essential to apply standard synthesis techniques with meticulous care, including structural and surface treatments to enhance intrinsic and extrinsic properties, as well as to employ novel in situ and ex situ technological approaches to assess the precise performance of full-cell LIBs under extreme temperatures. Additionally, understanding the physicochemical factors that influence LIB performance and preventing impurities that could cause internal short circuits are crucial. Therefore, a thorough investigation of electrode fabrication and cell assembly processes is necessary to achieve high-energy density and high-performance full-cell LIBs.

Author Contributions: Conceptualization, methodology, and software, F.G. and D.L.; writing—original draft preparation, F.G.; writing—review and editing, D.L. and K.A.; visualization, supervision, project administration, and funding acquisition, D.L. All authors have read and agreed to the published version of the manuscript.

Funding: This work was supported by the National R&D Program from Ministry of Trade, Industry and Energy of Korea (20021922).

Data Availability Statement: Not applicable.

Conflicts of Interest: The authors declare no conflicts of interest.

References

1. Behabtu, H.A.; Messagie, M.; Coosemans, T.; Berecibar, M.; Fante, K.A.; Kebede, A.A.; Mierlo, J.V. A review of energy storage technologies: Application potentials in renewable energy sources grid integration. *Sustainability* **2020**, *12*, 10511. [CrossRef]
2. China Energy Storage Alliance (CNESA). *CNESA Global Energy Storage Market Analysis-2019*; Q4 (Summary); China Energy Storage Alliance (CNESA): Beijing, China, 2020.
3. Whittingham, M.S. Electrical energy storage and intercalation chemistry. *Science* **1976**, *192*, 1126–1127. [CrossRef]
4. Whittingham, M.S.; Jacobson, A.J. *Intercalation Chemistry*; Academic Press: New York, NY, USA, 1982.
5. Nagaura, T.; Nagamine, M.; Tanabe, I.; Miyamoto, N. Solid-state batteries with sulfide-based electrolytes. *Prog. Batter. Sol. Cells* **1989**, *8*, 84–88.
6. Yazami, R.; Touzain, P. A reversible graphit-lithium negative electrode for electrochemical generators. *J. Power Sources* **1983**, *9*, 365–371. [CrossRef]

7. Sarre, G.; Blanchard, P.; Broussely, M. Aging of lithium-ion batteries. *J. Power Sources* **2004**, *127*, 65–71. [CrossRef]
8. Guo, J.; Li, Y.; Pedersen, K.; Stroe, D.-I. Lithium-ion battery, operation, degradation, and aging mechanism in electric vehicles: An overview. *Energies* **2021**, *14*, 5220. [CrossRef]
9. Zhang, X.; Li, P.; Huang, B.; Zhang, H. Numerical investigation on the thermal behavior of cylindrical lithium-ion batteries based on the electrochemical thermal coupling model. *Int. J. Heat Mass Transf.* **2022**, *199*, 123449. [CrossRef]
10. Román-Ramírez, L.; Marco, J. Design of Experiments Applied to Lithium-Ion Batteries: A Literature Review. *Appl. Energy* **2022**, *320*, 119305. [CrossRef]
11. Kim, H.-J.; Krishna, T.N.V.; Zeb, K.; Rajangam, V.; Gopi, C.V.V.M.; Sambasivam, S.; Raghavendra, K.V.G.; Obaidat, I.M. A comprehensive review of Li-ion battery materials and Their recycling techniques. *Electronics* **2020**, *9*, 1161. [CrossRef]
12. Barbosa, J.C.; Goncalves, R.; Costa, C.M.; Mendez, S.L. Recent advances on Materials for Lithium-ion batteries. *Energies* **2021**, *14*, 3145. [CrossRef]
13. Kim, T.; Song, W.; Son, D.-Y.; Ono, L.K.; Qi, Y. Lithium-ion batteries: Outlook on present, future, and hybridized technologies. *J. Mater. Chem. A* **2019**, *7*, 2942–2964. [CrossRef]
14. Chaudhary, M.; Tyagi, S.; Gupta, R.K.; Singh, B.P.; Singhal, R. Surface modification of cathode materials for energy storage devices: A review. *Surf. Coat. Technol.* **2021**, *412*, 127009. [CrossRef]
15. Wang, Y.; Xiao, X.; Li, Q.; Pang, H. Synthesis and Progress of New Oxygen vacant electrode materials for high-energy rechargeable battery applications. *Small* **2018**, *14*, 1802193. [CrossRef] [PubMed]
16. Ji, X.; Xia, Q.; Xu, Y.; Feng, H.; Wang, P.; Tan, Q. A review on progress of lithium-rich manganese-based cathodes for lithium-ion batteries. *J. Power Sources* **2021**, *487*, 229362. [CrossRef]
17. Fayaz, H.; Afzal, A.; Samee, A.D.M.; Soudagar, M.E.M.; Akram, N.; Mujtaba, M.A.; Jilte, M.; Islam, T.; Agbulut, U.; Saleel, C.A. Optimization of Thermal and Structural Design in Lithium-Ion Batteries to Obtain Energy Efficient Battery Thermal Management System (BTMS): A Critical Review. *Arch. Comput. Methods Eng.* **2022**, *29*, 129–194. [CrossRef]
18. Du, S.; Lai, Y.; Ai, L.; Ai, L.; Cheng, Y.; Tang, Y.; Jia, M. An investigation of irreversible heat generation in lithium-ion batteries based on a thermo-electrochemical coupling method. *Appl. Therm. Eng.* **2017**, *121*, 501–510. [CrossRef]
19. Goriparti, S.; Miele, E.; De Angelis, F.; Di Fabrizio, E.; Zaccaria, R.P.; Capiglia, C. Review on recent progress of nanostructured anode materials for Li-ion batteries. *J. Power Sources* **2014**, *257*, 421–443. [CrossRef]
20. Li, J.; Zhong, W.; Deng, Q.; Zhang, Q.; Yang, C. Recent progress in synthesis and surface modification of Nickle-rich layered oxide cathode materials for lithium-ion batteries. *Int. J. Extrem. Manuf.* **2022**, *4*, 042004. [CrossRef]
21. Zhang, J.; Qiao, J.; Sun, K.; Wang, Z. Balancing particle properties for practical lithium-ion batteries. *Particuology* **2022**, *61*, 18–29. [CrossRef]
22. Mayur, M.; DeCaluwe, S.C.; Kee, B.L.; Bessler, W.G. Modeling and simulation of the thermodynamics of lithium-ion battery intercalation materials in the open-source software Cantera. *Electrochim. Acta* **2019**, *323*, 134797. [CrossRef]
23. Kim, J.-S.; Lee, D.-C.; Lee, J.-J.; Kim, C.-W. Optimization for the maximum specific energy density of a lithium-ion battery using progressive quadratic response surface method and design of experiments. *Sci. Rep.* **2020**, *10*, 15586. [CrossRef] [PubMed]
24. Lain, M.J.; Brandon, J.; Kendrick, E. Design Strategies for High Power vs. High Energy Lithium-Ion Cells. *Batteries* **2019**, *5*, 64. [CrossRef]
25. Tikekar, M.D.; Choudhury, S.; Tu, Z.; Archer, L.A. Design principles for electrolytes and interfaces for stable lithium-metal batteries. *Nat. Energy* **2016**, *1*, 16114. [CrossRef]
26. Park, S.; Jeong, S.Y.; Lee, T.K.; Park, M.W.; Lim, H.Y.; Sung, J.; Cho, J.; Kwak, S.K.; Hong, S.Y.; Choi, N.-S. Replacing conventional battery electrolyte additives with dioxolone derivatives for high-energy-density lithium-ion batteries. *Nat. Commun.* **2021**, *12*, 838. [CrossRef] [PubMed]
27. Khan, F.N.U.; Rasul, M.G.; Sayem, A.S.M.; Mandal, N.K. Design and optimization of lithium-ion battery as an efficient energy storage device for electric vehicles: A comprehensive review. *J. Energy Storage* **2023**, *71*, 108033. [CrossRef]
28. XU, J.; Cai, X.; Cai, S.; Shao, Y.; Hu, C.; Lu, S.; Ding, S. High-energy lithium-ion batteries: Recent progress and a promising future in applications. *Energy Environ. Mater.* **2023**, *6*, e12450. [CrossRef]
29. Wang, Q.; O'Carroll, T.; Shi, F.; Huang, Y.; Chen, G.; Yang, X.; Nevar, A.; Dudko, N.; Tarasenko, N.; Xie, J.; et al. Designing Organic Material Electrodes for Lithium-Ion Batteries: Progress, Challenges, and Perspectives. *Electrochem. Energy Rev.* **2024**, *7*, 15. [CrossRef]
30. Chawla, N.; Bharti, N.; Singh, S. Recent Advances in Non-flammable electrolytes for safer lithium-ion batteries. *Batteries* **2019**, *5*, 19. [CrossRef]
31. Zhu, L.; Bao, C.; Xie, L.; Yang, X.; Cao, X. Review of synthesis and structural optimization of LiNi1/3Co1/3Mn1/3O2 cathode materials for lithium-ion batteries applications. *J. Alloys Comp.* **2020**, *831*, 154864. [CrossRef]
32. Ren, D.; Shen, Y.; Yang, Y.; Shen, L.; Levin, B.D.A.; Yu, Y.; Muller, D.A.; Arbuna, H.D. Systematic optimization of battery materials: Key parameter optimization for the scalable synthesis of uniform high-energy, and high-stability $LiNi_{0.6}Mn_{0.2}Co_{0.2}O_2$ cathode materials for Lithium-ion batteries. *ACS Appl. Mater. Interfaces* **2017**, *9*, 35811–35819. [CrossRef]
33. Julien, C.; Mauger, A.; Zaghib, K.; Groult, H. Optimization of layered cathode materials for Lithium-ion batteries. *Materials* **2016**, *9*, 595. [CrossRef] [PubMed]
34. Manthiram, A.; Chemelewski, K.; Lee, E.-S. A perspective on the high-voltage $LiMn_{1.5}Ni_{0.5}O_4$ spinel cathode for lithium-ion batteries. *Energy Environ. Sci.* **2014**, *7*, 1339–1350. [CrossRef]

35. Zheng, J.; Ye, Y.; Pan, F. Structure units as material genes in cathode materials for lithium-ion batteries. *Natl. Sci. Rev.* **2020**, *7*, 242–245. [CrossRef]

36. Zheng, J.; Xiao, J.; Zhang, J.-G. The roles of oxygen non-stoichiometry on the electrochemical properties of oxide-based cathode materials. *Nano Today* **2016**, *11*, 678–694. [CrossRef]

37. Kuganathan, N.; Ganeshalingham, S.; Chroneos, A. Defects, Dopants and Lithium mobility in $Li_9V_3(P_2O_7)_3(PO_4)_2$. *Sci. Rep.* **2018**, *8*, 8140. [CrossRef]

38. Kuganathan, N.; Solovjov, A.L.; Vovk, R.V.; Chroneos, A. Defects, diffusion, and dopants in Li_8SnO_6. *Heliyon* **2021**, *7*, e07460. [CrossRef]

39. Zhang, J.; Jiang, H.; Zeng, Y.; Zhang, Y.; Guo, H. Oxygen-defective Co_3O_4 for pseudo-capacitive lithium storage. *J. Power Sources* **2019**, *439*, 227026. [CrossRef]

40. Wang, F.; Wang, X.; Chang, Z.; Zhu, Y.; Fu, L.; Liu, X.; Wu, Y. Electrode materials with tailored facets for electrochemical energy storage. *Nanoscale Horiz.* **2016**, *1*, 272–289. [CrossRef]

41. Guo, B.; Ruan, H.; Zheng, C.; Fei, H.; Wei, M. Hierarchical $LiFePO_4$ with a controllable growth of the (010) facet for lithium-ion batteries. *Sci. Rep.* **2013**, *3*, 2788. [CrossRef]

42. Zhang, F.; Lou, S.; Li, S.; Yu, Z.; Liu, Q.; Dai, A.; Cao, C.; Toney, M.F.; Ge, M.; Xiao, X.; et al. Surface regulation enables high stability of single-crystal lithium-ion cathodes at high voltage. *Nat. Comm.* **2020**, *11*, 3050. [CrossRef]

43. Wang, Y.; Wang, E.; Zhang, X.; Yu, H. High-Voltage Single-Crystal Cathode materials for Lithium-Ion Batteries. *Energy Fuels* **2021**, *35*, 1918–1932. [CrossRef]

44. Langdon, J.; Manthiram, A. A perspective on single-crystal layered oxide cathodes for lithium-ion batteries. *Energy Storage Mater.* **2021**, *37*, 143–160. [CrossRef]

45. Qian, G.; Zhang, Y.; Li, L.; Zhang, R.; Xu, J.; Cheng, Z.; Xie, S.; Wang, H.; Rao, Q.; He, Y.; et al. Single crystal Nickle-rich layered oxide battery cathode materials: Synthesis, electrochemistry, and intra-granular fracture. *Energy Storage Mater.* **2020**, *27*, 140–149. [CrossRef]

46. Lu, C.-H.; Lin, S.-W. Influence of the particle size on the electrochemical properties of lithium manganese oxide. *J. Power Sources* **2001**, *97–98*, 458–460. [CrossRef]

47. Majdabadi, M.M.; Farhad, S.; Farkhondeh, M.; Fraser, R.A.; Fowler, M. Simplified electrochemical multi-particle model for $LiFePO_4$ cathodes in lithium-ion batteries. *J. Power Sources* **2015**, *275*, 633–643. [CrossRef]

48. Wu, S.; Yu, B.; Wu, Z.; Fang, S.; Shi, B.; Yang, J. Effect of particle size distribution on the electrochemical performance of micro-sized silicon-based negative materials. *RSC Adv.* **2018**, *8*, 8544–8551. [CrossRef]

49. Zhu, G.; Wang, Y.; Yang, S.; Qu, Q.; Zheng, H. Correlation between the physical parameters and the electrochemical performance of a silicon anode in lithium-ion batteries. *J. Mater.* **2019**, *5*, 164–175. [CrossRef]

50. Garcia, J.C.; Bareno, J.; Yan, J.; Chen, G.; Hauser, A.; Croy, J.R.; Iddir, H. Surface structure, morphology, and stability of $LiNi_{1/3}Mn_{1/3}Co_{1/3}O_2$ cathode material. *J. Phys. Chem. C* **2017**, *121*, 8290–8299. [CrossRef]

51. Li, H.; Ren, Y.; Yang, P.; Jian, Z.; Wang, W.; Xing, Y.; Zhang, S. Morphology, and size-controlled synthesis of the hierarchical structured $Li_{1.2}Mn_{0.54}Ni_{0.13}Co_{0.13}O_2$ cathode materials for lithium-ion batteries. *Electrochim. Acta* **2019**, *297*, 406–416. [CrossRef]

52. Wei, C.; He, W.; Zhang, X.; Xu, F.; Liu, Q.; Sun, C.; Song, X. Effect of morphology on the electrochemical performance $Li_3V_2(PO_4)_3$ cathode materials for lithium-ion batteries. *RSC Adv.* **2015**, *5*, 54225–54245. [CrossRef]

53. Wang, S.; Zhang, Z.; Jiang, Z.; Deb, A.; Yang, L.; Hirano, S.-I. Mesoporous $Li_3V_2(PO_4)_3$@CMK-3 nanocomposite cathode material for lithium-ion batteries. *J. Power Sources* **2014**, *253*, 294–299. [CrossRef]

54. Muller, M.; Schneider, L.; Bohn, N.; Binder, J.R.; Bauer, W. Effect of nanostructured and open-porous particle morphology on electrodes processing and electrochemical performance of Li-ion batteries. *ACS Appl. Energy Mater.* **2021**, *4*, 1993–2003. [CrossRef]

55. Seher, J.; Froba, M. Shape Matters: The effect of particle morphology on the fast-charging performance of $LifePO_4$/C nanoparticle composite electrode. *ACS Omega* **2021**, *6*, 24062–24069. [CrossRef] [PubMed]

56. Ghani, F.; Raza, A.; Kyung, D.; Kim, H.-S.; Lim, J.C.; Nah, I.W. Optimization of synthesis conditions of high-tap density $FeVO_4$ hollow microspheres via spray pyrolysis for lithium-ion batteries. *Appl. Surf. Sci.* **2019**, *497*, 143718. [CrossRef]

57. Birkl, C.R.; Roberts, M.R.; McTurk, E.; Bruce, P.G.; Howey, D.A. Degradation diagnostics for lithium-ion cells. *J. Power Sources* **2017**, *341*, 373–386. [CrossRef]

58. Liu, H.; Wolfman, M.; Karki, K.; Yu, Y.-S.; Stach, E.A.; Cabana, J.; Chapman, K.W.; Chupas, P.J. Intergranular cracking as a major cause of long-term capacity fading of layered cathodes. *Nano Lett.* **2017**, *17*, 3452–3457. [CrossRef]

59. Lei, Y.; Ni, J.; Hu, Z.; Wang, Z.; Gui, F.; Li, B.; Ming, P.; Zhang, C.; Elias, Y.; Aurbach, D.; et al. Surface modification of Li-rich Mn-based layered oxide cathodes: Challenges, materials, methods, and characterization. *Adv. Energy Mater.* **2020**, *10*, 2002506. [CrossRef]

60. Raza, A.; Ghani, F.; Lim, J.C.; Nah, I.W.; Kim, H.-S. Eco-friendly prepared mesoporous carbon encapsulated SnO_2 nanoparticles for high-reversible lithium-ion battery anodes. *Microporous Mesoporous Mater.* **2021**, *314*, 110853. [CrossRef]

61. Palacin, M.R. Recent Advances in rechargeable battery materials: A chemist's perspective. *Chem. Soc. Rev.* **2009**, *38*, 2565–2575. [CrossRef]

62. Cheng, H.; Shapter, J.G.; Li, Y.; Gao, G. Recent progress of advanced anode materials of lithium-ion batteries. *J. Energy Chem.* **2021**, *57*, 451–468. [CrossRef]

63. Bensalah, N.; Dawood, H. Review on synthesis, characterizations, and electrochemical properties of cathode materials for Lithium-ion batteries. *J. Mater. Sci. Eng.* **2016**, *5*, 4.

64. Lyu, Y.; Wu, X.; Wang, K.; Feng, Z.; Cheng, T.; Liu, Y.; Wang, M.; Chen, R.; Xu, L.; Zhou, J.; et al. An overview on the advances of $LiCoO_2$ cathodes for Lithium-Ion Batteries. *Adv. Energy Mater.* **2021**, *11*, 2000982. [CrossRef]

65. Liu, L.; Li, M.; Chu, L.; Jiang, B.; Lin, R.; Zhu, X.; Cao, G. Layered ternary metal oxides: Performance degradation mechanisms as cathodes, and design strategies for high-performance batteries. *Prog. Mater. Sci.* **2020**, *111*, 100655. [CrossRef]

66. Kim, K.H.; Jeong, H.; Lee, H.C.; Shon, J.K.; Park, J.; Park, H.-Y. Stable cycling of high-density three-dimensional sintered $LiCoO_2$ plate cathodes. *J. Power Sources* **2022**, *551*, 232223. [CrossRef]

67. Lv, Y.; Huang, S.; Zhao, Y.; Roy, S.; Lu, X.; Hou, Y.; Zhang, J. A review of nickel-rich layered oxides cathodes: Synthetic strategies, structural characteristics, failure mechanism, improvement approaches and prospects. *Appl. Energy* **2022**, *305*, 117849. [CrossRef]

68. Aishova, A.; Park, G.-T.; Yoon, C.S.; Sun, Y.-K. Cobalt-free high-capacity Ni-rich layered Li [$Ni_{0.9}Mn_{0.1}$] O_2 cathode. *Adv. Energy Mater.* **2020**, *10*, 1903179. [CrossRef]

69. Zhang, Q.; Su, Y.; Chen, L.; Lu, Y.; Bao, L.; He, T. Pre-oxidizing the precursors of Nickel-rich cathode materials to regulate their Li^+/Ni^{2+} cation ordering towards cyclability improvements. *J. Power Sources* **2018**, *396*, 734–741. [CrossRef]

70. Sun, J.; Cao, X.; Zhou, H. Advanced single-crystal layered Ni-rich cathode materials for next-generation high-energy density and long-life Li-ion batteries. *Phys. Rev. Mater.* **2022**, *6*, 070201. [CrossRef]

71. Zuo, W.; Luo, M.; Liu, X.; Wu, J.; Liu, H.; Li, J.; Winter, M.; Fu, R.; Yang, W.; Yang, Y. Li-rich cathodes for rechargeable Li-based batteries: Reaction mechanisms and advanced characterization techniques. *Energy Environ. Sci.* **2020**, *13*, 4450–4497. [CrossRef]

72. Zheng, H.; Han, X.; Guo, W.; Lin, L.; Xie, Q.; Liu, P.; He, W.; Wang, L.; Peng, D.-L. Recent developments and challenges of Li-rich Mn-based cathode materials for high-energy lithium-ion batteries. *Mater. Today Energy* **2020**, *18*, 100518. [CrossRef]

73. Li, Q.; Ning, D.; Wong, D.; An, K.; Tang, Y.; Zhou, D.; Schuck, G.; Chen, Z.; Zhang, N.; Liu, X. Improving the oxygen redox reversibility of Li-rich battery cathode materials via coulombic repulsive interactions strategy. *Nat. Commun.* **2022**, *13*, 1123. [CrossRef] [PubMed]

74. Cui, C.; Fan, X.; Zhou, X.; Chen, J.; Wang, Q.; Ma, L.; Yang, C.; Hu, E.; Yang, X.-Q.; Wang, C. Structure and interface design enable stable Li-rich cathode. *J. Am. Chem. Soc.* **2020**, *142*, 8918–8927. [CrossRef] [PubMed]

75. Tan, H.T.; Rui, X.; Sun, W.; Yan, Q.; Lim, T.M. Vanadium-based nanostructure materials for secondary lithium battery applications. *Nanoscale* **2015**, *7*, 14595. [CrossRef] [PubMed]

76. Huang, X.; Rui, X.; Hng, H.H.; Yan, Q. Vanadium pentaoxide-based cathode materials for Lithium-Ion Batteries: Morphology control, carbon hybridization, and cation doping. *Part. Part. Syst. Charact.* **2015**, *32*, 276–294. [CrossRef]

77. Chen, H.; Cheng, S.; Chen, D.; Jiang, Y.; Ang, E.H.; Liu, W.; Feng, Y.; Rui, X.; Yu, Y. Vanadate-based electrodes for rechargeable batteries. *Mater. Chem. Front.* **2021**, *5*, 1585. [CrossRef]

78. Radzi, Z.I.; Arifin, K.H.; Kufian, M.Z.; Balakrishnan, V.; Raihan, S.R.S.; Rahim, N.A.; Subramanium, R. Review of spinal $LiMn_2O_4$ cathode materials under high cut-off voltage in lithium-ion batteries: Challenges and strategies. *J. Electroanal. Chem.* **2022**, *920*, 116623. [CrossRef]

79. Ma, J.; Hu, P.; Cui, G.; Chen, L. Surface and Interface Issues in spinel $LiNi_{0.5}Mn_{1.5}O_4$: Insights into a potential cathode material for High-energy density lithium-ion batteries. *Chem. Mater.* **2016**, *28*, 3578–3606. [CrossRef]

80. Xu, X.; Deng, S.; Wang, H.; Liu, J.; Yan, H. Research progress in improving the cycling stability of high-voltage $LiNi_{0.5}Mn_{1.5}O_4$ cathode in Lithium-Ion Battery. *Nano-Micro Lett.* **2017**, *9*, 22. [CrossRef]

81. Ling, J.; Karuppiah, C.; Krishnan, S.G.; Reddy, M.V.; Misnon, I.I.; Rahim, M.H.A.; Yang, C.-C.; Jose, R. Phosphate polyanion materials as high-voltage lithium-ion battery cathodes: A Review. *Energy Fuels* **2021**, *35*, 10428–10450. [CrossRef]

82. Peng, Y.; Tan, R.; Ma, J.; Li, Q.; Wang, T.; Duan, X. Electrospun $Li_3V_2(PO_4)_3$ nanocubes/carbon nanofibers as free-standing cathodes for high-performance lithium-ion batteries. *J. Mater. Chem. A* **2019**, *7*, 14681–14688. [CrossRef]

83. Hu, J.; Huang, W.; Yang, L.; Pan, F. Structure and performance of the $LiFePO_4$ cathode material: From the bulk to the surface. *Nanoscale* **2020**, *12*, 15036–15044. [CrossRef] [PubMed]

84. Ramsubramanian, B.; Sundarrajan, S.; Chellappan, V.; Reddy, M.V.; Ramakrishna, S.; Zaghib, K. Recent development in carbon-$LiFePO_4$ cathode for Lithium-Ion Batteries: A Mini-Review. *Batteries* **2022**, *8*, 133. [CrossRef]

85. Islam, M.S.; Dominko, R.; Masquelier, C.; Sirisopanaporn, C.; Armstrong, A.R.; Bruce, P.G. Silicate cathodes for lithium batteries: Alternatives to phosphates? *J. Mater. Chem.* **2011**, *21*, 9811–9818. [CrossRef]

86. Bao, L.; Gao, W.; Su, Y.; Wang, Z.; Li, N.; Chen, S.; Wu, F. Progression of the silicate cathode materials used in lithium-ion batteries. *Chin. Sci. Bull.* **2013**, *58*, 575–584. [CrossRef]

87. Yang, S.-H.; Xue, H.; Guo, S.-P. Borates as promising electrode materials for rechargeable batteries. *Coord. Chem. Rev.* **2021**, *427*, 213551. [CrossRef]

88. Ma, T.; Muslim, A.; Su, Z. Microwave synthesis and electrochemical properties of lithium manganese borate as cathode for lithium-ion batteries. *J. Power Sources* **2015**, *282*, 95–99. [CrossRef]

89. Daniel, C.; Mohanty, D.; Li, J.; Wood, D.L. Review on Electrochemical Storage Materials and Technology. *AIP Conf. Proc.* **2014**, *1597*, 26–43.

90. Ghosh, S.; Bhattacharjee, U.; Bhowmik, S.; Martha, S.K. A Review on High-Capacity and High-Voltage Cathodes for Next-Generation Lithium-ion Batteries. *J. Energy Power Technol.* **2022**, *1*, 2. [CrossRef]

91. Asenbauer, J.; Eisenmann, T.; Kuenzel, M.; Kazzazi, A.; Chen, Z.; Bresser, D. The success story of graphite as a lithium-ion anode material-fundamentals, remaining challenges, and recent developments including silicon (oxide) composites. *Sustain. Energy Fuels* **2020**, *4*, 5387. [CrossRef]

92. Chang, H.; Wu, Y.-R.; Han, X.; Yi, T.-F. Recent developments in advanced anode materials for lithium-ion batteries. *Energy Mater.* **2021**, *1*, 100003. [CrossRef]

93. Zhao, W.; Choi, W.; Yoon, W.-S. Nanostructured Electrode materials for rechargeable Lithium-Ion batteries. *J. Electrochem. Sci. Technol.* **2020**, *11*, 195–219. [CrossRef]

94. Ma, D.; Cao, Z.; Hu, A. Si-based Anode materials for Li-ion batteries: A mini review. *Nano-Micro Lett.* **2014**, *6*, 347–358. [CrossRef] [PubMed]

95. Fang, S.; Bresser, D.; Passerini, S. Transition metal oxide anodes for electrochemical energy storage in Lithium- and Sodium-Ion Batteries. *Adv. Energy Mater.* **2020**, *10*, 1902485. [CrossRef]

96. Nzereogu, P.U.; Omah, A.D.; Ezema, F.I.; Iwuoha, E.I.; Nwanya, A.C. Anode materials for lithium-ion batteries: A review. *Appl. Surf. Sci. Adv.* **2022**, *9*, 100233. [CrossRef]

97. Parikh, P.; Sina, M.; Banerjee, A.; Wang, X.; D'Souza, M.S.; Doux, J.-M.; Wu, E.A.; Trieu, O.Y.; Gong, Y.; Zhou, Q.; et al. Role of polyacrylic acid (PAA) Binder on the solid electrolyte interphase in silicon anodes. *Chem. Mater.* **2019**, *31*, 2535–2544. [CrossRef]

98. Jiang, S.; Hu, B.; Shi, Z.; Chen, W.; Zhang, Z.; Zhang, L. Re-Engineering Poly (Acrylic Acid) Binder toward optimized electrochemical performance for silicon lithium-ion batteries: Branching Architecture leads to balanced properties of polymeric binders. *Adv. Funct. Mater.* **2020**, *30*, 1908558. [CrossRef]

99. Chen, J.; Liu, J.; Qi, Y.; Sun, T.; Li, X. Unveiling the roles of binder in the mechanical integrity of electrodes for Lithium-ion batteries. *J. Electrochem. Soc.* **2013**, *160*, A1502–A1509. [CrossRef]

100. Dong, T.; Mu, P.; Zhang, S.; Zhang, H.; Liu, W.; Cui, G. How Do Polymer Binders Assist Transition Metal Oxide Cathodes to Address the Challenge of High-Voltage Lithium Battery Applications? *Electrochem. Energy Rev.* **2021**, *4*, 545–565. [CrossRef]

101. Shi, Y.; Zhou, X.; Yu, G. Material and Structural Design of Novel Binder Systems for High-Energy, High-power Lithium-ion batteries. *Acc. Chem. Res.* **2017**, *50*, 2642–2652. [CrossRef]

102. Kazzazi, A.; Bresser, D.; Birrozzi, A.; van Zamory, J.; Hekmatfar, M.; Passerini, S. Comparative analysis of aqueous binders for high-energy Li-rich NMC as a lithium-ion cathode and the impact of adding phosphoric acid. *ACS Appl. Mater. Interfaces* **2018**, *10*, 17214–17222. [CrossRef]

103. He, J.; Wei, Y.; Hu, L.; Li, H.; Zhai, T. Aqueous binders enhanced high-performance GeP$_5$ anode for lithium-ion batteries. *Front. Chem.* **2018**, *6*, 21. [CrossRef] [PubMed]

104. Bresser, D.; Buchholz, D.; Moretti, A.; Varzi, A.; Passerini, S. Alternative binders for sustainable electrochemical energy storage-the transition to aqueous electrode processing and bio-derived polymers. *Energy Environ. Sci.* **2018**, *11*, 3096–3127. [CrossRef]

105. Fitz, O.; Ingenhoven, S.; Bischoff, C.; Gentischer, H.; Birke, K.P.; Saracsan, D.; Biro, D. Comparison of aqueous and non-aqueous based binder polymers and the mixing ratio of Zn//MnO$_2$ batteries with mildly acidic aqueous electrolytes. *Batteries* **2021**, *7*, 40. [CrossRef]

106. Wang, R.; Feng, L.; Yang, W.; Zhang, Y.; Zhang, Y.; Bai, W.; Liu, B.; Zhang, W.; Chuan, Y.; Zheng, Z.; et al. Effect of different binders on the electrochemical performance of metal oxide anode for lithium-ion batteries. *Nanoscale Res Lett.* **2017**, *12*, 575. [CrossRef]

107. Xing, J.; Bliznakov, S.; Bonville, L.; Oljace, M.; Maric, R. A review of non-aqueous electrolytes, binders, and separators for lithium-ion batteries. *Electrochem. Energy Rev.* **2022**, *5*, 14. [CrossRef]

108. Cholewinski, A.; Si, P.; Uceda, M.; Pope, M.; Zhao, B. Polymer Binders: Characterization and development towards aqueous electrode fabrication for sustainability. *Polymers* **2021**, *13*, 631. [CrossRef]

109. Yim, T.; Choi, S.J.; Jo, Y.N.; Kim, T.-H.; Kim, K.J.; Jeong, G.; Kim, Y.-J. Effect of binder properties on electrochemical performance for silicon-graphite anode: Method and application of binder screening. *Electrochim. Acta* **2014**, *136*, 112–120. [CrossRef]

110. Rapisarda, M.; Marken, F.; Meo, M. Graphene oxide and starch gel as a hybrid binder for environmentally friendly high-performance supercapacitors. *Commun. Chem.* **2021**, *4*, 169. [CrossRef] [PubMed]

111. Kannan, D.R.R.; Terala, P.K.; Moss, P.L.; Weatherspoon, M.H. Analysis of the separator thickness and porosity on the performance of lithium-ion batteries. *Intl. J. Electrochem.* **2018**, *2018*, 1925708.

112. Kim, P.J. Surface-functionalized separator for stable and reliable lithium-ion batteries: A review. *Nanomaterials* **2021**, *11*, 2275. [CrossRef]

113. Yuan, M.; Liu, K. Rational design on separators and liquid electrolytes for safer lithium-ion batteries. *J. Energy Chem.* **2020**, *43*, 58–70. [CrossRef]

114. Chen, K.; Li, Y.; Zhan, H. Advanced separators for lithium-ion batteries. *Earth Environ. Sci.* **2022**, *1011*, 012009. [CrossRef]

115. Ding, L.; Yan, N.; Zhang, S.; Xu, R.; Wu, T.; Wang, F.; Cao, Y.; Xiang, M. Facile manufacture technique for lithium-ion batteries composite separator via online construction of fumed SiO$_2$ coating. *Mater. Des.* **2022**, *215*, 110476. [CrossRef]

116. Kim, G.; Noh, J.H.; Lee, H.; Shin, J.; Lee, D. Roll-to-Roll Gravure Coating of PVDF on a Battery Separator for the Enhancement of Thermal Stability. *Polymers* **2023**, *15*, 4108. [CrossRef] [PubMed]

117. Liu, Z.; Jiang, Y.; Hu, Q.; Guo, S.; Yu, L.; Li, Q.; Liu, Q.; Hu, X. Safer lithium-ion batteries from separator aspect: Development and future perspective. *Energy Environ. Mater.* **2021**, *4*, 336–362. [CrossRef]

118. Huang, X.; He, R.; Li, M.; Chee, M.O.L.; Dong, P.; Lu, J. Functionalized separator for next generation batteries. *Mat. Today* **2020**, *41*, 143–155. [CrossRef]

119. Lagadec, M.F.; Zahn, R.; Wood, V. Designing Polyolefin separators to minimize the impact of local compressive stresses on lithium-ion battery performance. *J. Electrochem. Soc.* **2018**, *165*, A1829–A1836. [CrossRef]

120. Zhu, P.; Gastol, D.; Marshall, J.; Sommerville, R.; Goodship, V.; Kendrick, E. A review of current collectors for lithium-ion batteries. *J. Power Sources* **2021**, *485*, 229321. [CrossRef]

121. Yamada, M.; Watanbe, T.; Gunji, T.; Wu, J.; Matsumoto, F. Review of the design of current collectors for improving the battery performance in lithium-ion and post lithium-ion batteries. *Electrochem* **2020**, *1*, 124–159. [CrossRef]

122. Harper, G.; Sommerville, R.; Kendrick, E.; Driscoll, L.; Slater, P.; Stolkin, R.; Walton, A.; Christensen, P.; Heidrich, O.; Lambert, S. Recycling lithium-ion batteries from electric vehicles. *Nature* **2019**, *575*, 75–86. [CrossRef]

123. Zhang, S.; Jow, T. Aluminum corrosion in electrolyte of Li-ion battery. *J. Power Sources* **2002**, *109*, 458–464. [CrossRef]

124. Kumar, P.S.; Ayyasamy, S.; Tok, E.S.; Adams, S.; Reddy, M. Impact of electrical conductivity on the electrochemical performance of layered structured Lithium Trivanadate ($LiV_{3-x}M_xO_8$, M = Zn/Co/Fe/Sn/Ti/Zr/Nb/Mo, x = 0.01–0.1) as cathode material for energy storage. *ACS Omega* **2018**, *3*, 3036–3044. [CrossRef] [PubMed]

125. Nakamura, T.; Okano, S.; Yaguma, N.; Morinaga, Y.; Takahara, H.; Yamada, Y. Electrochemical performance of cathodes prepared on current collector with different surface morphologies. *J. Power Sources* **2013**, *244*, 532–537. [CrossRef]

126. Doberdo, I.; Loffler, N.; Laszczynski, N.; Cericola, D.; Penazzi, N.; Bodoardo, S.; Kim, G.-T.; Passerini, S. Enabling aqueous binder for lithium battery cathodes-carbon coating of aluminum current collector. *J. Power Sources* **2014**, *248*, 1000–1006. [CrossRef]

127. Boz, B.; Dev, T.; Salvadori, A.; Schaefer, J.L. Review-Electrolyte and Electrode Designs for enhanced ion transport properties to enable high performance lithium batteries. *J. Electrochem. Soc.* **2021**, *168*, 090501. [CrossRef]

128. Meutzner, F.; de Vivanco, M.U. Electrolytes-Technology review. *AIP Conf. Proc.* **2014**, *1597*, 185–195.

129. Chagnes, A.; Swiatowska, J. *Lithium Process Chemistry: Resources, Extractions, Batteries and Recycling*, 1st ed.; Elsevier: Amsterdam, The Netherland, 2015; Volume 5, pp. 167–189.

130. Zhang, H.; Liu, X.; Li, H.; Hasa, I.; Passerini, S. Challenges and Strategies for high-energy Aqueous Electrolytes rechargeable batteries. *Angew. Chem. Int. Ed.* **2021**, *60*, 598–616. [CrossRef]

131. Luo, J.-Y.; Cui, W.-J.; He, P.; Xia, Y.-Y. Raising the cycling stability of aqueous lithium-ion batteries by eliminating oxygen in the electrolyte. *Nat. Chem.* **2010**, *2*, 760–765. [CrossRef]

132. He, P.; Zhang, X.; Wang, Y.-G.; Cheng, L.; Xia, Y.-Y. Lithium-ion intercalation behavior of $LiFePO_4$ in aqueous and nonaqueous electrolyte solutions. *J. Electrochem. Soc.* **2007**, *155*, A144. [CrossRef]

133. Xu, K. Nonaqueous liquid electrolytes for lithium-based rechargeable batteries. *Chem. Rev.* **2004**, *104*, 4303–4418. [CrossRef]

134. Zhang, S.; Li, J.; Jiang, N.; Li, X.; Pasupath, S.; Fang, Y.; Liu, Q.; Dang, D. Rational Design of an Ionic Liquid-based electrolyte with High ionic conductivity towards safe Lithium/Lithium-Ion batteries. *Chem. Asian J.* **2019**, *14*, 2810–2814. [CrossRef] [PubMed]

135. Armand, M.; Endres, F.; MacFarlane, D.R.; Ohno, H.; Scrosati, B. Ionic-Liquid materials for the electrochemical challenges of the future. *Nat. Mater.* **2009**, *8*, 621–629. [CrossRef] [PubMed]

136. Lei, Z.; Chen, B.; Koo, Y.-M.; MacFarlane, D.R. Introduction: Ionic Liquids. *Chem. Rev.* **2017**, *117*, 6633–6635. [CrossRef] [PubMed]

137. Yu, Q.; Jiang, K.; Yu, G.; Chen, X.; Zhang, C.; Yao, Y.; Jiang, B.; Long, H. Recent progress of composite solid polymer electrolytes for all solid-state lithium metal batteries. *Chin. Chem. Lett.* **2021**, *32*, 2659–2678. [CrossRef]

138. Sasikumar, M.; Krishna, R.H.; Raja, M.; Therese, H.A.; Balakrishna, N.T.M.; Raghavan, P.; Sivakumar, P. Titanium dioxide nano-ceramic filler in solid polymer electrolytes: Strategy towards suppressed dendrite formation and enhanced electrochemical performance for safe lithium-ion batteries. *J. Alloys Compd.* **2021**, *882*, 160709. [CrossRef]

139. Suo, L.; Borodin, O.; Gao, T.; Olguin, M.; Ho, J.; Fan, X.; Luo, C.; Wang, C.; Xu, K. Water in salt electrolyte enables high-voltage aqueous lithium-ion chemistries. *Science* **2015**, *350*, 938–943. [CrossRef]

140. Chen, S.; Zhang, M.; Zou, P.; Sun, B.; Tao, S. Historical development, and novel concepts on electrolytes for aqueous rechargeable batteries. *Energy Environ. Sci.* **2022**, *15*, 1805–1839. [CrossRef]

141. Suo, L.; Oh, D.; Lin, Y.; Zhuo, Z.; Borodin, O.; Gao, T.; Wang, F.; Kushima, A.; Wang, Z.; Kim, H.-C.; et al. How solid-electrolyte interphase forms in aqueous electrolytes. *J. Am. Chem. Soc.* **2017**, *139*, 18670–18680. [CrossRef]

142. Yue, F.; Tie, Z.; Deng, S.; Wang, S.; Yang, M.; Niu, Z. An Ultralow Temperature Aqueous Battery with Proton Chemistry. *Angew. Chem. Int. Ed.* **2021**, *60*, 13882–13886. [CrossRef]

143. Sun, T.; Du, H.; Zheng, S.; Shi, J.; Tao, Z. High Power and Energy Density Aqueous Proton Battery Operated at $-90\,^\circ C$. *Adv. Funct. Mater.* **2021**, *31*, 2010127. [CrossRef]

144. Wang, F.; Borodin, O.; Ding, M.S.; Gobet, M.; Vatamanu, J.; Fan, X.; Gao, T.; Eidson, N.; Liang, Y.; Sun, W.; et al. Aqueous/Non-aqueous electrolyte for safe and high-energy Li-Ion batteries. *Joule* **2018**, *2*, 927–937. [CrossRef]

145. Chagnes, A.; Swiatowska, J. *New Developments in Lithium-Ion Batteries*, 1st ed.; Belharouak, I., Ed.; InTech: Vienna, Austria, 2012; Volume 6, pp. 145–172.

146. Hu, L.; Zhang, Z.; Amine, K. Electrochemical investigation of carbonate-based electrolytes for high voltage lithium-ion cells. *J. Power Sources* **2013**, *236*, 175–180. [CrossRef]

147. Yang, L.; Ravdel, B.; Lucht, B.L. Electrolyte Reaction with the surface of high voltage $LiNi_{0.5}Mn_{1.5}O_4$ cathodes for Lithium-ion batteries. *Electrochem. Solid State Lett.* **2010**, *13*, A95–A97. [CrossRef]

148. Flamme, B.; Garcia, G.R.; Weil, M.; Haddad, M.; Phansavath, P.; Vidal, V.R.; Chagnes, A. Guidelines to design organic electrolytes for lithium-ion batteries: Environmental impact, physicochemical and electrochemical properties. *Green Chem.* **2017**, *19*, 1828–1849. [CrossRef]

149. Abe, K.; Ushigoe, Y.; Yoshitake, H.; Yoshio, M. Functional electrolytes: Novel type additives for cathode materials, providing high cycleability performance. *J. Power Sources* **2006**, *153*, 328–335. [CrossRef]

150. Assary, R.S.; Curtiss, L.A.; Redfern, P.C.; Zhang, Z.; Amine, K. Computational Studies of Polysiloxanes: Oxidation Potentials and Decomposition Reactions. *J. Phys. Chem. C* **2011**, *115*, 12216–12223. [CrossRef]

151. Nanbu, N.; Suzuki, Y.; Ohtsuki, K.; Meguro, T.; Takehara, M.; Ue, M.; Sasaki, Y. Physical and Electrochemical Properties of Monofluorinated Ethyl Acetates for Lithium Rechargeable Batteries. *Electrochemistry* **2010**, *78*, 446–449. [CrossRef]

152. McElroy, P.J.; Gellen, A.T.; Kolahi, S.S. Excess second virial coefficients and critical temperatures of methyl acetate and diethyl sulfide. *J. Chem. Eng. Data* **1990**, *35*, 38. [CrossRef]

153. Laurence, C.; Nicolet, P.; Dalati, M.T.; Abboud, J.-L.M.; Notario, R. The empirical treatment of solvent-solute interactions: 15 years of π^*. *J. Phys. Chem.* **1994**, *98*, 5807–5816. [CrossRef]

154. Obama, M.; Oodera, Y.; Kohama, N.; Yanase, T.; Saito, Y.; Kusano, K. Densities, molar volumes, and cubic expansion coefficients of 78 aliphatic ethers. *J. Chem. Eng. Data* **1985**, *30*, 1–5. [CrossRef]

155. Dutt, H.D.P.H.; Jeewan, R. Dielectric relaxation studies of oligether of ethylene glycol at microwave frequencies. *Bull. Chem. Soc. Jpn.* **1991**, *64*, 2030–2031.

156. Tobishima, S.-I.; Okada, T. Lithium cycling efficiency and conductivity for high dielectric solvent/low viscosity solvent mixed systems. *Electrochim. Acta* **1985**, *30*, 17151722. [CrossRef]

157. Mozhzhukhina, N.; Méndez De Leo, L.P.; Calvo, E.J. Infrared spectroscopy studies on stability of dimethyl sulfoxide for application in a Li-Air Battery. *J. Phys. Chem. C* **2013**, *117*, 18375–18380. [CrossRef]

158. Bennion, D.N.; Tiedemann, W.H. Density, viscosity, and conductivity of lithium trifluoromethanesulfonate solutions in dimethyl-sulfite. *J. Chem. Eng. Data* **1971**, *16*, 368. [CrossRef]

159. Svirbely, W.J.; Lander, J.J. The dipole moments of diethyl sulfite, triethyl phosphate and tetraethyl silicate. *J. Am. Chem. Soc.* **1948**, *70*, 4121. [CrossRef] [PubMed]

160. Chen, J.; Vatamanu, J.; Xing, L.; Borodin, O.; Chen, H.; Guan, X.; Liu, X.; Xu, K.; Li, W. Improving electrochemical stability and low-temperature performance with water/acetonitrile hybrid electrolyte. *Adv. Energy Mater.* **2020**, *10*, 1902654. [CrossRef]

161. Li, Q.; Chen, J.; Fan, L.; Kong, X.; Lu, Y. Progress in electrolytes for rechargeable Li-based batteries and beyond. *Green Energy Environ.* **2016**, *1*, 18–42. [CrossRef]

162. Pan, S.; Yao, M.; Zhang, J.; Li, B.; Xing, C.; Song, X.; Su, P.; Zhang, H. Recognition of ionic liquids as high-voltage electrolytes for supercapacitors. *Front. Chem.* **2020**, *8*, 261. [CrossRef]

163. Ahmad, S. Polymer Electrolyte: Characteristics and peculiarities. *Ionics* **2009**, *15*, 309–321. [CrossRef]

164. Xi, G.; Xiao, M.; Wang, S.; Han, D.; Li, Y.; Meng, Y. Polymer-based Solid Electrolytes: Material Selection, design, and Application. *Adv. Funct. Mater.* **2021**, *31*, 2007598. [CrossRef]

165. Zhao, Y.; Wang, L.; Zhou, Y.; Liang, Z.; Tavajohi, N.; Li, B.; Li, T. Solid Polymer Electrolytes with High Conductivity, and transference number of Li ions for Li-based rechargeable Batteries. *Adv. Sci.* **2021**, *8*, 2003675. [CrossRef] [PubMed]

166. Mackanic, D.G.; Yan, X.; Zhang, Q.; Matsuhisa, N.; Yu, Z.; Jiang, Y.; Manika, T.; Lopez, J.; Yan, H.; Chen, X.; et al. Decoupling of mechanical properties and ionic conductivity in supramolecular lithium-ion conductors. *Nat. Commun.* **2019**, *10*, 5384. [CrossRef]

167. An, Y.; Han, X.; Liu, Y.; Azhar, A.; Na, J.; Nanjundan, A.K.; Wang, S.; Yu, J.; Yamauchi, Y. Progress in Solid polymer electrolytes for lithium-ion batteries and beyond. *Small* **2022**, *18*, 2103617. [CrossRef]

168. Aravindan, V.; Gnanaraj, J.; Madhavi, S.; Liu, H.-K. Lithium-ion conducting electrolyte salts for lithium batteries. *Chem.-Eur. J.* **2011**, *17*, 14326. [CrossRef] [PubMed]

169. Barbosa, J.C.; Goncalves, R.; Costa, C.M.; Bermudez, V.D.; Marijuan, A.F.; Zhang, Q.; Mendez, S.L. Metal-organic framework, and zeolite materials as active fillers for lithium-ion battery solid polymer electrolytes. *Mater. Adv.* **2021**, *2*, 3790–3805. [CrossRef]

170. Mazzapioda, L.; Sgambetterra, M.; Tsurumaki, A.; Navarra, M.A. Different approaches to obtain functionalized alumina as additive in polymer electrolyte membranes, Different approaches to obtain functionalized alumina as additive in polymer electrolyte membranes. *J. Solid State Electrochem.* **2022**, *26*, 17–27. [CrossRef]

171. Zhou, H.; Wang, Y.; Li, H.; He, P. The Development of a new type of rechargeable batteries based on Hybrid electrolytes. *ChemSusChem* **2010**, *3*, 1009–1019. [CrossRef]

172. Manthiram, A.; Yu, X.; Wang, S. Lithium battery chemistries enabled by solid-state electrolytes. *Nat. Rev. Mater.* **2017**, *2*, 16103. [CrossRef]

173. Han, L.; Lehmann, M.L.; Zhu, J.; Liu, T.; Zhou, Z.; Tang, X.; Heish, C.-T.; Sokolov, A.P.; Cao, P.; Chen, X.C.; et al. Recent developments, and challenges in hybrid solid electrolytes for Lithium-ion batteries. *Front. Energy Res.* **2020**, *8*, 202. [CrossRef]

174. Kwon, T.; Choi, I.; Park, M.J. Highly Conductive Solid-state hybrid electrolytes operating at subzero temperature. *ACS Appl. Mater. Interfaces* **2017**, *9*, 24250–24258. [CrossRef]

175. Janek, J.; Zeier, W.G. A solid future battery development. *Nat. Energy* **2016**, *1*, 16141. [CrossRef]

176. Liu, H.; Sun, Q.; Zhang, H.; Cheng, J.; Li, Y.; Zeng, Z.; Zhang, S.; Xu, X.; Ji, F.; Li, D.; et al. The application road of silicon-based anode in lithium-ion batteries: From liquid electrolyte to solid-state electrolyte. *Energy Storage Mater.* **2023**, *55*, 244–263. [CrossRef]

177. Murugan, R.; Thangadurai, V.; Weppner, W. Fast Lithium Ion Conduction in Garnet-Type $Li_7La_3Zr_2O_{12}$. *Angew. Chem. Int. Ed.* **2007**, *46*, 7778–7781. [CrossRef]

178. Kim, K.H.; Iriyama, Y.; Yamamoto, K.; Kumazaki, S.; Asaka, T.; Tanabe, K.; Fisher, C.A.J.; Hirayama, T.; Murugan, R.; Ogumi, Z. Characterization of the interface between $LiCoO_2$ and $Li_7La_3Zr_2O_{12}$ in an all-solid-state rechargeable lithium battery. *J. Power Sources* **2011**, *196*, 764–767. [CrossRef]

179. Ihrig, M.; Kuo, L.-Y.; Lobe, S.; Laptev, A.M.; Lin, C.-A.; Tu, C.-H.; Ye, R.; Kaghazchi, P.; Cressa, L.; Eswara, S.; et al. Thermal Recovery of the Electrochemically Degraded $LiCoO_2/Li_7La_3Zr_2O_{12}$:Al,Ta Interface in an All-Solid-State Lithium Battery. *ACS Appl. Mater. Interfaces* **2023**, *15*, 4101–4112. [CrossRef]

180. Teng, Y.; Liu, H.; Wang, Q.; He, Y.; Hua, Y.; Li, C.; Bai, J. In-doped $Li_7La_3Zr_2O_{12}$ nanofibers enhances electrochemical properties and conductivity of PEO-based composite electrolyte in all-solid-state lithium battery. *J. Energy Storage* **2024**, *76*, 109784. [CrossRef]

181. Morimoto, H.; Yamashita, H.; Tatsumisago, M.; Minami, T. Mechanochemical Synthesis of New Amorphous Materials of $60Li_2S\cdot40SiS_2$ with High Lithium Ion Conductivity. *J. Am. Ceram. Soc.* **1999**, *82*, 1352–1354. [CrossRef]

182. Wheaton, J.; Martin, S.W. Impact of impurities on the thermal properties of a $Li_2S–SiS_2–LiPO_3$ glass. *Appl. Glass Sci.* **2024**, *15*, 317–328. [CrossRef]

183. Olson, M.; Kmiec, S.; Riley, N.; Oldham, N.; Krupp, K.; Manthiram, A.; Martin, S.W. Structure and Properties of $Na_2S–SiS_2–P_2S_5–NaPO_3$ Glassy Solid Electrolytes. *Inorg. Chem.* **2024**, *63*, 9129–9144. [CrossRef]

184. Zhang, X.; Wang, S.; Xue, C.; Xin, C.; Lin, Y.; Shen, Y.; Li, L.; Nan, C.-W. Self-Suppression of Lithium Dendrite in All-Solid-State Lithium Metal Batteries with Poly(vinylidene difluoride)-Based Solid Electrolytes. *Adv. Mater.* **2019**, *31*, 186082. [CrossRef]

185. Tang, Y.; Xiong, Y.; Wu, L.; Xiong, X.; Me, T.; Wang, X. A Solid-State Lithium Battery with PVDF−HFP-Modified Fireproof Ionogel Polymer Electrolyte. *ACS Appl. Energy Mater.* **2023**, *6*, 4016–4026. [CrossRef]

186. Tran, H.K.; Truong, B.T.; Zhang, B.-R.; Jose, R.; Chang, J.-K.; Yang, C.-C. Sandwich-Structured Composite Polymer Electrolyte Based on PVDF-HFP/PPC/Al-Doped LLZO for High-Voltage Solid-State Lithium Batteries. *ACS Appl. Energy Mater.* **2023**, *6*, 1475–1487. [CrossRef]

187. Liu, Y.; Xu, X.; Jiao, X.; Kapitanova, O.O.; Song, Z.; Xiong, S. Role of Interfacial Defects on Electro–Chemo–Mechanical Failure of Solid-State Electrolyte. *Adv. Mater.* **2023**, *35*, 2301152. [CrossRef] [PubMed]

188. Xiong, S.; Xu, X.; Jiao, X.; Wang, Y.; Kapitanova, O.O.; Song, Z.; Liu, Y. Mechanical Failure of Solid-State Electrolyte Rooted in Synergy of Interfacial and Internal Defects. *Adv. Energy Mater.* **2023**, *13*, 2203614. [CrossRef]

189. Akhtar, S.; Lee, W.; Kim, M.; Park, M.S.; Yoon, W.-S. Conduction Mechanism of charge carriers in electrodes and design factors for the improvement of charge conduction in Li-Ion batteries. *J. Electrochem. Sci. Technol.* **2021**, *12*, 1–20. [CrossRef]

190. Zhang, Y.; Tao, L.; Xie, C.; Wang, D.; Zou, Y.; Chen, R.; Wang, Y.; Jia, C.; Wang, S. Defect Engineering on Electrode Materials for rechargeable batteries. *Adv. Mater.* **2020**, *32*, 1905923. [CrossRef]

191. Guo, W.; Meng, Y.; Hu, Y.; Wu, X.; Ju, Z.; Zhuang, Q. Surface, and Interface modification of electrode materials for lithium-ion batteries with organic liquid electrolytes. *Front. Energy Res.* **2020**, *8*, 170. [CrossRef]

192. Nasir, U.; Muralidharan, N.; Essehli, R.; Amin, R.; Belharouak, I. Valuation of Surface Coatings in High-Energy Density Lithium-Ion battery cathode Materials. *Energy Storage Mater.* **2021**, *38*, 309–328.

193. Vu, A.; Qian, Y.; Stein, A. Porous Electrode Materials for Lithium-Ion Batteries How to Prepare Them and What makes them Special. *Adv. Energy Mater.* **2012**, *2*, 1056–1085. [CrossRef]

194. Du, H.-L.; Jeong, M.-G.; Lee, Y.-S.; Choi, W.; Lee, J.K.; Oh, I.-H.; Jung, H.-G. Coating lithium titanate with nitrogen-doped carbon by simple refluxing for high-power lithium-ion batteries. *ACS Appl. Mater. Interfaces* **2015**, *7*, 10250–10257. [CrossRef]

195. Xu, K.; Wang, C. Batteries: Widening voltages windows. *Nat. Energy* **2016**, *1*, 16161. [CrossRef]

196. Gao, J.; Shi, S.-Q.; Li, H. Brief overview of electrochemical potential in lithium-ion batteries. *Chin. Phys. B* **2016**, *25*, 018210. [CrossRef]

197. Ven, A.V.D.; Bhattacharya, J.; Belak, A.A. Understanding Li Diffusion in Li-ion Intercalation compounds. *Acc. Chem. Res.* **2013**, *46*, 1216–1225.

198. Dong, C.; Xu, F.; Chen, L.; Chen, Z.; Cao, Y. Design Strategies for High-Voltage Aqueous batteries. *Small Struct.* **2021**, *2*, 2100001. [CrossRef]

199. Melot, B.C.; Tarascon, J.-M. Design and Preparation of Materials for Advanced Electrochemical Storage. *Acc. Chem. Res.* **2013**, *46*, 1226–1238. [CrossRef] [PubMed]

200. Masquelier, C.; Croguennec, L. Polyanionic (Phosphates, Silicates, Sulfates) Frameworks as Electrode materials for rechargeable Li (or Na) Batteries. *Chem. Rev.* **2013**, *113*, 6552–6591. [CrossRef]

201. Liu, C.; Neale, Z.G.; Cao, G. Understanding electrochemical potentials of cathode materials in rechargeable batteries. *Mat. Today* **2016**, *19*, 109–123. [CrossRef]

202. Yazami, R. *Nanomaterials for Lithium-Ion Batteries: Fundamentals and Applications*; CRC Press, Taylor & Francis Group: Boca Raton, FL, USA, 2013.

203. Park, J.-K. *Principles, and Applications of Lithium Secondary Batteries*; Wiley-VCH: Weinheim, Germany, 2012.

204. Hahn, B.P.; Long, J.W.; Rolison, D.R. Something from Nothing: Enhancing Electrochemical Charge Storage with Cation Vacancies. *Acc. Chem. Res.* **2013**, *46*, 1181–1191. [CrossRef]

205. Tang, Y.; Zhang, Y.; Li, W.; Ma, B.; Chen, X. Rational material design for ultrafast rechargeable lithium-ion batteries. *Chem. Soc. Rev.* **2015**, *44*, 5926–5940. [CrossRef]

206. Kim, I.G.; Ghani, F.; Lee, K.-Y.; Park, S.; Kwak, S.; Kim, H.-S.; Nah, I.W.; Lim, J.C. Electrochemical performance of Mn_2O_3 nanorods by N-doped reduced graphene oxide using ultrasonic spray pyrolysis for lithium storage. *Intl. J. Energy Res.* **2020**, *44*, 11171–11184. [CrossRef]

207. Cao, K.; Jin, T.; Yang, L.; Jiao, L. Recent progress in Conversion reaction metal oxide anodes for Li-ion batteries. *Mater. Chem. Front.* **2017**, *1*, 2213–2242. [CrossRef]

208. Ghani, F.; Nah, I.W.; Kim, H.-S.; Lim, J.C.; Marium, A.; Ijaz, M.F.; Rana, A.H.S. Facile one-step hydrothermal synthesis of $rGO@Ni_3V_2O_8$ interconnected hollow microspheres composite for lithium-ion batteries. *Nanomaterials* **2020**, *10*, 2389. [CrossRef] [PubMed]

209. He, L.-P.; Sun, S.-Y.; Song, X.-F.; Yu, J.-G. Recovery of cathode materials and Al from spent lithium-ion batteries by ultrasonic cleaning. *Waste Manag.* **2015**, *46*, 523–528. [CrossRef] [PubMed]

210. Liu, H.; Cheng, X.; Chong, Y.; Yuan, H.; Huang, J.-Q.; Zhang, Q. Advanced electrode processing of lithium-ion batteries: A review of powder technology in battery fabrication. *Particuology* **2021**, *57*, 56–71. [CrossRef]

211. Goncalves, R.; Mendez, S.L.; Costa, C.M. Electrode fabrication process and its influence in lithium-ion battery performance: State of the art and future trends. *Electrochem. Comm.* **2022**, *135*, 107210. [CrossRef]

212. Liu, Y.; Zhang, R.; Wang, J.; Wang, Y. Current and future lithium-ion battery manufacturing. *iScience* **2021**, *24*, 102332. [CrossRef]

213. Gonçalves, R.; Dias, P.; Hilliou, L.; Costa, P.; Silva, M.M.; Costa, C.M.; Galvan, S.C.; Mendez, S.L. Optimized printed cathode electrodes for high performance batteries. *Energy Technol.* **2021**, *9*, 2000805. [CrossRef]

214. Lalau, C.C.; Low, C.T.J. Electrophoretic deposition for lithium-ion battery electrode manufacture. *Batter. Supercaps* **2019**, *2*, 551–559. [CrossRef]

215. Goren, A.; Juarez, D.C.; Martins, P.; Ferdov, S.; Silva, M.M.; Tirado, J.L.; Costa, C.M.; Mendez, S.L. Influence of solvent evaporation rate in the preparation of carbon-coated lithium iron phosphate cathode films on battery performance. *Energy Technol.* **2016**, *4*, 573–582. [CrossRef]

216. Meyer, C.; Bockholt, H.; Haselrieder, W.; Kwade, A. Characterization of the calendaring process for compaction of electrodes for lithium-ion batteries. *J. Mater. Process. Technol.* **2017**, *249*, 172–178. [CrossRef]

217. Jeon, D.H. Wettability in electrodes and its impact on the performance of lithium-ion batteries. *Energy Storage Mater.* **2019**, *18*, 139–147. [CrossRef]

218. Robles, D.J.; Vyas, A.A.; Fear, C.; Jeevarajan, J.A.; Mukherjee, P.P. Overcharge and Aging Analytics of Li-Ion Cells. *J. Electrochem. Soc.* **2020**, *167*, 090547. [CrossRef]

219. Qian, J.; Liu, L.; Yang, J.; Li, S.; Wang, X.; Zhuang, H.L.; Lu, Y. Electrochemical surface passivation of $LiCoO_2$ particles at ultrahigh voltage and its applications in lithium-based batteries. *Nat. Commun.* **2018**, *9*, 4918. [CrossRef]

220. Yu, F.; Wang, Y.; Guo, C.; Liu, H.; Bao, W.; Li, J.; Zhang, P.; Wang, F. Spinal $LiMn_2O_4$ cathode materials in wide voltage window: Single-Crystalline versus Polycrystalline. *Crystals* **2022**, *12*, 317. [CrossRef]

221. Hu, L.-H.; Wu, F.-Y.; Lin, C.; Khlobystov, A.N.; Li, L.-J. Graphene-modified $LiFePO_4$ cathode for lithium-ion battery beyond theoretical capacity. *Nat. Commun.* **2013**, *4*, 1687. [CrossRef]

222. Song, H.; Li, J.; Luo, M.; Zhao, Q.; Liu, F. Ultra-Thin mesoporous LiV_3O_8 Nanosheet with exceptionally large specific area for fast and reversible Li storage in Lithium-ion battery cathode. *J. Electrochem. Soc.* **2021**, *168*, 050515. [CrossRef]

223. Wu, X.; Song, K.; Zhang, X.; Hu, N.; Li, L.; Li, W.; Zhang, L.; Zhang, H. Safety Issues in Lithium-Ion Batteries: Materials and cell Design. *Front. Energy Res.* **2019**, *7*, 65. [CrossRef]

224. Kim, C.-S.; Jeong, K.M.; Kim, K.; Yi, C.-W. Effects of Capacity ratios between anode and cathode on electrochemical properties for lithium polymer batteries. *Electrochim. Acta* **2015**, *155*, 431–436. [CrossRef]

225. Fanfeng, L.; Cheng, C.; Zhiyuan, Z.; Weikang, Z.; Zhengzhong, L.Y.U. The influence of N/P ratio on the performance of lithium iron phosphate batteries. *Energy Storage Sci. Technol.* **2021**, *10*, 1325–1329.

226. Angelopoulou, P.; Avgouropoulos, G. Effect of electrode loading on the electrochemical performance of $LiAl_{0.1}Mn_{1.9}O_4$ Cathode for lithium-ion batteries. *Mater. Res. Bull.* **2019**, *119*, 110562. [CrossRef]

batteries

Article

Enhanced Structural and Electrochemical Performance of LiNi$_{0.5}$Mn$_{1.5}$O$_4$ Cathode Material by PO$_4^{3-}$/Fe^{3+} Co-Doping

Yong Wang [1], Shaoxiong Fu [2], Xianzhen Du [1], Dong Wei [1], Jingpeng Zhang [1,*], Li Wang [1,2,*] and Guangchuan Liang [1,2]

[1] Shandong Goldencell Electronics Technology Co., Ltd., Zaozhuang 277100, China; tec@goldencell.biz (Y.W.)
[2] School of Materials Science and Engineering, Hebei University of Technology, Tianjin 300401, China; 13931100462@163.com
[*] Correspondence: tec03@goldencell.biz (J.Z.); wangli_hebut@163.com (L.W.)

Abstract: Series of PO$_4^{3-}$/Fe^{3+} co-doped samples of LiNi$_{0.5}$Mn$_{1.5-5/3x}$Fe$_x$P$_{2/3x}$O$_4$ (x = 0.01, 0.02, 0.03, 0.04, 0.05) have been synthesized by the coprecipitation–hydrothermal method, along with high-temperature calcination using FeSO$_4$ and NaH$_2$PO$_4$ as Fe^{3+} and PO$_4^{3-}$ sources, respectively. The effects of the PO$_4^{3-}$/Fe^{3+} co-doping amount on the crystal structure, particle morphology and electrochemical performance of LiNi$_{0.5}$Mn$_{1.5}$O$_4$ are intensively studied. The results show that the PO$_4^{3-}$/Fe^{3+} co-doping amount exerts a significant influence on the crystal structure and particle morphology, including increased crystallinity, lowered Mn^{3+} content, smaller primary particle size with decreased agglomeration and the exposure of high-energy (110) and (311) crystal surfaces in primary particles. The synergy of the above factors contributes to the obviously ameliorated electrochemical performance of the co-doped samples. The LiNi$_{0.5}$Mn$_{1.45}$Fe$_{0.03}$P$_{0.02}$O$_4$ sample exhibits the best cycling stability, and the LiNi$_{0.5}$Mn$_{1.4333}$Fe$_{0.04}$P$_{0.0267}$O$_4$ sample displays the best rate performance. The electrochemical properties of LiNi$_{0.5}$Mn$_{1.5-5/3x}$Fe$_x$P$_{2/3x}$O$_4$ can be regulated by adjusting the PO$_4^{3-}$/Fe^{3+} co-doping amount.

Keywords: lithium-ion battery; cathode material; LiNi$_{0.5}$Mn$_{1.5}$O$_4$; PO$_4^{3-}$/Fe^{3+} co-doping; electrochemical performance

Citation: Wang, Y.; Fu, S.; Du, X.; Wei, D.; Zhang, J.; Wang, L.; Liang, G. Enhanced Structural and Electrochemical Performance of LiNi$_{0.5}$Mn$_{1.5}$O$_4$ Cathode Material by PO$_4^{3-}$/Fe^{3+} Co-Doping. *Batteries* **2024**, *10*, 341. https://doi.org/10.3390/batteries10100341

Academic Editor: Yong-Joon Park

Received: 8 August 2024
Revised: 17 September 2024
Accepted: 24 September 2024
Published: 26 September 2024

1. Introduction

At present, lithium-ion batteries (LIBs) are widely adopted in portable electronic products, electric vehicles and energy storage equipment due to the advantages of high energy density, a long cycle life and high safety [1]. The capacity, energy density and cycling performance of LIBs mainly depend on the cathode material [2]. The current mainstream commercial cathode materials include LiCoO$_2$, LiMn$_2$O$_4$, LiFePO$_4$ and LiNi$_x$Co$_y$Mn$_z$O$_2$ (x + y + z = 1) ternary material. The low energy density of LiMn$_2$O$_4$, the complex preparation technology of LiFePO$_4$ and the high cost of Co for LiCoO$_2$ and ternary material limit their further application in electric vehicles and energy storage equipment. Therefore, there is an urgent need to develop a cathode material with a higher energy density, higher safety and lower cost. As is well known, the energy density of LIBs is directly affected by that of the cathode material, which is the product of its specific discharge capacity and operation voltage. Therefore, finding a cathode material with a higher specific capacity or operation voltage can effectively improve the energy density of LIBs.

Spinel LiNi$_{0.5}$Mn$_{1.5}$O$_4$ (LNMO) has been regarded as a promising next-generation cathode material due to its high operation voltage of 4.7 V (vs. Li/Li$^+$), high energy density of 650 Wh kg^{-1}, fast Li$^+$ insertion/extraction kinetics and abundant raw materials [3,4]. However, LNMO usually exhibits rapid capacity fading, mainly originating from irreversible structure change, transition metal ion dissolution and side reactions with liquid electrolytes [3]. Many endeavors have been undertaken to alleviate the above problems, such as surface coating and element doping. A surface coating using oxide [5], fluoride [6], phosphate [7] and other compounds (LaFeO$_3$ [8], Li$_{0.35}$La$_{0.55}$TiO$_3$ [9], etc.) has been applied to reduce side reactions with an electrolyte. Element doping is believed to be a cost-effective method to improve the structural stability and rate capability of a cathode material. In previous works, cations (Na$^+$ [10], Mg^{2+} [11], Cr^{3+} [12], Ru^{4+} [13], etc.) and anions (F$^-$ [14], S^{2-} [15], Cl$^-$ [16], etc.) have been used to dope into the LNMO lattice to improve the electrochemical performance. Presently, much attention has been paid to cation or anion single-doping, which can only improve the specific aspect of electrochemical performance. Therefore, in consideration of the integrated advantages of cation and anion doping, the cation and anion co-doping strategy has been proposed to regulate the spinel structure of LNMO material.

On the other hand, because the polyanion bonds are stronger than TM-O (TM = Ni, Mn) bonds, polyanion doping (PO$_4$$^{3-}$, BO$_4$$^{5-}$, SiO$_4$$^{4-}$, etc.) has been used to enhance the electrochemical performance of Ni-rich and Li-rich cathode materials [17–19]. However, until now, there are few reports on polyanion doping modification on LiNi$_{0.5}$Mn$_{1.5}$O$_4$ material. In our previous work [20], it was found that a PO$_4$$^{3-}$/Fe^{3+} co-doped sample exhibited better electrochemical performance than un-doped and single-doped samples. Herein, a series of PO$_4$$^{3-}$/Fe^{3+} co-doped samples of LiNi$_{0.5}$Mn$_{1.5-5/3x}$Fe$_x$P$_{2/3x}$O$_4$ (x = 0.01, 0.02, 0.03, 0.04, 0.05) were synthesized via the coprecipitation–hydrothermal method together with two-step calcinations using FeSO$_4$ as an Fe^{3+} source and NaH$_2$PO$_4$ as a PO$_4$$^{3-}$ source. The influence of the PO$_4$$^{3-}$/Fe^{3+} co-doping amount on the structure, particle morphology and electrochemical properties of LiNi$_{0.5}$Mn$_{1.5}$O$_4$ are intensively studied.

2. Materials and Methods

2.1. Material Synthesis

The preparation process of a PO$_4$$^{3-}$/Fe^{3+} co-doped sample is illustrated in Scheme 1. Firstly, a coprecipitation–hydrothermal method was adopted to prepare a PO$_4$$^{3-}$/Fe^{3+} co-doped carbonate precursor by using FeSO$_4$ and NaH$_2$PO$_4$ as Fe^{3+} and PO$_4$$^{3-}$ sources, respectively. To be specific, according to the formula LiNi$_{0.5}$Mn$_{1.5-5/3x}$Fe$_x$P$_{2/3x}$O$_4$ (x = 0.01, 0.02, 0.03, 0.04, 0.05), 0.6 mmol FeSO$_4$·7H$_2$O (99.0%, ShengAo, Tianjin, China) and 0.4 mmol NaH$_2$PO$_4$·2H$_2$O (99.0%, BoDi, Tianjin, China) were added to a mixture of deionized water (160 mL) and ethylene glycol (80 mL). After 30 min of stirring, NiSO$_4$·6H$_2$O (15 mmol, 99%, DaMao) and MnSO$_4$·H$_2$O (45 mmol, 99%, GuangFu, Tianjin, China) were added and stirred for 30 min to obtain a metal salt solution. NH$_4$HCO$_3$ (300 mmol, 99%, FuChen, Tianjin, China) was totally dissolved in deionized water (160 mL), and the obtained solution was added to the above metal salt solution via a peristaltic pump. After stirring for 30 min, the resulting suspension was placed into a Teflon-lined stainless-steel autoclave and maintained at 180 °C for 8 h in a blast oven. After cooling, repeated filtering and washing, the resultant PO$_4$$^{3-}$/Fe^{3+} co-doped carbonate powder was pre-sintered in a muffle furnace at 900 °C for 4 h to obtain a black oxide powder, which was uniformly blended with 5 wt.% excess Li$_2$CO$_3$ and sintered at 800 °C for 10 h. After sieving through a 325-mesh sieve the final LiNi$_{0.5}$Mn$_{1.5-5/3x}$Fe$_x$P$_{2/3x}$O$_4$ (x = 0.01, 0.02, 0.03, 0.04, 0.05) products were achieved and named as LNMO-FeP0.01, LNMO-FeP0.02, LNMO-FeP0.03, LNMO-FeP0.04, LNMO-FeP0.05, respectively. For comparison, the un-doped LNMO sample was synthesized based on the above process but with the absence of FeSO$_4$·7H$_2$O and NaH$_2$PO$_4$·2H$_2$O.

2.2. Material Characterization

The crystal structure was analyzed using an X-ray diffractometer (XRD, Smartlab 9KW, Rigaku, Japan) using CuKα radiation in the range of 2θ = 10–80° and Fourier transform infrared spectroscopy (FT-IR, V80, Bruker, Germany) in the range of 700–400 cm^{-1}. Scanning electron microscopy (SEM, JSM-7610F, Japan) was used to observe the particle morphology.

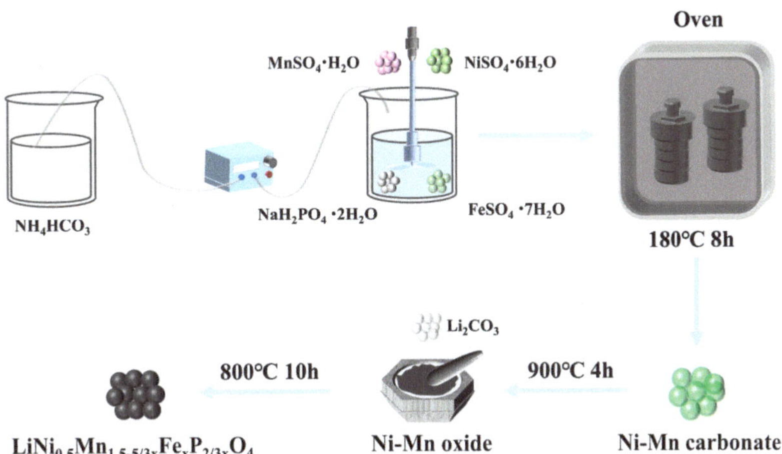

Scheme 1. Illustration of preparation process for PO_4^{3-}/Fe^{3+} co-doped sample.

2.3. Electrochemical Tests

CR2032 coin-type cells were assembled in an Ar-filled glove box, which consists of a cathode sheet, lithium metal anode, Celgard 2400 microporous membrane and commercial electrolyte purchased from Tinci Company. The cathode sheet was prepared as follows: LNMO powder, polyvinylidene fluoride (PVDF) and Super P (weight ratio 8:1:1) were mixed uniformly in N-methyl-2-pyrrolidone (NMP) to obtain a slurry, which was casted on aluminum foil using a doctor blade. After vacuum desiccation, the resultant Al foil was cut into 12 mm diameter round sheets. The constant-current charge/discharge tests were measured between 3.5 and 4.95 V at 25 °C on a Land battery test system (CT2001A, Wuhan, China). Cyclic voltammetry (CV) and electrochemical impedance spectroscopy (EIS) tests were conducted on the electrochemical workstation (CHI660E, Chenhua, Shanghai, China) using a two-electrode system. An EIS test was conducted in the frequency range of 100 kHz–100 mHz with a 5 mV amplitude.

3. Results and Discussion

Figure 1a shows the XRD patterns of the pristine LNMO and co-doped LiNi$_{0.5}$Mn$_{1.5-5/3x}$Fe$_x$P$_{2/3x}$O$_4$ samples. For all samples, the diffraction peaks can be ascribed to an *Fd3m* cubic spinel LiNi$_{0.5}$Mn$_{1.5}$O$_4$ (PDF #80-2162), and the sharp and narrow diffraction peaks suggest a well-crystallized spinel structure for all samples. The absence of impurity diffraction peaks also suggests that the Fe^{3+} and PO_4^{3-} ions have been successfully doped into the spinel lattice without changing the crystal structure of the LNMO material. The high phase purity contributes to less side reactions and the ameliorated electrochemical performance of the LNMO material. In contrast, the peak intensities of the co-doped samples are obviously increased in comparison with the un-doped sample, implying enhanced crystallinity after PO_4^{3-}/Fe^{3+} co-doping, which also benefits the electrochemical performance of the co-doped samples.

Jade 6.5 software was adopted to refine the XRD patterns, and the refined lattice constants *a* for the LNMO, LNMO-FeP0.01, LNMO-FeP0.02, LNMO-FeP0.03, LNMO-FeP0.04 and LNMO-FeP0.05 samples are 8.1804 Å, 8.1738 Å, 8.1640 Å, 8.1625 Å, 8.1597 Å and 8.1574 Å, respectively. It is evident that the lattice constants show a gradually decreasing trend with the PO_4^{3-}/Fe^{3+} co-doping amount increasing. This is primarily attributed to the P-O (410 kJ mol^{-1}, 298 K) and Fe-O (409 kJ mol^{-1}, 298 K) bonds being stronger than the Ni-O (392 kJ mol^{-1}, 298 K) and Mn-O (402 kJ mol^{-1}, 298 K) bonds, which can reduce oxygen evaporation during the calcination process, thus generating less Mn^{3+} ions to keep the charge neutrality. In consideration of the smaller ionic radius of Mn^{4+} (0.530 Å) compared with Mn^{3+} (0.645 Å), the increasing co-doping amount leads to the gradually decreased Mn^{3+} content, thereby inducing the gradually decreasing lattice constants. Figure 1b shows the magnified pattern of the (111) peak, where we observe that the (111) peak gradually shifts to higher angles along with the increase in the co-doping amount, in good correspondence with the gradually decreasing lattice constants. The shift of the (111) peak after PO_4^{3-}/Fe^{3+} co-doping also suggests their successful incorporation into the spinel lattice.

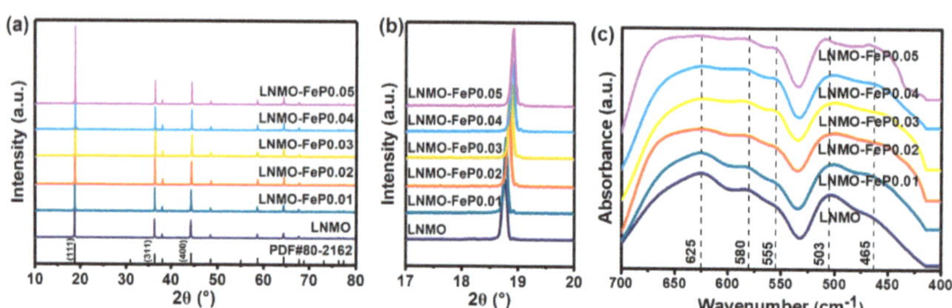

Figure 1. XRD patterns (**a**) and magnified pattern of (111) peak (**b**) and FT-IR spectra (**c**) of all samples.

As mentioned previously, for $LiNi_{0.5}Mn_{1.5}O_4$ materials, the higher the I_{311}/I_{400} peak intensity ratio is, the better the structural stability is [21]. According to Figure 1a, the calculated I_{311}/I_{400} intensity ratios for the LNMO, LNMO-Fe0.01, LNMO-FeP0.02, LNMO-FeP0.03, LNMO-FeP0.04 and LNMO-FeP0.05 samples are 1.071, 1.104, 1.170, 1.181, 1.142 and 1.084, respectively. That is, the enhanced I_{311}/I_{400} ratios of the co-doped samples imply improved structural stability and cycling stability after PO_4^{3-}/Fe^{3+} co-doping, mainly due to the higher bonding strength of the Fe-O and P-O bonds. Among the co-doped samples, the LNMO-FeP0.03 sample has the highest I_{311}/I_{400} ratio, suggesting that it has the best cycling capability However, further increasing the co-doping amount to $x = 0.05$ leads to a decrease in the I_{311}/I_{400} ratio, maybe due to the lattice distortion resulting from the excessive doping of PO_4^{3-}/Fe^{3+} ions.

In addition, the relative peak intensity in the XRD pattern can reflect the relative exposure of crystal planes [22]. In order to analyze the changes in the exposed crystal planes of samples, the relevant diffraction peaks were normalized to the (111) peak to explore the effect of PO_4^{3-}/Fe^{3+} co-doping on the selective growth of the (440) and (311) planes, and the obtained I_{440}/I_{111} and I_{311}/I_{111} intensity ratios are listed in Table 1. It is found that appropriate PO_4^{3-}/Fe^{3+} co-doping ($x \leq 0.04$) can enhance the I_{440}/I_{111} and I_{311}/I_{111} intensity ratios, whereas overmuch co-doping ($x = 0.05$) decreases the ratios again. That is, appropriate PO_4^{3-}/Fe^{3+} co-doping can increase the proportion of exposed (110) and (311) crystal planes. The changes in exposed crystal planes after PO_4^{3-}/Fe^{3+} co-doping can also be observed from Figure 2.

Table 1. I_{440}/I_{111} and I_{311}/I_{111} intensity ratios for un-doped and co-doped samples.

Sample	I_{440}/I_{111}	I_{311}/I_{111}
LNMO	0.0857	0.312
LNMO-FeP0.01	0.0862	0.313
LNMO-FeP0.02	0.0871	0.314
LNMO-FeP0.03	0.0895	0.321
LNMO-FeP0.04	0.0967	0.328
LNMO-FeP0.05	0.0870	0.309

$LiNi_{0.5}Mn_{1.5}O_4$ material generally displays two different crystal structures, including a disordered *Fd3m* structure and an ordered *P4₃32* structure [23], which can be distinguished by means of FT-IR, and the corresponding spectra are shown in Figure 1c. Generally, the ordered structure usually has eight absorption peaks, while the disordered one only has five broadened absorption peaks [24]. From Figure 1c, all samples exhibit five broadened absorption peaks at 625, 580, 555, 503 and 465 cm^{-1}, manifesting the dominant disordered *Fd3m* structure. As is widely accepted, the disordered structure usually exhibits higher electronic conductivity than the ordered one because of the presence of Mn^{3+} [25]. In addition, from Figure 1c, the 625 cm^{-1} absorption peak is stronger than the 580 cm^{-1} absorption peak, also suggesting a dominant disordered *Fd3m* structure [21]. At the same time, the Ni/Mn disordering degree can be evaluated by the ratio of the 625 cm^{-1} peak to the 580 cm^{-1} peak (I_{625}/I_{580}), and a higher I_{625}/I_{580} ratio usually means a higher degree of Ni/Mn disordering [26]. According to Figure 1c, the I_{625}/I_{580} intensity ratios of the LNMO, LNMO-Fe0.01, LNMO-FeP0.02, LNMO-FeP0.03, LNMO-FeP0.04 and LNMO-FeP0.05 samples are 1.193, 1.143, 1.054, 1.021, 1.020 and 1.018, respectively. In other words, the degree of Ni/Mn disordering (Mn^{3+} content) gradually decreases with the co-doping amount. In our previous work [20], the XPS analysis results of un-doped LNMO and co-doped LNMO-FeP0.02 samples also verify the decreased Mn^{3+} content in the LNMO-FeP0.02 sample. In addition, the XPS results imply that Fe^{3+} and PO$_4^{3-}$ ions have been successfully doped into the LNMO-FeP0.02 lattice.

Figure 2 shows SEM images of the un-doped and co-doped samples. From Figure 2a,g, we observe that un-doped LNMO displays a secondary microspherical structure constituted by truncated octahedral primary particles with poor crystallinity and severe agglomeration, which may affect the Li$^+$ ion insertion/extraction process, which is detrimental to the electrochemical properties of the LNMO material.

From Figure 2b,h, LNMO-FeP0.01 shows a similar particle morphology to un-doped LNMO but with improved crystallinity and reduced agglomeration. At a co-doping amount of $x = 0.02$ (Figure 2c,i), the particle morphology and size of the LNMO-FeP0.02 sample change greatly, and most microsphere particles are replaced by dispersed small particles with greatly decreased primary particle sizes. Besides the (111) and (100) planes, a (110) plane also appears. The emergence of small dispersed particles with high-energy (110) crystal planes is mainly attributed to the enhanced crystal stability caused by the stronger Fe-O and P-O bonds, which allows the particles to exist as small dispersed particles with additional (110) planes.

When $x = 0.03$ and 0.04, the primary particle size becomes smaller, as shown in Figure 2j,k. However, differently from LNMO-FeP0.02, the primary particles of LNMO-FeP0.03 and LNMO-FeP0.04 agglomerate together to reduce the total surface energy. And as seen from Figure 2j,k, the primary particles even exhibit a (311) plane besides the (111), (100) and (110) planes, which is induced by the synergistic effect of PO$_4^{3-}$/Fe^{3+} co-doping. When x further increases to 0.05, as shown in Figure 2f,l, primary LNMO-FeP0.05 particles agglomerate into the secondary microsphere structure, and the (311) and (110) planes decrease or even disappear. This may be due to the fact that excessive PO$_4^{3-}$/Fe^{3+} doping may cause lattice distortion, which in turn destroys the stability of the crystal structure.

Figure 3a shows the first-cycle charge/discharge curves of the un-doped and co-doped samples at a 0.2 C rate. The first specific discharge capacities at the 0.2 C rate are 120.0, 131.4, 134.5, 134.6, 145.2 and 130.6 mAh g^{-1}, respectively, for LNMO, LNMO-Fe0.01, LNMO-FeP0.02, LNMO-FeP0.03, LNMO-FeP0.04 and LNMO-FeP0.05. That is, the first specific discharge capacities are increased after co-doping because of the improved crystallinity, decreased primary particle size and agglomeration, as well as the appearance of high-energy crystal planes. The emergence of high-energy (110) and (311) crystal planes contribute to LNMO-FeP0.04 having the highest first discharge capacity.

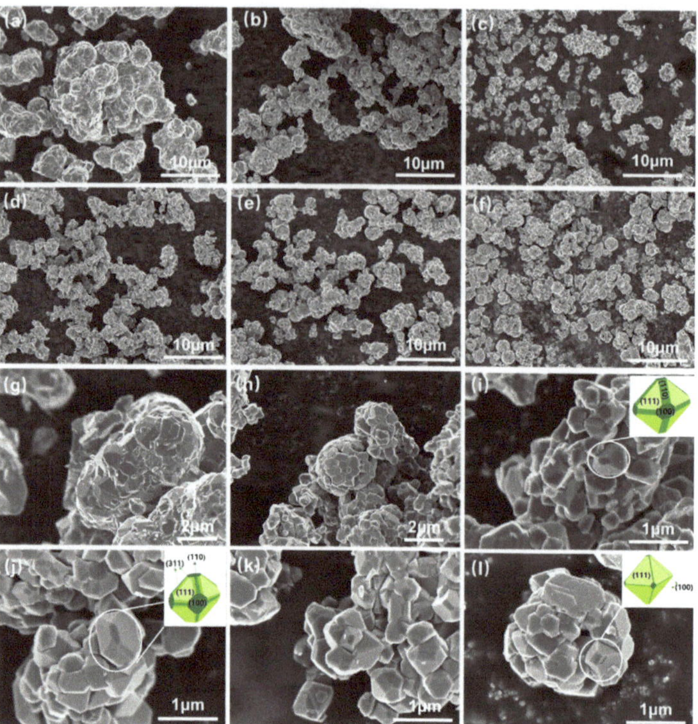

Figure 2. SEM images of all samples: (**a,g**) LNMO, (**b,h**) LNMO-FeP0.01, (**c,i**) LNMO-FeP0.02, (**d,j**) LNMO-FeP0.03, (**e,k**) LNMO-FeP0.04, (**f,l**) LNMO-FeP0.05.

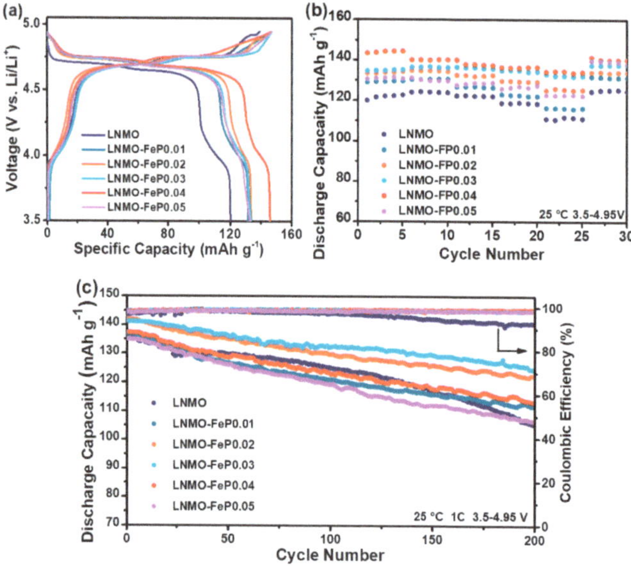

Figure 3. First-cycle charge/discharge curves at 0.2 C rate (**a**), rate capability curves (**b**) and cycling and coulombic efficiency curves (**c**) for all samples.

In addition, the ~4.7 V and ~4.0 V plateaus in the charge/discharge curves of all samples can be separately ascribed to Ni^{2+}/Ni^{4+} and Mn^{3+}/Mn^{4+} redox couples. The presence of a ~4.0 V plateau also confirms the dominant disordered structure, consistent with the above FT-IR results. In addition, the relative Mn^{3+} contents can be evaluated by the discharge capacity between 3.8 and 4.25 V divided by the total discharge capacity according to the first-cycle discharge curves [27], which are 14.83%, 11.47%, 9.37%, 9.21%, 8.92% and 8.03%, respectively, for LNMO, LNMO-Fe0.01, LNMO-FeP0.02, LNMO-FeP0.03, LNMO-FeP0.04 and LNMO-FeP0.05. That is, the Mn^{3+} content gradually decreases with the co-doping amount, due to the fact that that the stronger Fe-O and P-O bonds reduce oxygen loss during calcination and then result in the formation of fewer Mn^{3+} ions.

Figure 3b shows the rate capability curves of the un-doped and co-doped samples, with five cycles at 0.2 C, 1 C, 2 C, 5 C and 10 C rates and then back to 0.2 C. The discharge capacities at each rate are enhanced after co-doping, suggesting that the rate performance is improved after co-doping. The 10 C discharge capacities for LNMO, LNMO-Fe0.01, LNMO-FeP0.02, LNMO-FeP0.03, LNMO-FeP0.04 and LNMO-FeP0.05 are 111.7, 116.4, 125.2, 132.3, 134.6 and 122.8 mAh g^{-1}. That is, LNMO-FeP0.04 displays the optimal rate capability among the co-doped samples. When $x \leq 0.04$, the gradually improved rate capability is mainly induced by the smaller primary particles, higher crystallinity, lower agglomeration and exposure of high-energy (110) and (311) crystal planes. The smaller primary particles may enlarge the electrode/electrolyte contact area and make the Li^+ ions' diffusion path shorter. The increased exposure of high-energy (110) and (311) crystal planes is more conductive to Li^+ ion diffusion [12]. The optimal rate capability of the LNMO-FeP0.04 sample is mainly attributed to the greater exposure of the (110) and (311) crystal planes. As x further increases to 0.05, the rate performance deteriorates adversely due to the lower Mn^{3+} content and the reduction or disappearance of the (311) and (110) crystal planes.

The cycling performance test was conducted on the un-doped and co-doped samples at 1 C and 25 °C, as shown in Figure 3c. The gradual decrease in discharge capacities with the cycle number is mainly ascribed to interfacial side reactions between the electrode and electrolyte and the accompanying continual growth of a CEI (Cathode–Electrolyte Interphase) layer on the electrode surface. The capacity retention rates are 76.7%, 80.9%, 85.9%, 88.8%, 83.1% and 78.6%, respectively, for LNMO, LNMO-Fe0.01, LNMO-FeP0.02, LNMO-FeP0.03, LNMO-FeP0.04 and LNMO-FeP0.05 after 200 cycles. It is evident that the cycling performance is obviously improved after $PO_4{}^{3-}/Fe^{3+}$ co-doping. The LNMO-FeP0.03 sample exhibits the optimal cycling performance. When the co-doping amount $x \leq 0.03$, the cycling performance is gradually enhanced because of the reduced Mn^{3+} content and the decreased (111) crystal plane. A lower Mn^{3+} content may reduce the Jahn–Teller effect and Mn^{2+} dissolution during cycling [22]. In addition, the crystal planes contacting the electrolyte greatly affect the cycling capability of the LNMO material [28]. The (111) crystal plane has been reported to accelerate Mn^{2+} dissolution and demonstrate unstable interface behaviors at a high voltage, thereby adversely affecting the cycling capability of the LNMO material [28,29]. Therefore, increased exposure of the (111) crystal plane is disadvantageous to the cycling performance of the LNMO material. On the other hand, stronger Fe-O and P-O bonds can make the crystal structure more stable and reduce lattice stress during cycling, which is also conductive to the cycling performance of the LNMO material. However, a further increase in the co-doping amount leads to a decline in the cycling capability of LNMO-FeP0.04 and LNMO-FeP0.05, probably because of the lattice distortion caused by excessive doping disrupting the crystal structure stability. Additionally, the increased agglomeration and decreased Mn^{3+} content are also detrimental to the cycling performance of the LNMO-FeP0.04 and LNMO-FeP0.05 samples.

Figure 4 shows the CV curves obtained at different scan rates. Due to the limitations of Li^+ diffusion dynamics, the oxidation/reduction peaks of all electrodes shift towards higher and lower potentials, respectively, as the scan rate increases. As shown in the insets of Figure 4a–f, the peak current (i_p) of the oxidation/reduction peaks exhibits a linear relation with the square root of the scan rate ($v^{1/2}$). The Li^+ diffusion coefficient (D_{Li}) for each redox peak can be obtained based on the Randles–Sevcik equation [30]:

$$i_p = (2.69 \times 10^5)n^{3/2}AD_{Li}^{1/2}v^{1/2}C_0 \tag{1}$$

where i_p is peak current (mA), n is 1, A is the electrode surface area (~1.13 cm^2), v is the voltage scan rate (mV s^{-1}), and C_0 is the initial Li^+ ion concentration in the cathode (mol cm^{-3}), which can be calculated according to the equation $C_0 = 8/(N_A \cdot V)$, where N_A is 6.02×10^{23}, and V is the refined lattice volume from the XRD pattern. The average Li^+ diffusion coefficient (D_a) was obtained by averaging the four D_{Li} values, as shown in Table 2. The D_a values are 1.084×10^{-10}, 2.251×10^{-10}, 3.609×10^{-10}, 4.011×10^{-10}, 4.582×10^{-10} and 2.594×10^{-10} cm^2 s^{-1}, respectively, for LNMO, LNMO-Fe0.01, LNMO-FeP0.02, LNMO-FeP0.03, LNMO-FeP0.04 and LNMO-FeP0.05. Evidently, all the co-doped samples exhibit higher Li^+ diffusion coefficients than the un-doped ones, which is attributable to improved crystallinity, reduced agglomeration, smaller primary particle size and a higher proportion of exposed (110) and (311) crystal planes. Compared to the (111) and (110) planes, the (110) and (311) planes expedite Li^+ ion diffusion during the cycling process [28], leading to a better rate capability of LNMO materials with more (110) and (311) planes, which accounts for the largest Li^+ ion diffusion coefficient of LNMO-FeP0.04. However, when $x = 0.05$, the Li^+ diffusion coefficient decreases significantly, possibly due to the disappearance of the (110) and (311) planes and increased agglomeration.

Table 2. Li^+ ions diffusion coefficients (D_{Li}) for un-doped and co-doped samples.

Sample	D_{Li} ($\times 10^{-10}$ cm^2 s^{-1})				D_a ($\times 10^{-10}$ cm^2 s^{-1})
	O_1	R_1	O_2	R_2	
LNMO	0.497	1.694	1.327	0.817	1.084
LNMO-FeP0.01	1.947	3.933	2.027	1.098	2.251
LNMO-FeP0.02	2.039	2.959	6.136	3.302	3.609
LNMO-FeP0.03	3.665	6.502	2.606	3.269	4.011
LNMO-FeP0.04	1.694	6.058	6.850	3.725	4.582
LNMO-FeP0.05	1.565	4.745	1.410	2.657	2.594

To further determine the improved cycling capability of the co-doped samples, the EIS was tested to observe the impedance change after 50, 100, 150 and 200 cycles. Figure 5 shows the obtained Nyquist plots and fitting curves. All plots comprise two semicircles and a sloping line, representing the resistance of Li^+ ions passing through the electrode surface film (R_{sf}), charge transfer resistance (R_{ct}) and Warburg resistance associated with Li^+ ion diffusion in the electrode bulk (Z_w). Manthiram et al. [31] pointed out that R_{sf} and R_{ct} include surface film and charge transfer resistances from the cathode and Li anode. Because the Li anode goes through the same electrochemical process before the EIS test, the obtained R_{sf} and R_{ct} values can roughly represent the resistances from the cathode. Table 3 lists the fitted R_{sf} and R_{ct} values, obtained using ZView2 software, according to the equivalent circuit in the inset of Figure 5. For all electrodes, the gradual increase in R_{sf} with the cycle number may be caused by the continual side reactions and CEI growth. However, after different cycle numbers, the R_{sf} of co-doped samples is decreased compared to the un-doped sample, implying the effectively inhibited side reactions and CEI growth after co-doping. On one hand, stronger Fe-O and P-O bonds have a strong inhibitory effect on lattice distortion during cycling. On the other hand, the exposed crystal planes also affect interfacial side reactions. The exposure of more (110) and (311) crystal planes could reduce interfacial side reactions, thereby decreasing interfacial impedance. Additionally, when the co-doping amount $x = 0.03$, the LNMO-FeP0.03 sample exhibits the lowest R_{sf} values due to its higher proportion of exposed (110) and (311) planes and moderate agglomeration. However, when $x \geq 0.04$, the particle agglomeration rates of the samples increase, and the (110) and (311) crystal planes gradually decrease until their disappearance at $x = 0.05$, thus leading to increased side reactions. Consequently, LNMO-FeP0.04 and LNMO-FeP0.05 exhibit higher R_{sf} values than LNMO-FeP0.03.

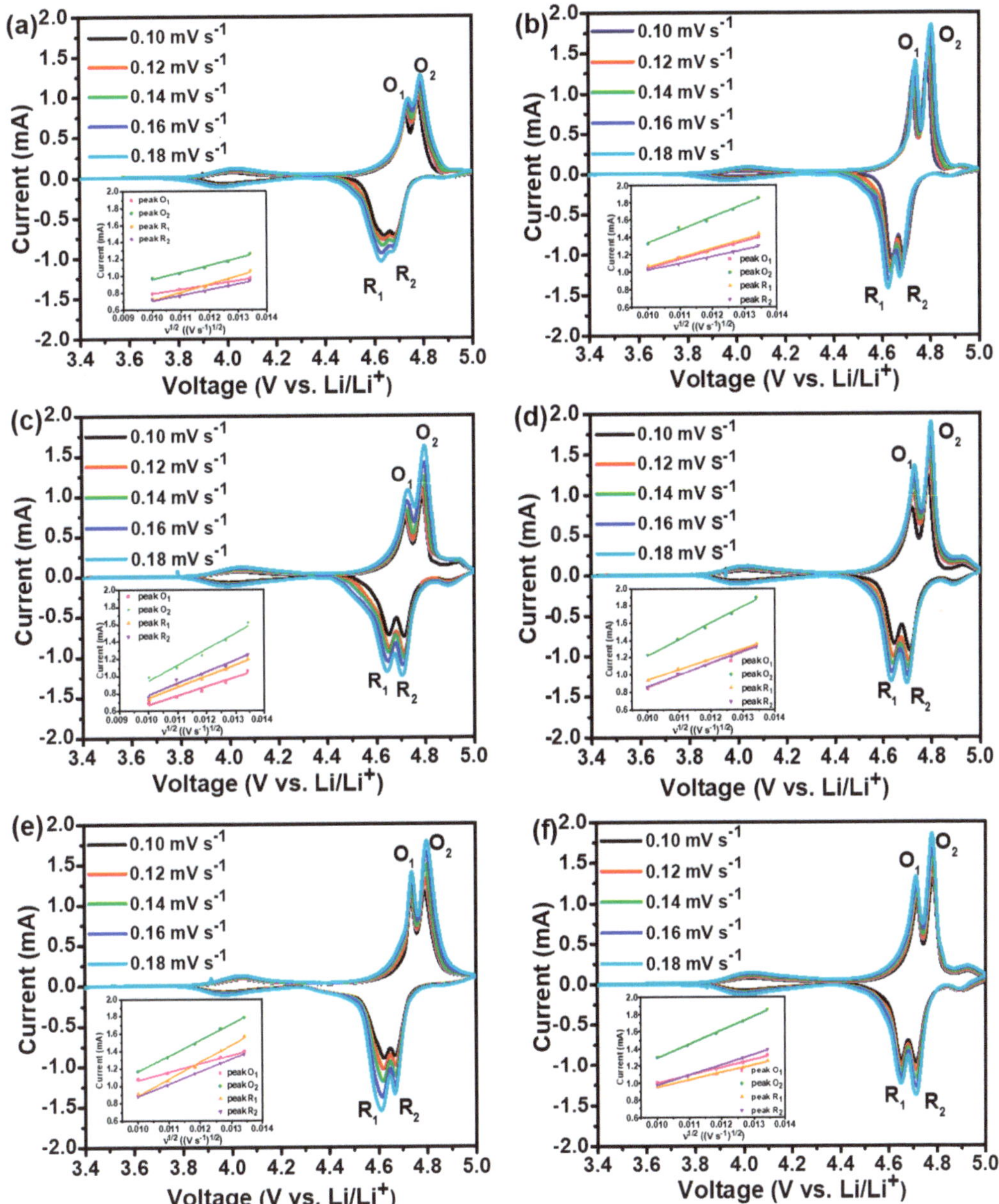

Figure 4. CV curves obtained at different scan rates and linear relationship between peak current and square root of scan rates for un-doped and co-doped samples: (**a**) LNMO, (**b**) LNMO-FeP0.01, (**c**) LNMO-FeP0.02, (**d**) LNMO-FeP0.03, (**e**) LNMO-FeP0.04, (**f**) LNMO-FeP0.05.

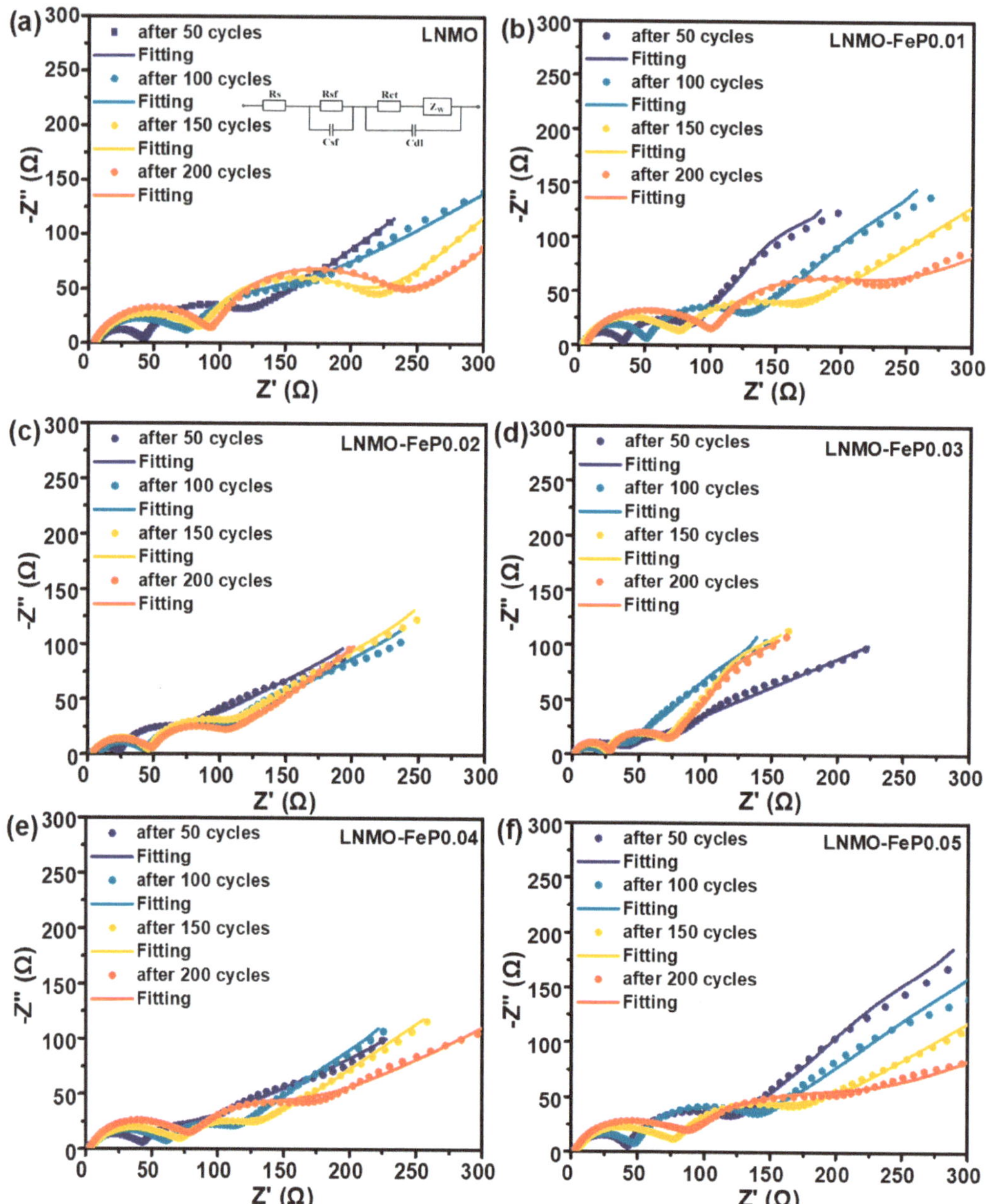

Figure 5. Nyquist plots and fitting lines after different cycle numbers for un-doped and co-doped samples: (**a**) LNMO, (**b**) LNMO-FeP0.01, (**c**) LNMO-FeP0.02, (**d**) LNMO-FeP0.03, (**e**) LNMO-FeP0.04, (**f**) LNMO-FeP0.05.

In addition, the R_{ct} values after 50, 100, 150 and 200 cycles are also compared in Table 3, showing a gradually increasing trend with cycling. Among them, the LNMO-FeP0.03 electrode displays smaller R_{ct} values during the whole cycle process, suggesting its faster electrochemical kinetics, maybe due to the fewer interfacial side reactions.

Table 3. Fitted R_{sf} and R_{ct} values according to Figure 5a–f.

Sample	Resistance (Ω)	After 50 Cycles	After 100 Cycles	After 150 Cycles	After 200 Cycles
LNMO	R_{sf}	38.56	68.16	79.70	91.11
	R_{ct}	65.22	72.16	123.70	136.20
LNMO-FeP0.01	R_{sf}	30.60	49.26	72.25	97.14
	R_{ct}	49.37	68.34	78.01	92.70
LNMO-FeP0.02	R_{sf}	21.75	36.36	44.34	46.70
	R_{ct}	34.54	36.04	43.86	48.64
LNMO-FeP0.03	R_{sf}	12.48	18.84	24.09	25.71
	R_{ct}	15.60	24.38	41.57	43.21
LNMO-FeP0.04	R_{sf}	38.74	57.52	66.39	74.19
	R_{ct}	26.65	45.79	52.78	65.52
LNMO-FeP0.05	R_{sf}	39.10	44.73	72.28	78.66
	R_{ct}	70.32	77.62	84.31	114.50

4. Conclusions

$LiNi_{0.5}Mn_{1.5}O_4$ materials with different PO_4^{3-}/Fe^{3+} co-doping amounts have been synthesized using a combined coprecipitation–hydrothermal method along with high-temperature calcination. The results reveal that the Ni/Mn disordering degree and Mn^{3+} content gradually decrease with the co-doping amounts. SEM observations show that the particle morphology of the samples is influenced significantly by the co-doping amounts. The aggregation degree first decreases and then increases with the co-doping amounts. Additionally, when the co-doping amount $x \leq 0.04$, high-energy (110) and (311) crystal planes gradually increase with increasing co-doping amounts. However, when the co-doping amount increases to $x = 0.05$, the (311) and (110) crystal planes gradually decrease or even disappear. Electrochemical tests demonstrate that the $LiNi_{0.5}Mn_{1.45}Fe_{0.03}P_{0.02}O_4$ sample ($x = 0.03$) exhibits the best cycling stability, and the $LiNi_{0.5}Mn_{1.4333}Fe_{0.04}P_{0.0267}O_4$ ($x = 0.04$) sample exhibits the optimal rate performance. This co-doping strategy can also be adopted to other cathode materials to achieve an intriguing electrochemical performance.

Author Contributions: Y.W. and S.F. contributed equally to this work. Conceptualization, methodology and software, S.F.; formal analysis, Y.W.; data curation, X.D.; resources, D.W.; writing—review and editing, J.Z. and L.W.; funding acquisition, L.W.; supervision, G.L. All authors have read and agreed to the published version of the manuscript.

Funding: This research was funded by the National Natural Science Foundation of China, grant number [No. 51802074].

Data Availability Statement: The raw data supporting the conclusions of this article will be made available by the authors on request.

Conflicts of Interest: Authors Yong Wang, Xianzhen Du, Dong Wei, Jingpeng Zhang and Guangchuan Liang were employed by the company Shandong Goldencell Electronics Technology Co., Ltd. The remaining authors declare that the research was conducted in the absence of any commercial or financial relationships that could be construed as a potential conflict of interest.

References

1. Qian, H.; Ren, H.; Zhang, Y.; He, X.; Li, W.; Wang, J.; Hu, J.; Yang, H.; Sari, H.M.K.; Chen, Y.; et al. Surface doping vs. bulk doping of cathode materials for lithium-ion batteries: A review. *Electrochem. Energy Rev.* **2022**, *5*, 2. [CrossRef]
2. Yi, H.; Liang, Y.; Qian, Y.; Feng, Y.; Li, Z.; Zhang, X. Low-cost Mn-based cathode materials for lithium-ion batteries. *Batteries* **2023**, *9*, 246. [CrossRef]
3. Wu, W.; Zuo, S.; Zhang, X.; Feng, X. Two-step solid state synthesis of medium entropy $LiNi_{0.5}Mn_{1.5}O_4$ cathode with enhanced electrochemical performance. *Batteries* **2023**, *9*, 91. [CrossRef]

4. Wang, B.; Son, S.-B.; Badami, P.; Trask, S.E.; Abraham, D.; Qin, Y.; Yang, Z.; Wu, X.; Jansen, A.; Liao, C. Understanding and mitigating the dissolution and delamination issues encountered with high-voltage $LiNi_{0.5}Mn_{1.5}O_4$. *Batteries* **2023**, *9*, 435. [CrossRef]
5. Ma, C.; Wen, Y.; Qiao, Q.; He, P.; Ren, S.; Li, M.; Zhao, P.; Qiu, J.; Tang, G. Improving electrochemical performance of high-voltage spinel $LiNi_{0.5}Mn_{1.5}O_4$ cathodes by silicon oxide surface modification. *ACS Appl. Energy Mater.* **2021**, *4*, 12201–12210. [CrossRef]
6. Yu, C.; Dong, L.; Zhang, Y.; Du, K.; Gao, M.; Zhao, H.; Bai, Y. Promoting electrochemical performances of $LiNi_{0.5}Mn_{1.5}O_4$ cathodes via YF_3 surface coating. *Solid State Ion.* **2020**, *357*, 115464. [CrossRef]
7. Wu, Y.; Ben, L.; Zhan, Y.; Yu, H.; Qi, W.; Zhao, W.; Huang, X. Binding Li_3PO_4 to spinel $LiNi_{0.5}Mn_{1.5}O_4$ via a surface Co-containing bridging layer to improve the electrochemical performance. *Energy Technol.* **2021**, *9*, 2100147. [CrossRef]
8. Mou, J.R.; Deng, Y.L.; He, L.H.; Zheng, Q.J.; Jiang, N.; Lin, D.M. Critical roles of semi-conductive $LaFeO_3$ coating in enhancing cycling stability and rate capability of 5 V $LiNi_{0.5}Mn_{1.5}O_4$ cathode materials. *Electrochim. Acta* **2018**, *260*, 101–111. [CrossRef]
9. Mereacre, V.; Stüble, P.; Trouillet, V.; Ahmed, S.; Volz, K.; Binder, J.R. Coating versus doping: Understanding the enhanced performance of high-voltage batteries by the coating of spinel $LiNi_{0.5}Mn_{1.5}O_4$ with $Li_{0.35}La_{0.55}TiO_3$. *Adv. Mater. Interfaces* **2023**, *10*, 2201324. [CrossRef]
10. Wang, J.; Chen, D.; Wu, W.; Wang, L.; Liang, G. Effects of Na^+ doping on crystalline structure and electrochemical performances of $LiNi_{0.5}Mn_{1.5}O_4$ cathode material. *Trans. Nonferrous Met. Soc. China* **2017**, *27*, 2239–2248. [CrossRef]
11. Liang, G.; Wu, Z.; Didier, C.; Zhang, W.; Cuan, J.; Li, B.; Ko, K.-Y.; Hung, P.-Y.; Lu, C.-Z.; Chen, Y.; et al. A long cycle-life high-voltage spinel lithium-ion battery electrode achieved by site-selective doping. *Angew. Chem. Int. Ed.* **2020**, *59*, 10594–10602. [CrossRef] [PubMed]
12. Hsiao, Y.-S.; Huang, J.-H.; Cheng, T.-H.; Hu, C.-W.; Wu, N.-J.; Yen, C.-Y.; Hsu, S.-C.; Weng, H.C.; Chen, C.-P. Cr-doped $LiNi_{0.5}Mn_{1.5}O_4$ derived from bimetallic Ni/Mn metal-organic framework as high-performance cathode for lithium-ion batteries. *J. Energy Storage* **2023**, *68*, 107686. [CrossRef]
13. Zhou, D.; Li, J.; Chen, C.; Lin, F.; Wu, H.; Guo, J. A hydrothermal synthesis of Ru-doped $LiMn_{1.5}Ni_{0.5}O_4$ cathode materials for enhanced electrochemical performance. *RSC Adv.* **2021**, *11*, 12549. [CrossRef]
14. Luo, Y.; Li, H.; Lu, T.; Zhang, Y.; Mao, S.S.; Liu, Z.; Wen, W.; Xie, J.; Yan, L. Fluorine gradient-doped $LiNi_{0.5}Mn_{1.5}O_4$ spinel with improved high voltage stability for Li-ion batteries. *Electrochim. Acta* **2017**, *238*, 237–245. [CrossRef]
15. Kim, D.-W.; Zettsu, N.; Shiiba, H.; Sánchez-Santolino, G.; Ishikawa, R.; Ikuhara, Y.; Teshima, K. Metastable oxysulfide surface formation on $LiNi_{0.5}Mn_{1.5}O_4$ single crystal particles by carbothermal reaction with sulfur-doped heterocarbon nanoparticles: New insight into their structural and electrochemical characteristics, and their potential applications. *J. Mater. Chem. A* **2020**, *8*, 22302–22314.
16. Kim, W.-K.; Han, D.-W.; Ryu, W.-H.; Lim, S.-J.; Eom, J.-Y.; Kwon, H.-S. Effects of Cl doping on the structural and electrochemical properties of high voltage $LiMn_{1.5}Ni_{0.5}O_4$ cathode materials for Li-ion batteries. *J. Alloys Compd.* **2014**, *592*, 48–52. [CrossRef]
17. Chen, T.; Li, X.; Wang, H.; Yan, X.; Wang, L.; Deng, B.; Ge, W.; Qu, M. The effect of gradient boracic polyanion-doping on structure, morphology, and cycling performance of Ni-rich $LiNi_{0.8}Co_{0.15}Al_{0.05}O_2$ cathode material. *J. Power Sources* **2018**, *374*, 1–11. [CrossRef]
18. Yang, G.; Pan, K.; Lai, F.; Wang, Z.; Chu, Y.; Yang, S.; Han, J.; Wang, H.; Zhang, X.; Li, Q. Integrated co-modification of PO_4^{3-} polyanion doping and Li_2TiO_3 coating for Ni-rich layered $LiNi_{0.6}Co_{0.2}Mn_{0.2}O_2$ cathode material of lithium-ion batteries. *Chem. Eng. J.* **2021**, *421*, 129964. [CrossRef]
19. Zhang, H.-Z.; Li, F.; Pan, G.-L.; Li, G.-R.; Gao, X.-P. The effect of polyanion-doping on the structure and electrochemical performance of Li-rich layered oxides as cathode for lithium-ion batteries. *J. Electrochem. Soc.* **2015**, *162*, A1899–A1904. [CrossRef]
20. Fu, S.; Zhang, Y.; Bian, Y.; Xu, J.; Wang, L.; Liang, G. Effect of Fe^{3+} and/or PO_4^{3-} doping on the electrochemical performance of $LiNi_{0.5}Mn_{1.5}O_4$ cathode material for Li-ion batteries. *Ind. Eng. Chem. Res.* **2023**, *62*, 1016–1028. [CrossRef]
21. Lan, L.; Li, S.; Li, J.; Lu, L.; Lu, Y.; Huang, S.; Xu, S.; Pan, C.; Zhao, F. Enhancement of the electrochemical performance of the spinel structure $LiNi_{0.5-x}Ga_xMn_{1.5}O_4$ cathode material by Ga doping. *Nanoscale Res. Lett.* **2018**, *13*, 251. [CrossRef] [PubMed]
22. Gong, J.; Fu, S.; Zhang, Y.; Yan, S.; Lang, Y.; Guo, J.; Wang, L.; Liang, G. Enhanced electrochemical performance of 5V $LiNi_{0.5}Mn_{1.5-x}Zr_xO_4$ cathode material for lithium-ion batteries. *ChemistrySelect* **2021**, *6*, 7202–7212. [CrossRef]
23. Liu, D.; Hamel-Paquet, J.; Trottier, J.; Barray, F.; Gariépy, V.; Hovington, P.; Guerfi, A.; Mauger, A.; Julien, C.M.; Goodenough, J.B.; et al. Synthesis of pure phase disordered $LiMn_{1.45}Cr_{0.1}Ni_{0.45}O_4$ by a post-annealing method. *J. Power Sources* **2012**, *217*, 400–406. [CrossRef]
24. Kunduraci, M.; Amatucci, G.G. Synthesis and characterization of nanostructured 4.7 V $Li_xMn_{1.5}Ni_{0.5}O_4$ spinels for high-power lithium-ion batteries. *J. Electrochem. Soc.* **2006**, *153*, A1345–A1352. [CrossRef]
25. Kim, J.-H.; Myung, S.-T.; Yoon, C.S.; Kang, S.G.; Sun, Y.-K. Comparative study of $LiNi_{0.5}Mn_{1.5}O_{4-\delta}$ and $LiNi_{0.5}Mn_{1.5}O_4$ cathodes having two crystallographic structures: $Fd3m$ and $P4_332$. *Chem. Mater.* **2004**, *16*, 906–914. [CrossRef]
26. Zheng, X.; Liu, W.; Qu, Q.; Zheng, H.; Huang, Y. Bi-functions of titanium and lanthanum co-doping to enhance the electrochemical performance of spinel $LiNi_{0.5}Mn_{1.5}O_4$ cathode. *J. Mater.* **2019**, *5*, 156–163. [CrossRef]
27. Zeng, Y.-P.; Wu, X.-l.; Mei, P.; Cong, L.-N.; Yao, C.; Wang, R.-S.; Xie, H.-M.; Sun, L.-Q. Effect of cationic and anionic substitutions on the electrochemical properties of $LiNi_{0.5}Mn_{1.5}O_4$ spinel cathode materials. *Electrochim. Acta* **2014**, *138*, 493–500. [CrossRef]

28. Liu, H.; Wang, J.; Zhang, X.; Zhou, D.; Qi, X.; Qiu, B.; Fang, J.; Kloepsch, R.; Schumacher, G.; Liu, Z.; et al. Morphological evolution of high-voltage spinel LiNi$_{0.5}$Mn$_{1.5}$O$_4$ cathode materials for lithium-ion batteries: The critical effects of surface orientations and particle size. *ACS Appl. Mater. Interfaces* **2016**, *8*, 4661–4675. [CrossRef]
29. Chen, Z.; Zhao, R.; Du, P.; Hu, H.; Wang, T.; Zhu, L.; Chen, H. Polyhedral LiNi$_{0.5}$Mn$_{1.5}$O$_4$ with excellent electrochemical properties for lithium-ion batteries. *J. Mater. Chem. A* **2014**, *2*, 12835–12848. [CrossRef]
30. Liu, W.; Li, X.; Xiong, D.; Hao, Y.; Li, J.; Kou, H.; Yan, B.; Li, D.; Lu, S.; Koo, A.; et al. Significantly improving cycling performance of cathodes in lithium ion batteries: The effect of Al$_2$O$_3$ and LiAlO$_2$ coatings on LiNi$_{0.6}$Co$_{0.2}$Mn$_{0.2}$O$_2$. *Nano Energy* **2018**, *44*, 111–120. [CrossRef]
31. Liu, J.; Manthiram, A. Understanding the improved electrochemical performances of Fe-substituted 5 V spinel cathode LiMn$_{1.5}$Ni$_{0.5}$O$_4$. *J. Phys. Chem. C* **2009**, *113*, 15073–15079. [CrossRef]

MDPI AG

Grosspeteranlage 5

4052 Basel

Switzerland

Tel.: +41 61 683 77 34

Batteries Editorial Office

E-mail: batteries@mdpi.com

www.mdpi.com/journal/batteries